Karolina Wieszczycka, Bartosz Tylkowski, Katarzyna Staszak (Eds.)
Metals in Wastes

Also of interest

Chemical Synergies.
From the Lab to In Silico Modeling
Bandeira, Tylkowski (Eds.), 2018
ISBN 978-3-11-048135-8, e-ISBN 978-3-11-048206-5

Environmental Pollution Control.
Liu, Zhang, Liu, 2017
ISBN 978-3-11-053789-5, e-ISBN 978-3-11-053831-1

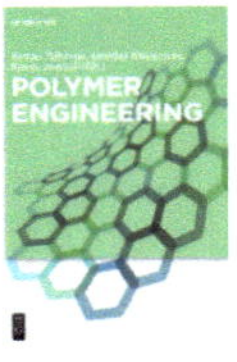

Polymer Engineering.
Tylkowski, Wieszczycka, Jastrzab (Eds.), 2017
ISBN 978-3-11-046828-1, e-ISBN 978-3-11-046974-5

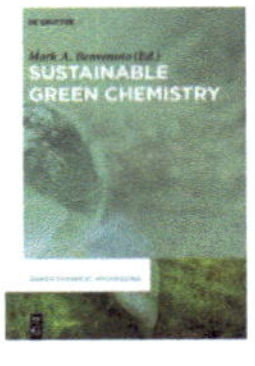

Sustainable Green Chemistry.
Benvenuto (Ed.), 2017
ISBN 978-3-11-044189-5, e-ISBN 978-3-11-043585-6

Physical Sciences Reviews.
e-ISSN 2365-659X

Metals in Wastes

Edited by
Karolina Wieszczycka, Bartosz Tylkowski,
Katarzyna Staszak

DE GRUYTER

Editors
Dr. habil. Karolina Wieszczycka
Poznań University of Technology
Institute of Chemical Technology and Engineering
Berdychowo 4
60-965 Poznań
Poland
karolina.wieszczycka@put.poznan.pl

Dr. Bartosz Tylkowski
Centre Tecnològic de la Química de Catalunya,
Carrer Marcelli Domingo s/n
Tarragona, 43007
Spain

and

Department of Chemical Engineering
Rovira i Virgili University
Av. Paisos Catalans 26
Tarragona, 43307
Spain
bartosz.tylkowski@urv.cat

Dr. habil. Katarzyna Staszak
Poznań University of Technology
Institute of Chemical Technology and Engineering
Berdychowo 4
60-965 Poznań
Poland
katarzyna.staszak@put.poznan.pl

ISBN 978-3-11-054628-6
e-ISBN (PDF) 978-3-11-054706-1
e-ISBN (EPUB) 978-3-11-054652-1

Library of Congress Control Number: 2018945147

Bibliographic information published by the Deutsche Nationalbibliothek
The Deutsche Nationalbibliothek lists this publication in the Deutsche Nationalbibliografie; detailed bibliographic data are available on the Internet at http://dnb.dnb.de.

Typesetting: Integra Software Services Pvt. Ltd.
Printing and binding: CPI books GmbH, Leck
Cover image: Yuji Kotani / DigitalVision / Getty Images Plus

www.degruyter.com

Preface

The idea for this book came up during our continuous searching of potential secondary sources of metals, as well as the methods of their recycling. Despite many works on this subject, there is a lack of compact guide on this topic. The goal was to provide an overview of the main areas of the field of sources of metals in industrial wastes, the composition of wastes, and innovative methods for recovery of critical and valuable metals. The choice of topic is not accidental. Nowadays, the recovery of raw materials from waste sources is not only an ecological idea, but it is the overarching goal of the zero waste economy. The book is addressed to materials scientists, engineers, chemists, and industrial chemists who work in this area. It could be also an inspiration for those working outside of the field.

The book aims to review the potential source and composition of metals in industrial wastes and to present the readers methods for their recovery. The discussed issues were divided in line with the main industrial sources of waste. Chapter 1 provides an overview of metals contents and their recovery from waste electrical and electronic equipment (WEEE), knows as e-waste. Chapter 2 shows huge interest in recovery of valuable metals in automotive industry, mainly from spent automotive catalysts, while Chapter 3 from rubbery industry. Mineral extraction industries, where at each mining step large volume of waste are generated and most of them represent potentially sources of crucial metals are highlighted in Chapter 4. Chapter 5 describes wastes from electroplating and other metal finishing operations which are considered as hazardous and may represent a loss of raw materials. The potential sources of various metals in chemical and petrochemical processes are discussed in Chapter 6. Special emphasis is put on the catalysts used in the industry. Chapter 7 focuses on the presentation of economic aspects of metals recovery from various sources.

We would like to express our gratitude to the contributing authors in making this project a success, and to Lena Stoll from DeGruyter Publisher, Germany for her assistance and encouragement in this venture. Moreover, we would like to gratefully acknowledge a financial support from the Polish National Centre of Science funds according to decision No. DEC-2017/27/B/ST8/01484 and from Polish Ministry of Science and Higher Education grant 03/32/DSPB/0810 as well as from the People Programme (Marie Curie Actions) of the Seventh Framework Programme of the European Union (FP7/2007-2013) under REA grant agreement no. 600388 (TECNIOspring programme), and from the Agency for Business Competitiveness of the Government of Catalonia, ACCIÓ.

Karolina Wieszczycka
Bartosz Tylkowski
Katarzyna Staszak

https://doi.org/10.1515/9783110547061-201

Contents

List of contributing authors

Maria Rosinda Costa Ismael
Universidade de Lisboa
Instituto Superior Técnico
Chemical Engineering Department
Av. Rovisco Pais 1
1049-001 Lisboa
Portugal
rosinda.ismael@tecnico.ulisboa.pt

Dobrawa Kwaśniewska
Poznań University of Economics and Business
Department of Technology and Instrumental Analysis
Faculty of Commodity Science
Aleja Niepodległości 10
61-875 Poznań
Poland
dobrawa.kwasniewska@ue.poznan.pl

Maria Teresa Angelino Reis
Universidade de Lisboa
Instituto Superior Técnico
Chemical Engineering Department
Av. Rovisco Pais 1
1049-001 Lisboa
Portugal
teresareis@tecnico.ulisboa.pt

Magdalena Regel-Rosocka
Poznań University of Technology
Institute of Chemical Technology and Engineering
Berdychowo 4
60-965 Poznań
Poland
magdalena.regel-rosocka@put.poznan.pl

Martyna Rzelewska
Poznań University of Technology
Institute of Chemical Technology and Engineering
Berdychowo 4
60-965 Poznań
Poland
martyna.rzelewska@put.poznan.pl

Katarzyna Staszak
Poznań University of Technology
Institute of Chemical Technology & Engineering
Berdychowo 4
60-965 Poznań
Poland
katarzyna.staszak@put.poznan.pl

Maciej Staszak
Poznań University of Technology
Institute of Chemical Technology and Engineering
Berdychowo 4
60-965 Poznań
Poland
maciej.staszak@put.poznan.pl

Daria Wieczorek
Poznań University of Economics and Business
Department of Technology and Instrumental Analysis
Faculty of Commodity Science
Aleja Niepodległości 10
61-875 Poznań
Poland
d.wieczorek@ue.poznan.pl

Karolina Wieszczycka
Poznań University of Technology
Institute of Chemical Technology & Engineering
Berdychowo 4
61-875 Poznań
Poland
karolina.wieszczycka@put.poznan.pl

https://doi.org/10.1515/9783110547061-202

Magdalena Regel-Rosocka

1 Electronic wastes

Abstract: *E-waste* amount is growing at about 4% annually, and has become the fastest growing waste stream in the industrialized world. Over 50 million tons of e-waste are produced globally each year, and some of them end up in landfills causing danger of toxic chemicals leakage over time. E-waste is also sent to developing countries where informal processing of waste electrical and electronic equipment (WEEE) causes serious health and pollution problems. A huge interest in recovery of valuable metals from WEEE is clearly visible in a great number of scientific, popular scientific publications or government and industrial reports.

Keywords: waste electrical and electronic equipment, metal content, metal recovery

1.1 Electronic wastes – definition, categories

In the era of technical and technological development, the production of electrical and electronic equipment is the fastest growing field in the world [1, 2]. At the end of their life cycle, these devices become waste, often hazardous to the environment, containing a mixture of various metals, plastics, rubber, glass and ceramics. *Waste electrical and electronic equipment* (*WEEE*) or *e-waste* means electrical or electronic equipment which is waste including all components, subassemblies and consumables which are part of the product at the time of discarding. As WEEE is the fastest growing waste stream in the European Union (EU) with some "9 million tonnes generated in 2005, and expected to grow to more than 12 million tonnes by 2020" [3], two EU directives have been put in place: the *WEEE Directive* (*Directive 2012/19/EU*) [4] and the Directive on the restriction of the use of certain hazardous substances in electrical and electronic equipment (RoHS recast Directive 2011/65/EU) [5]. According to the WEEE Directive, these wastes are categorized as shown in Table 1.1.

Though batteries and accumulators often are part of EEE and some authors of scientific articles consider them as e-waste, generally, they are excluded from WEEE and their collecting and recycling are regulated by separate legal acts, e.g. the EU Directive 2006/66/EC on batteries and accumulators and waste batteries and accumulators. Thus, recycling of batteries and metal recovery from them do not cover the subject of this chapter.

This article has previously been published in the journal *Physical Sciences Reviews*. Please cite as: Regel-Rosocka, M. Electronic wastes. *Physical Sciences Reviews* [Online] **2018,** *3* (5).
DOI: 10.1515/psr-2018-0020

https://doi.org/10.1515/9783110547061-001

Table 1.1: Categories of EEE covered by Directive 2012/19/EU [4].

No.	Category	Examples
1	Small household appliances	Toasters, vacuum cleaners, electric knives
2	Large household appliances	Refrigerators, washing machines, microwave ovens, air conditioner appliances
3	IT and telecommunications equipment	Personal computers, laptops – CPU, mouse, screen and keyboard included, printers, telephones (cordless, pay, cellular)
4	Consumer equipment and photovoltaic panels	Radio, television, video, hi-fi sets, photovoltaic panels
5	Lighting equipment	Straight and compact fluorescent lamps, sodium lamps and metal halide lamps
6	Electrical and electronic tools (with the exception of large-scale stationary industrial tools)	Drills; saws; equipment for turning, milling, sanding, grinding, sawing, cutting, shearing, drilling, making holes, punching, folding, bending or similar processing of wood; metal; and other materials
7	Toys, leisure and sports equipment	Electric trains or car racing sets, game consoles, video games, sports equipment with electric or electronic components, coin slot machines
8	Medical devices (with the exception of all implanted and infected products)	Radiotherapy, cardiology, dialysis, nuclear medicine equipment, analyzers, freezers, fertilization tests
9	Monitoring and control instruments	Smoke detectors, heating regulators, thermostats, measuring, weighing or adjusting appliances for household or as laboratory equipment
10	Automatic dispensers	Automatic dispensers for drinks, bottles, cans, food, money

1.2 Economic and ecological aspect of WEEE recycling

Continuous technological development causes constantly increasing demand for various metals, which are called strategic or critical metals. Generally, they are obtained from natural (primary) ores, which are minerals containing metals in the form of oxides, sulfides, carbonates and other substances such as sand. Recovery of metals from the ores is a costly and environmentally expensive method. It generates large amounts of harmful gases (mainly SO_2 and CO_2) and solid waste. Often the concentration of metal in the ore is very small (e.g. platinum group metals, PGM), which forces a huge amount of raw material to be processed, causing degradation of the areas where mining takes place. Comparing WEEE to natural ores, e-waste can be considered as a relatively concentrated source of a variety of metals, i.e. precious, rare earth elements (REEs), cobalt, etc. Thus, production of metals from scrap can be less energy and water-intensive than ore processing. Average values of energy

Table 1.2: Average energy consumption and water use for metal production from scrap (secondary sources) and ores (primary sources) [6].

Metals	Energy use in metal extraction, MJ/kg		Water use, m^3/t	
	Scrap	Ores	Scrap	Ores
Cr	6	36.5	12	72
Fe	6	60	14	325
Pb	9	38.5	40	62.5
Mg	10	197.5	2	8.5
Al	10	580	2	165.5
Zn	11	47.5	20	87.5
Cu	14	1,035	15	120
Sn	15	1,330	5	102.5
Ni	20	250	20	190
Co	80	1,120	65	1,020
Ag	130	2,380	30	130
Au	185	32,800	30	270,000
Zr	230	1,410	260	12,800
PGM	2,400	136,680	4,500	650,000
REE	3,000	6,350	750	1,540

consumption and water use for metal production from scrap and ores are presented in Table 1.2. Of course, these values will differ depending on ore grade, availability and processing technology, e.g. PGM production from high-grade ores consumes near 19,000 MJ/kg energy and 100,000 m^3 of water/ton, while from low-grade ores energy consumption increases even up to 255,000 MJ/kg, and water use to 1,200,000 m^3/t [6]. However, for all metals presented in Table 1.2, scrap processing gains advantage over traditional metal production from primary sources.

Some metals, e.g. REEs, are abundant only in very limited area which creates monopoly and increases prices. That is the reason for many governments to indicate so called "critical materials" that are economically important but are a subject of high supply risk (Figure 1.1) [6–8]. For instance, EU strongly supports development of different ways to guarantee continuous supplies of critical materials. In 2014 European Commission announced a list of *20 critical raw materials* (on the base of high supply risk and economic importance) which includes [6, 8] antimony (Sb), beryllium (Be), borates, chromium (Cr), cobalt (Co), coking coal, fluorspar, gallium (Ga), germanium (Ge), indium (In), magnesite, magnesium (Mg), natural graphite, niobium (Nb), PGMs, phosphate rock, REEs (heavy), REEs (light), silicon metal (Si) and tungsten (W). Most of these materials are metals, and they are considered critical because the European industry consumes 25–30% of the world's production, while only 3% of global metal production is placed in Europe, and many important metals are not produced in Europe at all [6, 9]. Thus, the following actions are undertaken to provide a stable and affordable supply of critical materials [6]:

1. sustainable supply of raw materials from European countries,
2. secure access to raw materials by ensuring undistorted world market conditions (e.g. diplomacy with resource-rich countries such as China and resource-dependent countries such as the US and Japan for cooperation),
3. more efficient recycling of raw materials and products containing critical materials through promoting collections, stopping illegal exports of end-of-life (EoL) products and promoting research on system optimization (e.g. improving product design, the Eco-Design Directive) and on tackling technical challenges.

Particularly the last action is an important aspect of the Europe's transition toward a *circular economy*, "*where the value of products, materials and resources is maintained in the economy for as long as possible, and the generation of waste minimised, and is an essential contribution to the EU's efforts to develop a sustainable, low carbon, resource efficient and competitive economy*" [10].

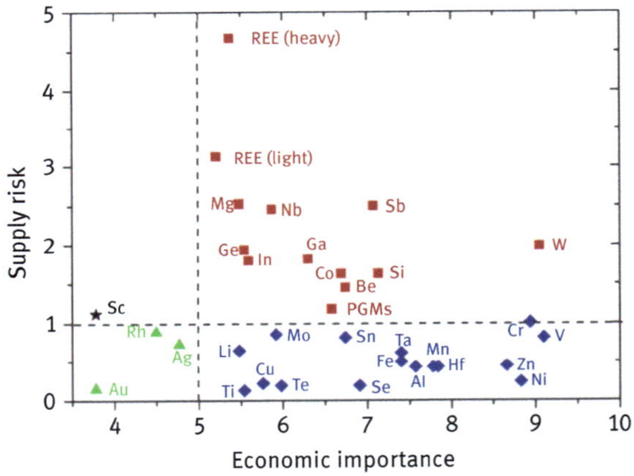

Figure 1.1: Critical metals for EU according to JCR economic importance and supply risk assessment in 2014 (according to data from [8]).

Thus, the idea of WEEE Directive, implemented in 2003, lies in introducing a closed circuit of metals in Europe, so that recovered materials could partially cover the increasing demand for metals. This would have the effect of limiting gas emissions as well as solving the problem of storing huge amounts of electrical and electronic waste, which are an abundant source of metals. The minimum targets of WEEE recovery and recycling are attributed to the categories listed in Table 1.1, and for example, until August 2018 waste small household appliances (category 1) and automatic dispensers (category 10) should be recovered (collected) at minimum 85% and recycled at 80% level [4]. However, recycling alone cannot cover all demand for raw materials because some losses during collection of material for recycle at EoL are inevitable, and also some limitations in quality of the recycled products, e.g. critical metals. Many applications require both high-quality grades and the absence of specific impurities, and it is difficult or even impossible to produce such pure metals or alloys from mixed sources that emerge from EoL waste. Thus, recovery of metals from WEEE has some limitations, and their application will always need the use of some primary sources. However, efficient recycling of metals is predicted to extend the lifetime of natural ore supply, e.g. if Nb is recycled at 50%, its lifetime of supply will be extended for 72 years, if at 70% the supply is extended for 360 years [9]. Au recycling at 70% would extend its supply for 70 years but now the overall efficiency of gold and palladium recycling from WEEE in Europe is estimated to be below 20%. For that reason, it is still a lot of work to do to improve the target recovery of metals, particularly critical metals, not only in Europe [11–13].

Moreover, recycling of WEEE protects the environment from harmful impact of hazardous metals that are present in these devices and should be recovered. Hazardous effects to the human beings and other life forms are listed in Table 1.3.

Table 1.3: Toxicity of metals typically applied as WEEE components.

Components	Toxicity	Type of WEEE	Ref.
Arsenic (As)	Human carcinogen, skin diseases	CRT screens, LCDs, doping material for silicons	[14–16]
Cadmium (Cd)	Human carcinogen, neurotoxic, damages kidneys, accumulates in bones leading to osteomalacia, osteoporosis, and pathologic fractures, inhalation of fumes causes fume fever	PCB, switches, semiconductor chips, CRT	[14, 15, 17, 18]
Chromium (Cr)	Human carcinogen (Cr(VI)), eye and skin irritation	ICT waste fines, floppy discs, CRT, plastic computer housing	[14–16, 19]
Copper (Cu)	Fumes and dust cause irritation of the upper respiratory tract, nausea, gastric pain, diarrhea	Wiring, ICT waste fines, portable audio scrap, PCB, keyboards	[14–16, 20]
Lead (Pb)	Human probable carcinogen, abdominal pain, nausea, vomiting, anemia (usually hemolytic) neurological/neurophysiological effects, affects nervous system, red blood cells and kidneys	ICT waste fines, PCB, CRT	[15–21]
Mercury (Hg)	Toxic to humans, affects nervous system	LCDs	[22–24]

(continued)

Table 1.3: (*continued*)

Components	Toxicity	Type of WEEE	Ref.
Nickel (Ni)	Human carcinogen, contact allergen, pulmonary irritation, asthma	CRT, ICT waste fines, PCB, printed wiring boards	[14, 15, 18, 20]
Zinc (Zn)	Nausea, vomiting, abdominal pain, diarrhea, respiratory irritation	CRT screens, ICT waste fines, PCB, keyboards	[14, 15, 20]

Note: CRT: cathode ray tube; ICT: information and communication technologies; PCB: printed circuit board; LCD: liquid crystal display.

It was reported that about 50–80% (in 2005) of the e-waste collected in developed countries is shipped to recycling centers in China, India, Pakistan, Vietnam, the Philippines, and much smaller amounts to African countries (e.g. Nigeria) because of low labor costs, and less stringent environmental regulations in these countries [1, 11, 13, 18, 21, 25]. Stopping illegal shipments and making sure that WEEE is collected and properly treated within recycling infrastructure is a challenge in e-waste management for the developed economies. Uncontrolled conditions of recycling expose workers and the environment to hazardous impact of WEEE components, particularly heavy metals or also nonmetallic compounds, such as polychlorinated biphenyls, dioxins/furans, polycyclic aromatic hydrocarbons or flame retardants [1, 17, 19]. Recycling of e-waste is a relatively recent source of exposure to lead which dust or fumes are inhaled by workers. There are many reports on exposure of people working at disassembling and recycling of WEEE to hazardous substances through inhalation, dust ingestion, dermal exposure, and dietary intake and health consequences [1, 17]. A typical example of the hazards of e-waste recycling exported from developed countries is the Guiyu town in China, which claims to be the largest e-waste site in the world [1, 25, 26]. As it has been shown in this paragraph, the idea of recovery of metals from secondary sources is of great importance not only economic but also environmental.

1.3 Metal content in WEEE

Only smartphones' sale is increasing year by year with global revenues projected to reach 768 billion USD in 2017, up from USD 544 billion in 2015 [27]. It means that also e-waste amount will grow exponentially in the future. However, no one asks what exactly is in a smartphone. A typical iPhone is estimated to contain around 0.034 g of Au, 0.34 g of Ag, 0.015 g of Pd and less than 0.001 g of Pt. It is also composed of 25 g of Al and around 15 g of Cu [26]. It was calculated that 1 ton of iPhones would deliver 300 times more Au than a ton of gold ore and 6.5 times more Ag than a ton of silver ore

[26]. Smartphones are also a valuable source of REE (yttrium, lanthanum, terbium, neodymium, gadolinium and praseodymium), scarce in Europe and difficult to mine in other countries. As 2 billion smartphone users change phones every 11 months, it means that phone users store "treasure" of various metals because 1 million mobile phones contain 16 tons of Cu, 350 kg of Ag, 34 kg of Au and 15 kg of Pd [26].

Every year about 300 tons of Au are used for production of electronic devices around the world (293 tons in 2008 and 246 tons in 2009 – this decline is explained by the economic crisis), which accounts for 12% of annual gold production. Thus, electronic equipment produced each year is a potential source of significant quantities of this valuable metal [28, 29], and is regarded even as "a high-grade ore" [21].

The metals presented in Table 1.4 account for a significant part of the value of the equipment produced, and therefore their recovery becomes a viable undertaking, both in terms of environmental and financial considerations. In the case of PC-boards or mobile phones, over 80% of their value comes from the presence of precious metals such as *Ag, Au, Pd.*

Table 1.4: Contribution of metals in value of typical electronic components [29, 30].

Components	Fe, %	Al, %	Cu, %	Ag, %	Au, %	Pd, %
TV-PCB	4	14	35	7	33	7
PC-PCB	0	1	13	5	69	12
Mobile phone	0	0	6	11	71	11
Portable audio	3	1	73	4	16	3
DVD-player	15	3	30	5	42	5
Calculator	1	4	10	6	76	3

As it was said before, WEEE is a mixture of various metals and alloys, mainly steel, copper, aluminum, zinc and lead, precious metals and nonmetallic components (glass, ceramics, rubber, wood, plastics). It can also change rapidly within one category over time. The content of metals in e-waste varies depending on the type of equipment and its age. Old devices contain considerable amount of hazardous elements, such as lead or mercury. WEEE composition changes with technological development and legislative regulations. For example, as a consequence of replacement of cathode ray tube (CRT) monitors with liquid crystal displays (*LCDs*), concentration of Pb in e-waste will be reduced because each CRT tube contained 0.5–5 kg of Pb [17, 19, 31]. However, as the majority of LCD screens is illuminated by mercury-containing backlights, waste LCDs contain toxic Hg [22–24], and other metals such as In, Sn, Al and Zn [22, 32–39]. About 95% of laptops and notebooks use cold cathode fluorescent discharge lamps (CCFL) with a low vapor pressure that contain mercury as backlight units [23]. It is important to emphasize that nowadays

CCFLs are replaced by light-emitting diode (LED) backlights which are safer due to Hg absence [24]. For LCD televisions, the LED penetration rate exceeded 50% in the last quarter of 2011 [24]. However, in e-waste LCDs with CCFL will be still present for some years. As in the future, new products which we cannot even think of today, will be on sale, e-waste recycling is continuously challenging responsibility.

After mechanical processing and separation of WEEE components, iron usually dominates (from 54% in TV sets to 85% in PCs) then copper and aluminum (from a few to 10%) and precious metals (usually less than 1%) in the metallic fraction [30]. Also, the availability of the metals differs depending on the part of a device, e.g. printed circuit boards (PCBs) or mobile phones are considered to be rather mono-streams and relatively "clean" WEEE, which processing is not as complicated as of complex electronic equipment or mixtures of all kind of EoL products or waste from different streams [20, 29]. Table 1.5 shows content of metals in the six exemplary components or electronic devices.

Table 1.5: Weight contribution of components (wt% or ppm) in exemplary electronic devices [29].

Devices	Fe, %	Al, %	Cu, %	Plastics, %	Ag, ppm	Au, ppm	Pd, ppm
Monitor board	30	15	12	28	280	20	10
PC-board	7	5	18	23	900	200	80
Mobile phone	7	3	13	43	3,000	320	120
Portable audio	23	1	21	47	150	12	4
DVD-player	62	2	5	24	115	15	4
Calculator	4	5	3	61	260	50	5

The most valuable metals, and thus, the most cost effective in processing, are components of computer hardware and mobile phones, for example PCBs, all kinds of plugs and switches, memory slots, graphics card slots, jumpers or transistors [30]. Precious metals and copper are embedded mainly in PCBs a key component of WEEE. PCB weight contribution in some electronics is as follows: 7% in CRT TVs (old type), 18.8% in computers, 21% in laptops, 21.3% in mobile phones, 11.9% in LCDs [19, 40]. Metals in PCB are entrapped in nonconductive plastic, glass fiber reinforced plastic or ceramic (mainly silica or alumina) substrate which must be shredded prior metal recovery to make metals accessible. A typical PCB is comprised of 30% plastics, 30% refractory oxides (ceramic compounds) and 40% metals [41]. Generally, the following examples of e-waste categories according to value of the material are proposed [42]:

1. *low value*: TV-boards, monitor boards, printer boards, cordless phones, calculators, shredded bulk material after Al-/Fe-separation,
2. *medium value*: PC-boards, laptop-and handheld-computer circuit boards,
3. *high value*: PCBs from mainframes, mobile phones, integrated circuit (IC) chips, multilayer ceramic capacitors.

Metal contents in PCBs from different devices are presented in Table 1.6. Umicore – one of the companies that recycle e-waste – proposes to categorize PCBs into different qualities, depending on their gold content [42]. It is schematically presented in Figure 1.2.

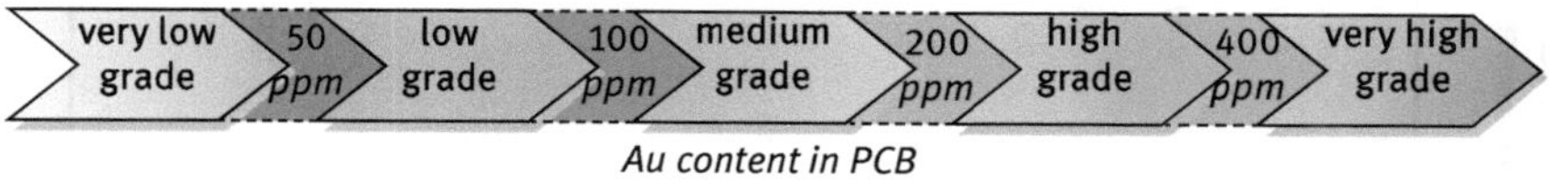

Figure 1.2: Value categories of PCB waste depending on Au content [42].

As it is presented in Table 1.6, metal content depends on the type of device and differs also for various fractions separated from one type of WEEE. To some extent also method of metal analysis can determine the results of metal content. Most authors used inductively coupled plasma atomic emission spectrometry to measure metal contents after digestion of the milled WEEE in mixtures of mineral acids [43–50], also microwave plasma atomic emission spectroscopy was applied [51] and/or X-ray fluorescence [48].

In some reports, PCBs are considered as "bare" (dismantled), in some as printed circuit board assembly (PCBA), which is a PCB mounted with electronic components of a range of sizes and compositions [40, 52]. PCBA consists of various components such as capacitors, resistors, IC chips, thermistors and transformers mounted onto a nonconductive plastic, glass fiber reinforced plastic or ceramic substrate and linked with copper or aluminum wires. These components contain metals, among them *Cu* (wires) of the highest content, followed by *Sn, Fe, Cd* and *Pb*, used in soldering and lead frames, Ta used in capacitors, *Ga, In, Ti, Si, Ge, As, Sb, Se* and *Te* found in ICs, or precious metals *Au, Ag* and *Pd*, applied as contact materials or plating layers components of ICs [40, 52]. However, legislation restrictions constrain the electronics industry and force it to replace, for example Sn–Pb solders with Pb-free alternatives, such as Sn/Ag/Cu alloy [21].

Thus, Pb containing WEEE is limited to older devices, in some years lead content in e-waste should decrease. Nowadays, PCBs account for the majority of e-scrap recycled both in industrial plants (smelters) and in small workshops (mainly in Asia or Africa) because they are relatively rich source of *Cu* and precious metals, particularly of gold.

Table 1.6: Metal content in PCBs in wt% unless stated otherwise.

Devices							Computers	Laptops					Various WEEE	TV				Mobile phones	Printer*
Metals																			
Cu	10–26.8	4.35	17.3	16.4	2.7	19.7	10.9	17.6	24.8	16.0	6–27	28.7	25.24	11.2	23.0	32.7	35.5	26	32.5
Al	1.33–4.78	1.90	0.2	1.9	–	82 ppm	1.7	2.0	2.5	5.0	2.0–7.2	1.7	0.69	0.3	1.0	2.2	0.26	–	3.73
Pb	0.99–4.19	1.82	1.25	0.09	–	684 ppm	1.5	0.22	0.57	2.0	1–4.2	1.3	2.22	–	0.12	0.6	1.87	–	0.00
Zn	0.16–2.17	0.02	0.17	0.029	–	554 ppm	1.1	0.45	–	1.0	0.2–2.2	–	2.02	0.15	0.3	–	5.92	–	0.64
Ni	0.28–2.35	0.06	0.72	0.026	7.4	61 ppm	2.5	0.57	–	1.0	0.3–5.4	–	0.93	0.02	1.1	1.9	3.41	13.49	0.34
Fe	1.22–8.0	2.80	0.21	2.33	8.6	194 ppm	7.7	3.8	0.18	5.0	1.2–8.0	0.6	0.98	0.0043	3.8	1.4	12.5	–	1.4
Sn	1.0–5.28	4.51	–	–	–	–	–	–	–	3.0	1–5.6	3.8	3.17	–	–	–	3.39	3.08	0.96
Sb	0.06–0.4	1.23	–	–	–	–	3.9	–	–	–	0.1–0.4	–	–	–	–	<0.05	–	–	0.11
Bi	–	0.02	–	–	–	–	–	–	–	–	–	–	–	–	–	–	–	–	–
Ca	–	1.97	–	–	–	–	–	–	7.1	–	–	–	–	0.4	–	3.4	0.09	–	1.13
Ba	–	–	0.32	–	–	–	–	–	–	–	–	–	–	–	–	–	0.16	–	0.27
Cr	–	–	0.013	200 ppm	–	–	–	0.01	–	–	–	–	–	91 ppm	0.2	<0.01	–	–	–
Co	–	–	–	–	–	–	–	–	–	–	1–4,000 ppm	–	–	0.04	–	–	–	–	–
As	–	–	–	–	–	–	–	–	–	–	–	–	–	2.2 ppm	–	–	–	–	–
Cd	–	–	–	–	–	–	–	–	–	–	–	–	–	157 ppm	–	–	–	–	–
Au	80–1,000 ppm	–	–	2 ppm	0.0105	–	0.005	0.0029	–	0.025	250–2,050 ppm	80 ppm	890 ppm	0.14 ppm	0.032	–	–	–	0.004
Ag	110–3,301 ppm	–	–	–	<0.0001	–	0.008	–	–	0.100	110–4,500 ppm	70 ppm	1,907 ppm	48 ppm	–	–	0.21	0.26	0.31
Pt	4.6–30 ppm	–	–	–	–	–	–	–	–	–	5–30 ppm	–	17 ppm	–	–	–	–	0.053	–

(continued)

Table 1.6: (*continued*)

Devices / Metals							Computers	Laptops					Various WEEE	TV				Mobile phones	Printer*
Pd	10–294 ppm	–	–	–	–	64 ppm	0.002	–	–	0.010	50–4,000 ppm	30 ppm	47 ppm	–	–	–	–	0.027	–
REE (Pr, Nd, Dy)	–	–	–	–	–	–	–	–	–	–	–	–	–	–	–	–	–	0.1905	–
Nonmetallic:		80	80	89	~80	~80	70	75	65	66.8		64		~88	70	58	36	58	58
Ceramic	30										30		22.14						
Plastics	30										30		32.41						
Ref.	[40, 52]	[44]	[45]	[47]	[51]	[50]	[21]	[47]	[46]	[43]	[19]	[53]	[54]	[55]	[47]	[48]	[49]	[56]	[49]

Note: *Magnetic fraction.

As the variety of EEE is really wide, in this chapter only some exemplary e-wastes will be considered as a secondary source of metals. PCBs (mentioned in Table 1.6) are present in all waste devices but also other parts of EEE contain valuable metals. For example, nowadays, indium recovery from LCDs becomes more important because there are no natural ores of this metal (it is associated with Pb and Zn ores) and its concentration in primary ores ranges from 10 to 20 ppm [20]. Over 70% of In consumption (as indium-tin-oxide – *ITO*) is applied for transparent conductive coatings in flat panel displays such as LCD-TVs, smartphones, laptops and tablets [38, 42]; this metal is also applied in high efficiency photovoltaic devices, infrared detectors and high speed transistors [20]. Content of various metals in LCD panels is presented in Table 1.7.

LCD display is built from several layers of various materials that contain 25 or more components, including a transparent conductive coating made up of a thin

Table 1.7: Metal content in LCD panels.

Unit **Metals**	**ppm**	**ppm**	**ppm**	**%**	**ppm**	**ppm**	**%**	**ppm**
In	53 ± 6	130 ± 30	110 ± 20	7.95	200 ± 50	530	0.06	30 ± 5
Sn	260 ± 30	–	–	1.01	20 ± 3	–	0.01	72 ± 4
Ga	–	<10	<10	–	–	–	–	–
Al	3.3% ± 0.7	5% ± 2	–	–	65–825	–	14.4	1,030 ± 50
Fe	3,100 ± 600	600 ± 400	–	0.22	70–500	–	0.34	2,200 ± 800
Ca	–	–	–	0.77	–	–	9.6	–
Pb	–	–	–	0.03	–	–	–	–
Cr	–	–	–	0.46	4 ± 0.4	–	0.01	–
Zn	–	–	–	0.10	20–270	–	0.18	370 ± 85
Ag	–	–	–	–	4 ± 3	–	–	–
Co	–	–	–	–	4 ± 0.4	–	–	2 ± 0.4
Cu	–	–	–	–	10–250	–	0.02	370 ± 22
Mg	–	–	–	–	6 ± 1	–	–	–
Mn	–	–	–	–	2 ± 0.5	–	–	–
Ni	–	–	–	–	17 ± 4	–	–	41 ± 9
As	–	–	–	–	–	346	0.9	–
Sb	–	–	–	–	–	24	–	–
Sr	–	–	–	–	–	–	3.43	–
Ba	–	–	–	–	–	–	0.85	–
Ti	–	–	–	–	–	–	0.13	–
K	–	–	–	–	–	–	0.34	–
Eu	–	–	–	–	–	–	–	31 ± 20
Mo	–	–	–	–	–	–	–	11 ± 1
Nd	–	–	–	–	–	–	–	19 ± 4
Y	–	–	–	–	–	–	–	319 ± 35
Ref.	[36, 37]	[36, 37]	[36, 37]	[33]	[20]	[57]	[22]	[58]

layer of indium and tin oxide and their alloys [34, 59, 60]. A coupled indium(III) oxide and tin(IV) oxide (ITO) used in LCDs contain 90% of In_2O_3 and 10% of SnO_2. The content of indium in the ITO glass is approximately 102 mg/kg, however, in some devices it reaches even up to 530 ppm. It means that indium content in the waste LCDs is much higher than its grade in the primary ores. Other layers include glass, foil and liquid crystals. Liquid crystals are a mixture of 10–25 organic compounds that only differ in alkyl or alkoxy side chains by varying the number of carbon atoms.

1.4 Mechanical processing of WEEE

Generally, processing of WEEE covers some stages of mechanical operations such as dismantling, classification, size reduction (crushing/shredding), and finally separation for metals and nonmetals (magnetic, gravity, electrostatic separation, flotation). When separated, *upgraded e-waste*, known as *e-scrap*, is treated by pyrometallurgy, hydrometallurgy or biometallurgy to recover metals that can be valuable product of the WEEE recycling. Schematic representation of the main stages of processing of WEEE is shown in Figure 1.3.

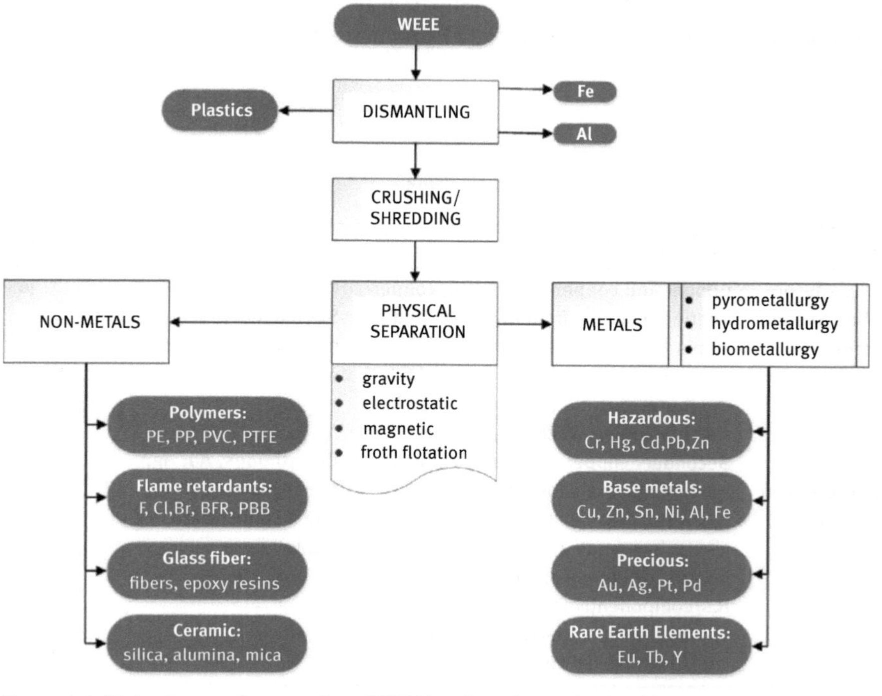

Figure 1.3: Main stages of processing of WEEE and products obtained [19, 63, 64].

Dismantling aims at recovery of useful elements such as plastics or steel elements, and preliminary separation of different materials and removal of hazardous components [20, 61, 62].

1.5 Recovery of metals from WEEE

Further processing of metallic fraction applies chemical treatment by pyrometallurgy, hydrometallurgy or biometallurgy. Sometimes term "hydrochemistry" is used instead of "hydrometallurgy" because the former covers a broader range of operations and reactions [20]. In practice, these different techniques are combined to design the best technology for efficient recovery of metals from e-waste.

1.5.1 Pyrometallurgy

Pyrometallurgy means thermal treatment of materials (minerals, ores, concentrates or wastes) and includes burning, smelting in a plasma arc furnace or blast furnace, drossing, sintering, melting and reactions in a gas phase at high temperatures (>1,200°C). Recovery of nonferrous and precious metals from WEEE with pyrometallurgy has been processed for over 20 years [62]. As pyrometallurgical processes are high energy consuming, and thus, generate high-operational costs, they demand continuous supply of feed materials of stable composition with high precious metal content. Some industrial pyrometallurgical processes for WEEE treatment are presented in Table 1.8.

Table 1.8: Pyrometallurgical plants for WEEE processing.

Process	Feed materials	Products	Ref.
Aurubis, Kayser Recycling System (Germany)	Copper and copper alloy scrap, copper-bearing residues from foundries and semis fabricators, shredder materials, galvanic slimes, slags, ashes and filter dust, precious metal-bearing raw materials, including *WEEEscrap*	Cu cathodes, $NiSO_4$, SnPb alloy, precious metals, Zn-bearing oxide	[65]
Noranda, Horne Smelter (Glencore Xstrata, Canada)	Secondary copper and precious metal-bearing materials (ashes, router dust, catalysts, muds, sludges), *electronic waste*: low-grade and high-grade circuit boards, cell phones, pins/punchings/ICs/components, lead frames/trims/bare boards, sweeps, insulated consumer wire/degaussing wire, copper yokes	Cu cathodes, Au, Ag, Se, Te, $CuSO_4$, $NiSO_4$ and a concentrate of PGMs	[62, 66]

(*continued*)

Table 1.8: (*continued*)

Process	Feed materials	Products	Ref.
Boliden, Rönnskär smelter (Sweden)	Copper and lead concentrates, *electronic waste*: circuit boards from computers and mobile phones	Cu, Pb, Au and Ag, Se, copper telluride, PGM concentrates, crude nickel sulfate, nickel matte, Zn clinker, Sb, In or Cd	[2, 62, 67]
Umicore (Belgium)	Residues from the various manufacturing (e.g. anode slimes), *electronic waste*: low, medium and high value	Precious metals: Ag, Au, Pt, Pd, Rh, Ir, Ru; base metals: Pb, Cu; special metals: In, Te, Se; chemicals: $NaSbO_3$	[42, 62, 64, 68]

1.5.1.1 Aurubis process (Kayser Recycling System, KRS)

The key step of pyrometallurgical Aurubis process takes place in the KRS equipped with a submerged lance furnace (13 m high). A submerged lance is immersed into the furnace and supplies with heating oil, oxygen and air necessary for the combustion (reduction) process. Metals, such as Cu, Ni, Sn, Pb and precious metals (mainly Au and Ag), contained in the raw materials are enriched in the furnace, and Cu content amounts to 80%. It is further enriched in a top blown rotary converter up to 95% and separated from Sn and Pb. Sn-Pb alloy is produced from a slag, while Zn is enriched in KRS oxide. Recycling materials rich in precious metals and intermediate smelter products originating both from Aurubis' production plants and from external metal smelters or precious metal separating plants are further processed to enrich and recover precious metals and separate various by-metals. Pyrometallurgical separation of Pb and Cu and enrichment of precious metals is carried out in an electric furnace in various smelting campaigns. Such by-products, as Pb, Bi, Sb and Te, are separated in the lead refinery and sold as lead bullion, Pb-Bi alloy, Sb concentrates and Te concentrates. The precious metals accumulate in "rich lead" and comprise even 70% of it [65]. Schematic representation of Aurubis process in presented in Figure 1.4.

1.5.1.2 Noranda process (Horne Smelter)

Noranda process (now owned by Glencore Xstrata) has been operating in Canada for over 30 years [62, 66]. The Horne Smelter, located in Rouyn-Noranda, processes electronic scrap containing copper and precious metals, being now the world's largest e-scrap processor with the annual capacity of 840,000 tons of copper and precious metal-bearing materials. The smelter uses both copper concentrates and precious metal-bearing recyclable materials to produce a 99.1% copper anode, which is shipped to the Canadian Copper Refinery in Montreal to be converted into final

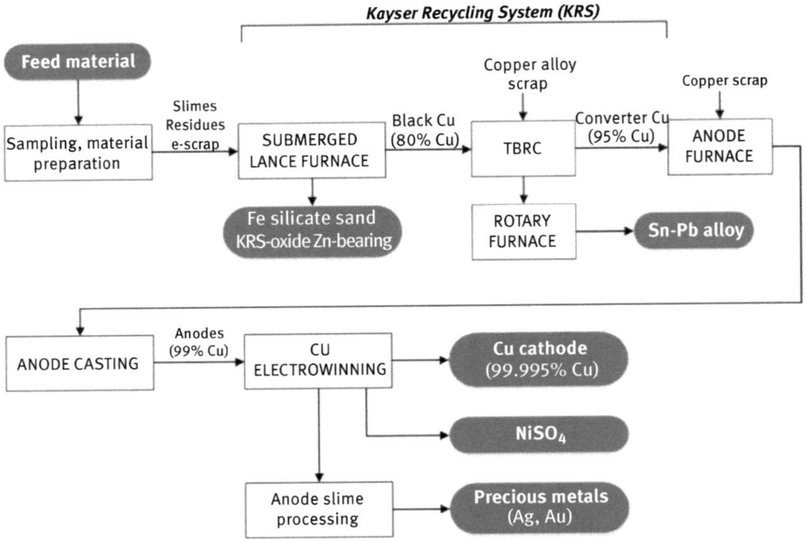

Figure 1.4: The Aurubis process of Cu and precious metals recovery from secondary materials [65].

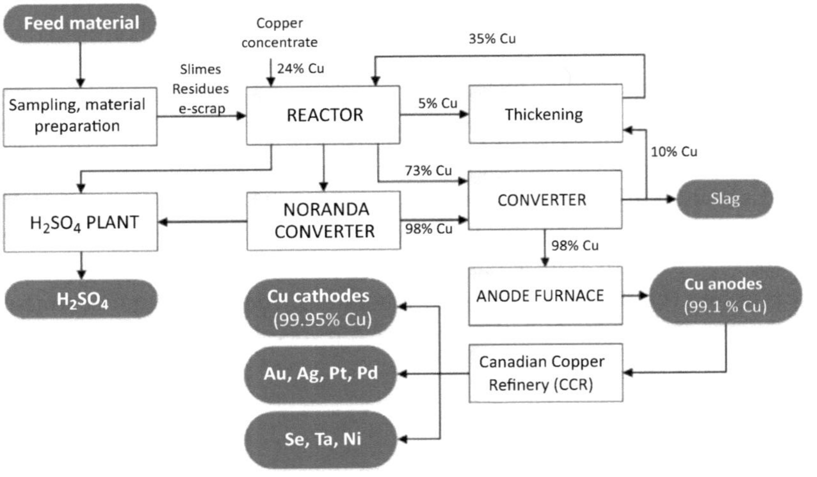

Figure 1.5: Scheme of the Noranda process [62].

product – 99.95% copper cathodes. Approximately, 15% of the feed material for smelting operation is from recyclable materials [66]. The process scheme is shown in Figure 1.5.

The raw material feeding the reactor is immersed in a molten metal bath at 1,250°C, which is mixed by an oxygen-enriched air mixture (over 39% O_2). The aim of this step is to oxidize iron, lead and zinc to oxides that will be removed along with silica slag. After cooling, the slag is ground to recover metal residues therefrom. The copper matte containing precious metals is directed to a converter where copper is enriched, and then it is further refined in an anode furnace and cast into anodes with purity of 99.1%. The remaining 0.9% is comprised of precious metals: gold, silver, platinum, palladium and other metals such as selenium, tallium and nickel. In the presented pyrometallurgical process, plastics and other combustible materials from WEEE are used as a "fuel" to reduce energy costs.

1.5.1.3 Boliden process (Rönnskär Smelter)

Boliden's Rönnskär smelter in northern Sweden is one of the world's largest recycler of metal from electronic material with annual capacity of 120,000 tons, including PCBs from computers and mobile phones (sourced primarily from Europe) [2]. Generally, the process is similar to that presented in Figure 1.5 because the smelted e-scrap is integrated with the smelter's main copper flow for further refining. However, the shredded electronic material, before being sent to the main copper flow, is processed in a Kaldo furnace, specially adapted by Boliden for e-scrap smelting [2, 62, 67]. Secondary copper raw materials that are high grade can be fed directly to the smelter, while these low grade (e-scrap) are upgraded in a Kaldo furnace. The furnace consists of a leaning cylinder that rotates during the smelting process to ensure an even heat distribution. Black copper – product of Kaldo furnace – is directed to main smelter flow (to converter) for further refining to produce copper and precious metals. Fifty percent of Ag and 30% of Cu produced in this process are from recycled materials. Anode slime is processed to recover Ag, Au, Pd, Ni, Se and Zn, and slag from Kaldo furnace is directed to Pb refinery, where also some amount of Sb, In or Cd can be obtained [2, 62].

Similarly to the Noranda process, here, also plastics and other combustible materials which melt during e-scrap smelting are a source of energy for the smelting process, and, additionally, generate steam that is converted into electricity or district heating [2].

1.5.1.4 Umicore process (IsaSmelt)

Umicore (Hoboken, Belgium) carries out a unique process of metals' recovery based on copper, lead and nickel as collector metals, focusing particularly on recovery of precious metals. Various industrial wastes and by-products from the nonferrous metal industry, such as slag or anode slimes, precious metals containing materials, spent industrial and automotive catalysts, and WEEE are used in the process as the feed material. E-waste represents 10% of the 250,000 tons of the waste processed in Hoboken/year. The Hoboken plant is one of the world's largest precious metals recycling facilities with a capacity of over 50 tons of PGMs (25 tons of each Pt and

Pd, 5 tons of Rh), over 100 tons of Au and 2,400 tons of Ag [42, 64]. Apart from the main products such as precious metals and base metals, Umicore Precious Metals Refining has also an annual capacity of 50 tons of indium (from LCD feed).

As this process uses also spent automotive catalysts as a feed material, it will be presented in detail in *Chapter Wastes generated by automotive industry – spent automotive catalysts.*

1.5.1.5 Summary of pyrometallurgical processes for WEEE recycling

Pyrometallurgy is a traditional method for precious metals recovery from WEEE for many years, but this method has some limitations. One of them is emission into the environment of gases such as SO_2, NO_x or CO_2. Only partial recovery of metals is possible on the pyrometallurgy path, and electrochemical (electrolysis) or hydrometallurgical (leaching) methods are required to obtain high-grade metals. Aluminum and iron cannot be recovered in metallic form in integrated furnaces, and these metals pass to slags. On the one hand, glass and ceramic e-waste components increase also the amount of slag in the blast furnaces, thereby increasing the loss of precious and non-precious metals. On the other hand, plastics or organic residues contained in the feed materials decrease operational cost partly replacing in the furnaces traditional reducing agent and energy source – coke. However, furnaces for e-scrap treatment must be specially designed as integrated metal smelters (special installations of off-gas treatment) to protect the environment from emission of dioxins. These hazardous compounds are released as a consequence of presence of halogenated flame retardants in plastics (e.g. PCB).

Generally, all the pyrometallurgical processes presented here consist of three stages of metal recovery. The first is smelting, the next conversion process, and the final – refining to produce pure metals. Aurubis, Boliden and Noranda processes are focused on copper recovery, while Umicore seems to be the most flexible regarding to feed material and aims at the recovery of precious and special metals.

1.5.2 Hydrometallurgy

Hydrometallurgy, called also “wet techniques”, means application of solutions of various chemical compounds to dissolve, extract, recover elements from different materials. Hydrometallurgy seems to be a perspective for the future e-waste treatment as it is easier to control and operate, more predictable, applies simpler and cheaper equipment and operates in lower temperatures than pyrometallurgy. However, such drawbacks as generation of harmful solutions or sludges that must be managed or recycled make hydrometallurgy of e-waste still not as advanced as pyrometallurgy. Now, on the industrial scale, hydrometallurgy is used as auxiliary operations (leaching or electrolysis) to purify metals smelted in pyrometallurgical operations. Therefore, the examples of hydrometallurgical treatment of WEEE presented here are rather small-scale operations or research proposals that can be developed on bigger scale in the future. However, 4 years ago an interesting prototype of small portable plant for hydrometallurgical recovery of metals from WEEE was investigated and proposed [54, 69]. Facing domination of pyrometallurgy in the e-waste recycling on industrial scale, such small portable plant is an alternative to pyrometallurgical process, and can be readily applied in small plants (small- and medium-sized enterprises – SME) without their own recycling plant.

Hydrometallurgical treatment of e-waste consists of many steps of leaching with acids or bases to dissolve metallic components. To increase the yield of metal leaching, it is important to crush and ground the material as much as possible. Leaching means solid–liquid extraction of metals from e-scrap to solutions of leaching agents. The pregnant leaching/leach solutions are then purified and the required metals are separated from them with liquid–liquid extraction, adsorption, ion exchange or cementation.

1.5.2.1 Gold/precious metals recovery

Hydrometallurgical methods are frequently studied to recover gold from e-waste because of high price of this metal and relatively high content in electronic components (Table 1.6). Au is almost always present as thin coating over copper, iron or silver substrates in PCBs [70], and the amount of gold in PCB is an order of magnitude higher than in the natural ores (average 10 mg Au/kg ore) [71]. Some examples of hydrometallurgical proposals for recovery of precious metals are given in Table 1.9.

Table 1.9: Hydrometallurgical proposals for precious metals recovery from WEEE (recoveries given in brackets).

Components	Operations	Products	Ref.
PCB	– Ag *leaching* with aqua regia – Formation of red *precipitate* of $Pd(NH_4)_2Cl_6$. – Au *liquid–liquid extraction* with toluene – Formation of Au nanoparticles (φ 20 nm) with dodecanethiol and sodium borohydride solutions	Ag (98%) Pd (93%) Au (97%)	[43]
PCB	– *Leaching* in sulfuric media at 75°C – *Leaching* in an oxidant media at 75°C	– Al (90%), Zn (40%), Sn (8.6%) – Cu (98.5%), Zn (60%), Al (10%)	[49]
PCB	– *Enrichment* of Pd during Cu recovery – Pd dissolution in a special solution of $CuSO_4$ and NaCl – Pd recovered during the *dissolution-extraction-stripping* process – Regeneration and recycling of extractant (diisoamyl sulfide, S201)	Pd (96.9%)	[50]

(continued)

Table 1.9: *(continued)*

Components	Operations	Products	Ref.
PCB (portable plant)	– Cu *leaching* with H_2SO_4 at 70°C – Au leaching with thiourea in the presence of ($Fe_2(SO_4)_3$) as an oxidant in H_2SO_4 at 40°C – *Electrolysis* to win metallic copper from leaching solution	Cu, Au, Ag	[54]
PCB (portable plant)	– First, *leaching* with H_2SO_4 and H_2O_2 – Leaching with thiourea to dissolve Au and Ag – Selective *precipitation* to recover metals from leach liquor	Cu (90%) Au (90%) Ag (90%)	[69]
PCB	– HNO_3 *leaching* at 70°C to dissolve base metals – Mechanical crushing of PC chips – HNO_3 leaching of crushed e-scrap – Au *leaching* with aqua regia from solid residue – Au *precipitation* with $FeSO_4$	Au	[72]
PCB	– Base metal *leaching* with sodium citrate solution – *Direct electrowinning* of base metals from citrate solution – *Recycling* of the barren citric solution to the leaching stage	Cu (71%) Fe (83%) Pb (94%)	[73]
PCB	– Cu *leaching* with H_2SO_4, 17% (vol.) H_2O_2 – Cu *cementation* with Zn powder – Ag, Au leaching with thiourea and Fe^{3+} ions (oxidizing agent) in diluted H_2SO_4 – *Neutralization* of thiourea solution with NaOH – Ag, Au *cementation* with Zn powder	In one step: Au (90%), Ag (75%) In four steps: Cu (>95%)	[74]
PCB	– Two consecutive H_2SO_4 *leaching* steps in the presence of H_2O_2 as an oxidizing agents to dissolve Cu – Au *leaching* from solid residue with acidic thiourea in the presence of Fe^{3+} – Pd and remained Au leaching from the solid residue with NaClO-HCl-H_2O_2 – Pd and Au *precipitation* by sodium borohydride	Cu (99%), Au (86%), Ag (71%)	[75]

(*continued*)

Table 1.9: (*continued*)

Components	Operations	Products	Ref.
PCBs from mobile phones	– Au *leaching* with a commercial reagent (cyanide) – Ag leaching with HNO_3 – Au and Ag leaching with alternative reagent with sodium thiosulfate $Na_2S_2O_3$ is more efficient when $CuSO_4$ was added	Au (60%) Ag (100%) Au (15% – in alternative agent) Ag (3% – in alternative agent)	[76]
E-waste	– *Etching* of Ag with $FeCl_3$ or $CuCl_2$ solutions – *Regeneration* of spent etching solutions by oxidizing with atmospheric oxygen	Detached Au particles, ceramic and plastic supports	[70]
E-scrap of at least 80% Au content	– *Chlorination* of e-scrap at 300–700°C – HCl *washing* to dissolve impurities-metal chlorides – NH_4OH and HNO_3 washing to dissolve AgCl	99.9% purity Au	[77]
Processor scrap	– Without shredding or grinding stages to reduce secondary dust pollution, reaction time and contaminant by-products – Environmentally friendly Au recovery from scrap by *oxidizing* of the scrap substrate with ammonium $(NH_4)_2S_2O_8$ or potassium persulfate $K_2S_2O_8$	Au (>98%)	[51, 78]
Coatings of Cu–based e-waste scrap	– Selective *extraction* of Ag in the form of diamminesilver(I) complex using base-activated persulfate – ammonia system – Easy *reduction* of the complex to metallic silver	Ag (50% recovery for silver plates and up to 100% for silver powder with $K_2S_2O_8$)	[79]
Printed wiring boards and integrated circuits (ATMI process)	– Selective *chemical desoldering* in an acidic solution at 30°C and 40°C – Collectin salts or oxides of solders (Pb/Sn and Sn/Ag) – ICs grinding and nontoxic, environmentally benign *leaching* at 30°C to remove all precious metals – *Chemical reduction or electrowinning* to collect gold, silver, palladium	Sn, Pb and Ag (>95%) Precious metals, mainly Au (>99% recovery, >99.9% purity)	[80]

Solutions of such complexing agents as cyanides, thiourea, halides and some strong acids (e.g. aqua regia – 3:1 mixture of HCl and HNO_3) are used as typical leaching agents not only for Au dissolution but also for other metals present in the e-scrap. Aqua regia is a very effective leachant because it makes possible to dissolve Ag, Pd, and Au at the same time. For over 100 years, cyanide leaching has been playing an important role in industrial Au and Ag recovery, particularly from natural ores, due to high gold recoveries, robustness and relatively low costs [81–83]. Gold leaching with cyanides relies on an electrochemical dissolution of Au in the alkaline cyanide solution and formation of the $Au(CN)_2^-$ complex according to the general reaction:

$$4Au + 8CN^- + O_2 + H_2O \rightarrow 4Au(CN)_2^- + 4OH^- \tag{1.1}$$

The cyanide leaching is not considered an eco-friendly process [81], hence, to replace cyanides, other complexing/leaching agents are proposed:
- thiourea – less toxic, faster rates of Au and Ag dissolving; disadvantage – carcinogenic and sensitive to Cu, Pb, Zn, As, Sb presence [78, 84],
- thiosulfate – nontoxic and economical compared to cyanide leaching; disadvantage – low Au recovery [81],
- thiocyanate – effective, forms stable and soluble complexes of aurothiocyanate and aurithiocyanate [84],
- mineral acids with different oxidants such as H_2O_2, O_2, Fe^{3+} or addition of NaCl.

Thus, recently some novel approaches to Au leaching have been proposed, e.g. non-cyanide leaching processes [69, 81, 82], use of more environmental friendly leaching agents, such as ammonium persulfate $(NH_4)_2S_2O_8$ or copper chloride to oxidize (etch) non-noble metals to detach solid gold and ceramic support [51, 70, 78].

Recently, a portable/mobile plant for hydrometallurgical metal recovery from WEEE was designed, and a prototype HydroWEEE was installed in a container lorry [54, 69]. It can treat four different e-wastes, i.e. fluorescent lamps (Y recovery), CRTs (Y, Zn), Li-ion batteries (Li, Co), PCBs (Au). Such a small mobile plant is addressed to SME without their own recycling plant. It was estimated that the portable process is economically feasible for the amount of gold exceeding the limit value of 500 ppm, and investment return is expected after up to 7 years of plant operation, assuming processing of 50–100 kg of WEEE [54]. A scheme of metal recovery in the portable plant is shown in Figure 1.6.

The process comprises several steps of leaching, filtration and cementation. Copper is extracted from the waste material with 2 M sulfuric acid as a leaching agent. Solid residue after Cu leaching step is leached by thiourea in the presence of ferric ion ($Fe_2(SO_4)_3$) as an oxidant in sulfuric acid solution to extract gold. Thiourea with Fe^{3+} ions (oxidizing agent) in diluted H_2SO_4 is often proposed as an efficient solution for Au or Ag leaching [54, 69, 76, 81, 85]. In the next step, Cu is electrowon

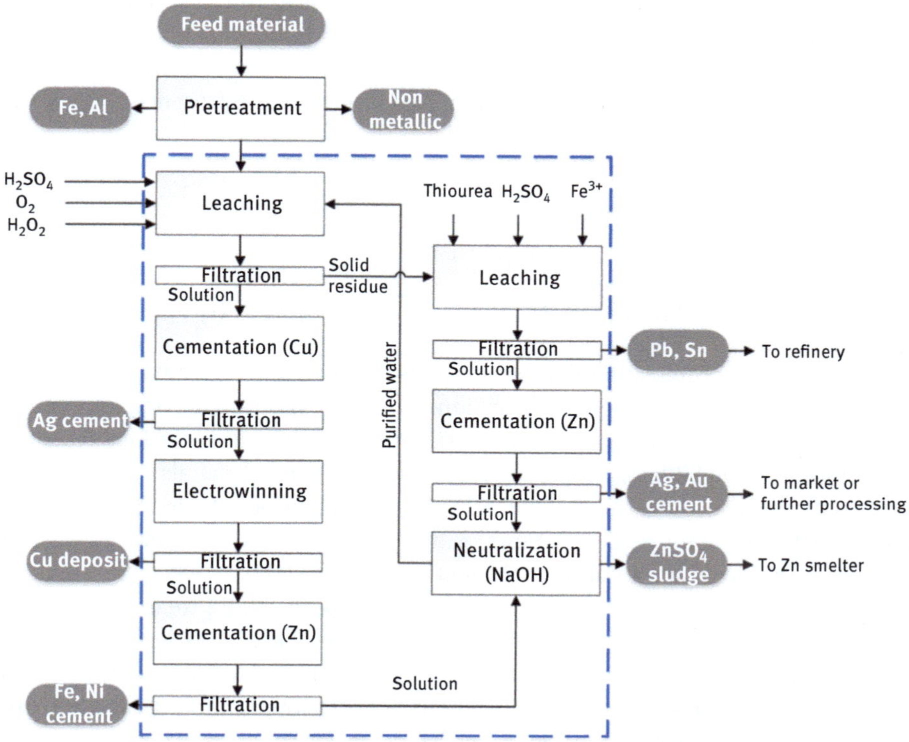

Figure 1.6: Scheme for hydrometallurgical recovery of base and precious metals from WEEE in a portable plant HydroWEEE [54].

from sulfate solution as a cathode Cu. Au and Ag produced as a cement powder can be sold to refinery or internally refined up to commercial purity metal powders.

Thiourea leaching in the presence of Fe^{3+} ions is also proposed for selective recovery of Cu, Ag, Au and Pd from waste PCBs [75]. High Cu recovery (99%) is obtained in two steps of sulfuric acid leaching in the presence of oxidizing agent (H_2O_2). In the next step, Au and Ag are leached from solid residue by thiourea and ferric ions, achieving 85% and about 70% recoveries, respectively. The remaining Au and Pd are dissolved in a NaClO-HCl-H_2O_2 leaching system. Finally, they are quantitatively precipitated as metals with sodium borohydride (reducing agent) according to the following reactions:

$$7PdCl_4^{2-} + 2BH_4 + 20OH^- \leftrightarrow 2BO_3^{3-} + 14H_2O + 7Pd + 28Cl^- \tag{1.2}$$

$$7AuCl_4^{2-} + 2BH_4 + 20OH^- \leftrightarrow 2BO_3^{3-} + 14H_2O + 7Au + 28Cl^- \tag{1.3}$$

To increase Au recovery not only in thiourea but also in thiosulfate leaching, cupric (Cu^{2+}) or ferric (Fe^{3+}) ions are added into leaching solutions as catalysts. However, thiosulfate can be decomposed by the presence of copper as well as other metals [83]. The electrochemical reactions of Au leaching in the presence of Cu are as follows:

$$Au + 2S_2O_3^{2-} \rightarrow Au(S_2O_3)_2^{3-} + e^- \tag{1.4}$$

$$Cu(NH_3)_4^{2+} + 3S_2O_3^{2-} + e^- \rightarrow 2Cu(S_2O_3)_3^{5-} + 4NH_3 \tag{1.5}$$

Generally, the use of thiosulfate-based solutions for Au and Ag leaching from PCBs is not efficient even in the presence of Cu^{2+} ions [76]. Also, because of complex composition of e-waste, for the recovery of metals present in the PCB, various leaching solutions are needed. For example, high recoveries of different metals (84% Cu, 82% Fe, 77% Al, 76% Zn, 70% Ni, 90% Pd, 88% Pb, 83% Sn) were obtained using H_2O_2, NaCl, H_2SO_4 and aqua regia solutions. Ninety-eight percent gold and 93% silver recovery with thiosulfate solution took as many as 48 h [81].

ATMI (Advanced Technology Materials, Inc.) proposes WEEE treatment without shredding or grinding which is advantageous because there is no loss (even up to 40% in case of shredding) of precious metals and/or no hazard of formation of dangerous metal fines, dust containing brominated flame retardants and dioxins [80]. After selective chemical desoldering of feed e-waste (PWBs and ICs) in an acidic solution at 30°C and 40°C, solders (Pb/Sn and Sn/Ag) are collected as salts (>95% recovery of Sn, Pb and Ag), while gold connectors containing copper, gold and base metals remain intact on the board. The desoldered ICs are still operational and may be reused which contributes substantially to the sustainable development because manufacturing 2 g of new chips requires 1.6 kg of fossil fuels, 72 g of chemicals, 32 kg of water and 700 g of gases. ICs may be also ground to recover precious metals from them by nontoxic, environmentally benign leaching (authors do not reveal chemical composition of the leaching bath) at 30°C, which selectively removes all precious metals (gold, silver, palladium) that further are collected by chemical reduction or electrowinning. The process was scaled up and applied as a recycling pilot line to treat about 200 kg/h of PWBs.

Some laboratory scale proposals, based on etching (selective oxidation) of non-noble metals such as Ni, Fe, and Cu, also seem interesting and perspective. Chemical etching to remove the excess of metals from copper-plated boards is well known in electronics as a step of PCB production. Thus, etching of non-noble metals from e-scrap to produce solid Au eliminates gold leaching step, and reduces reaction time and contaminant by-products. Focusing on recyclable etching agents (oxidizers), easily renewable $CuCl_2$ solution was proposed for recovery of precious metals, in particular gold. $CuCl_2$ can be completely regenerated by oxidation of the cuprous chloride, formed during the etching, using the atmospheric oxygen. This $CuCl_2$-based process is considered to be a closed circuit and environmentally friendly process because it applies only clean reagents (energy and the air) leading to effective separation of gold from non-noble metals and substrate (support materials, i.e. plastic, ceramic), and finally, the etching solution is easily and completely regenerated by oxidation in the air atmosphere, and is recycled to the process [70].

Another example of etching solution is ammonium ($(NH_4)_2S_2O_8$) or potassium persulfate ($K_2S_2O_8$). During reaction of $(NH_4)_2S_2O_8$ solutions and solid scraps of processors in oxygen flow and at 95°C, iron, nickel and copper were partially oxidized and 99% of gold was released (as a solid) from the substrate [51].

It is noteworthy that base-activated persulfate – ammonia system can be used for efficient silver recovery from coatings of Cu-based e-waste scrap [79]. Moreover, presence of Cu either in the metallic or ionic form is very important because it is a source of the persulfate activation and increase in the oxidizing strength of the solution. Each gram of base-activated potassium persulfate under specified conditions may leach almost 100 mg of silver coatings in a form of well soluble diamminesilver(I) complex. one gram of the silver complex is obtained from about 40 g of the scrap. Then, the silver complex can be reduced to metallic silver.

Completely different approach is represented by authors of a dissolution-extraction-stripping process (Figure 1.7) [50], who proposed use of liquid–liquid extraction to recover Pd(II) from chloride leaching solution. Diisoamyl sulfide (S201, represented in the equation as R_2S) in dodecane was used as an organic phase to extract palladium(II) chloro-complexes according to the following reaction:

$$2R_2S + PdCl_4^{2-} \leftrightarrow 2PdCl_2 \cdot 2R_2S + 2Cl^- \tag{1.6}$$

The extracted Pd(II) was stripped with ammonia solution from the organic phase, and then, the organic phase was washed with NaCl solution and reused in subsequent extraction. However, there is no information on further treatment of Pd(II) ammonia solution to recover metallic Pd.

These several examples are focused on leaching of Au or other precious metals, and only some operations of further treatment are shown. However, it is important step of each recovery process to purify the leaching solution and selectively extract metals from a mixture. Among these, purification and separation operations are not only abovementioned liquid–liquid extraction, cementation, neutralization or electrolysis but also ion exchange or adsorption [81].

1.5.2.2 Indium recovery

Indium recovery from LCD panels is a quite new problem in WEEE recycling, and is mainly based on hydrometallurgy. A typical process of LCD treatment includes pretreatment (crushing and grinding) or preleaching, e.g. with acetone, to remove

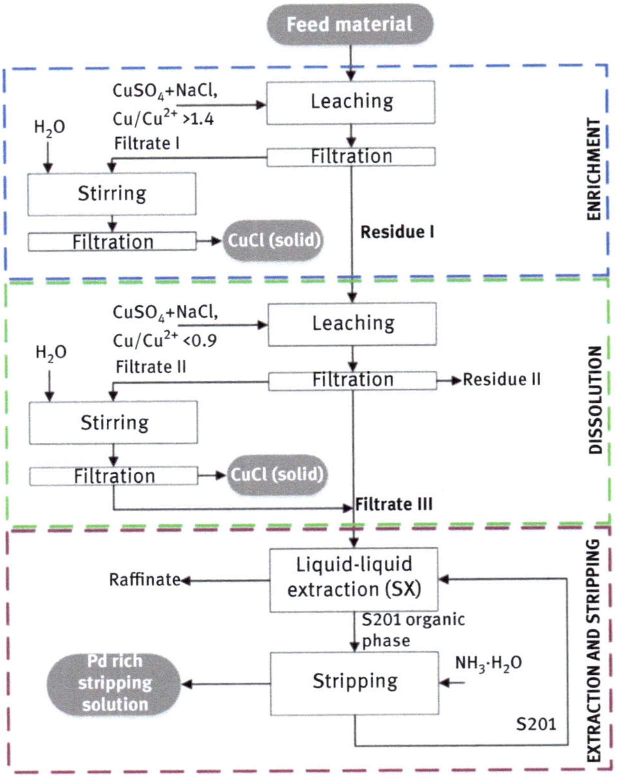

Figure 1.7: Scheme of dissolution-extraction-stripping process [50].

plastic or other organic materials. Then, separation of LCD polarizing film by thermal shock method is proposed, further, liquid crystals (considered as harmful because they are a mixture of 10–25 various compounds, including aromatic-based polymers with benzene, cyano-group, fluorine, bromine and chlorine) are removed, e.g. by ultrasonic cleaning. To maximize the recovery of the main fractions, plastic, glass and In concentrate, different pretreatment operations are proposed. For example, two physico-mechanical processes were developed by Ferella et al. [59], i.e. wet process, in which around 40% of the original LCD panel weight was recovered as In concentrate, 20% and 15% of the weight were plastic and glass fractions, respectively. A dry process is less advantageous because In is recovered as a mixture of glass and indium (85%), while 11% are plastics. The plastic fraction can be easily recycled, and In can be recovered from the concentrate by hydrometallurgical techniques. Scheme of LCD recycling process including dismantling, leaching and precipitation of In from leach solutions is shown in Figure 1.8 (see also Table 1.10). In this process, Silveira et al. [86] recovered 0.2986 kg In from 1,000 kg of LCD screens (corresponding to indium content of 0.0299 wt% in the cell phone screens).

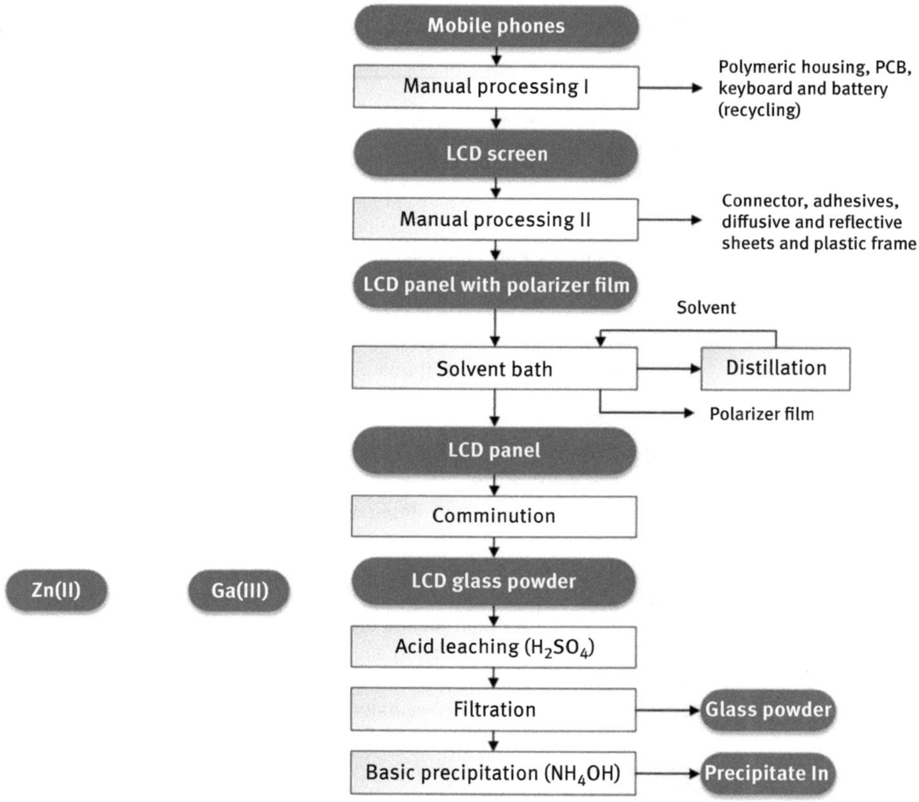

Figure 1.8: Scheme of LCD recycling process [86].

Indium contained in thin film conductive layers or indium concentrate (depending on the pretreatment operations) is dissolved by acid leaching (with mineral acids), according to the following reaction:

$$In_2O_3SnO_2 + 10HCl \xrightarrow{\text{conc. HCl}} 2InCl_3 + SnCl_4 + 5H_2O \tag{1.7}$$

Strong oxidative acids mixed with strong acid not only speedup In dissolution but also prevent Sn^{4+} from being reduced to Sn^{2+}, (black precipitation of SnO which has a negative influence on the indium recovery). Moreover, obviously not only the required metal is leached from the waste but also many accompanying ones (base metals), therefore further purification is necessary. Some examples of hydrometallurgical processes for In recovery are presented in Table 1.10.

Though, indium precipitation from pregnant leach liquors is considered environmentally friendly and simple, in many cases this operation is not selective enough. Thus, liquid–liquid extraction is often proposed to recover In selectively from the

Table 1.10: Hydrometallurgical proposals for indium recovery from waste LCD (recoveries given in brackets).

Components	Operations	Products	Ref.
LCD	*Sharp Corporation process* – In *leaching* with HCl from crushed LCD panels – Removal of glass and film by *filtration* – Purification of leaching solution containing In and other metal ions at *anion-exchange resin* – Precipitation of In(III) sludge from the concentrated In solution after sorption of accompanying metal ions with NaOH at pH 4.5–5.5	In (90%)	[22]
LCD	– In and Sn *leaching* with 5 M HCl and 10% H_2O_2 at 75°C for 120 min	In	[34]
LCD	– *Cross-current leaching* (6–10 steps) with 2 M H_2SO_4 at 80°C for 10 min – *Cementation* of In with zinc powder (5 g/dm^3, pH 3, 55°C)	In (>90%), Al and Sn (>85%)	[35–37]
LCD	– In *leaching* with 6 N HCl at room temperature which allowed solubilization of approximately 90% In – In *extraction* through the PEG–ammonium sulfate–water system (aqueous biphasic systems)	In (30%)	[38]
LCD	– 100% of indium *leaching* with 1 M H_2SO_4, 10% wt/vol S/L, 80°C, 3 h, without oxidant or, alternatively, with 0.75 M H_2SO_4, 10% wt/vol S/L, 80°C, 3 h and 10% vol. H_2O_2 – In recovery from pregnant solution by *ion-exchange* (5% wt/vol. Amberlite IRC748 resin, at pH 3, 24 h) – In *stripping* from the resin by 2 M H_2SO_4 in 1 h	In	[59]
LCD	– Three-stage process from nitric acid media in the presence of SCN^- as a complexing agent. – Pb removal from 0.1 M HNO_3 by *electrodeposition* (8 h, in the absence of SCN^-) – Sn removal by *electrodeposition* from the residual electrolyte from stage 1 with addition of 0.02 M SCN^- (8 h) – In recovery at the anode of the cylindrical mesh electrode cell as an oxyhydroxide phase by increasing SCN^- concentration to 0.1 M (24 h)	Pb (97%), Sn (94%), In (98%)	[87]

(*continued*)

Table 1.10: (*continued*)

Components	Operations	Products	Ref.
Gallium zinc oxides (GZO), indium gallium zinc oxides (IGZO) from thin film solar industry and display panel manufacturing	– *Leaching* by HNO_3 – Two steps of liquid–liquid *extraction* to separate In, Ga and Zn – Zn remains in the raffinate, In and Ga stripped with HCl – *Precipitation* of the metal hydroxides with NaOH – Formation of metal oxides by *calcination*	Zn, Ga, In (99.9%)	[88]
Mobile phones LCD	– Atmospheric pressure *leaching*: 1 M citric acid as leaching agents, 90°C, 5 vol.% H_2O_2, 120 min – *Supercritical CO_2 leaching*: co-solvents – citric acid and 5 vol.% H_2O_2, 100°C, 15 MPa, 30 min	In (76.5% – traditional leaching) In (94.6% – supercritical CO_2)	[89]
Mobile phones LCD	– *Leaching* with 2.4 M HCl – Separation by *homogeneous liquid-liquid extraction* (HoLLE) with a solution of Zonyl FSA ($CF_3(CF_2)_nCH_2CH_2SCH_2CH_2CO_2H$ (n = 6–8 in the presence of chelating agent (1,10-phenanthroline)	In (97%), Al (91%), Fe (95%), Sn (91%), Si (67%), Sr (10%)	[90]
Flat panel displays, LCD	– *Leaching* by HCl (formation of soluble metal chloride complexes) – In *extraction* with 0.25 M Cyanex 923 in kerosene (2 stages) and *stripping* with 1 M HNO_3 (2 stages) – Y extraction with 0.2 M DEHPA in kerosene (3 stages) and stripping with 2 M HCl	Y, 95% purity In	[58, 91–93]
LCD	– Separation of LCD polarizing film by *thermal shock* method at 230–240°C, – Removal of liquid crystals between the glass substrates by the *ultrasonic cleaning* in industrial detergents at a frequency of 40 KHz (*P* = 40 W) for 10 min – In recovery from glass by *leaching* in a mixture of concentrated HCl and HNO_3 (HCl:HNO_3:H_2O = 45:5:50, vol.), 30 min at 60°C	In (92%)	[94]
Sulfate solution (model)	– Preparation of resin: formation of PVA–boric acid protective layer on solvent impregnated resins impregnated with EHEHPA – In *adsorption* from H_2SO_4 (pH 1.5) – In *desorption* from the resin with 0.5–2 M HCl	In (98.9%)	[95]

(*continued*)

Table 1.10: (*continued*)

Components	Operations	Products	Ref.
LCD leaching solution	– Preparation of HZ830 resin: impregnation with Cyanex 923 – In *adsorption* from HCl solution – In *desorption* with H_2SO_4	In (93%)	[96]
In(III) industrial wastewater	– Preparation of phosphorylated sawdust beads – In *adsorption* from industrial wastewater, pH 3.5 – In *desorption* with 6 bed volumes of 0.5 M HCl	In (95%)	[97]
LCD	– Manual removal of screens from cell phones – *Pretreatment* for removal of the polarizing film from the glass of LCD panels – In *leaching* with 1.0 M H_2SO_4, 1:50 S/L, 90°C, 1 h, 500 rpm – In *precipitation* with NH_4OH at pH 7.4	In (96.4% in leaching, 99.8% in precipitation step)	[86]

leaching solutions. Acidic [58, 91–93, 98] (bis(2-ethylhexyl)phsosphoric acid – DEHPA, 2-ethylhexyl-2-ethylhexylphosphonic acid – EHEHPA, bis(2,4,4-trimethylpentyl)phosphinic acid – Cyanex 272), basic [98] (trioctylamine, mixture of tertiary amines – Adogen 364, tetrabutylammonium chloride) or solvating [91, 92, 98–101] extractants (tributyl phosphate, trioctylphosphine oxide, mixture of trialkylphosphine oxides – Cyanex 923, Cyanex 925, methyl isobutyl ketone) are proposed as indium extractants.

Yang et al. [91, 92] proposed DEHPA for In recovery from acidic leaching solutions but, finally, a solvating extractant Cyanex 923 was proposed to separate selectively In from Y, Eu, Nd, Cu, while DEHPA was applied to separate Y from Cu, Nd, Eu and partly from Fe [58]. The major impurities such as Al(III), Cu(II), Fe(III), and Zn(II) were limited by properly selected conditions of extraction and stripping.

DEHPA was also shown to extract both indium and tin from H_2SO_4 media after leaching of LCD wastes [93]. The proposed solvent system contains two countercurrent stages (A/O = 8) of extraction and two countercurrent stages (A/O = 1/2) of stripping (Figure 1.9). In this process, indium is concentrated from 0.74 to 12 g/dm^3 and is almost completely recovered from the leaching solution. Use of HCl as a stripping solution leads to high In/Sn selectivity in stripping steps, and as a part of the whole process provides with high concentration and separation of In from Sn from ITO.

DEHPA was also used as an extractant for recovery of indium, gallium and zinc from spent gallium zinc oxide (GZO) and indium gallium zinc oxide (IGZO) elements by a process comprised of leaching, solvent extraction, stripping and chemical precipitation. Scheme of the whole recovery process is shown in Figure 1.10.

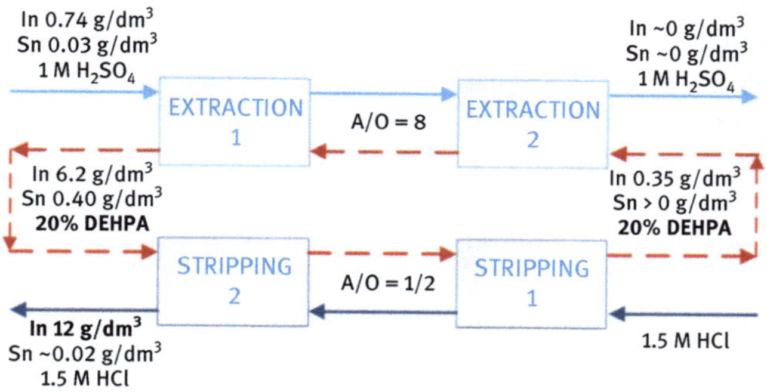

Figure 1.9: Scheme of countercurrent extraction-stripping process for separation of In from Sn [93].

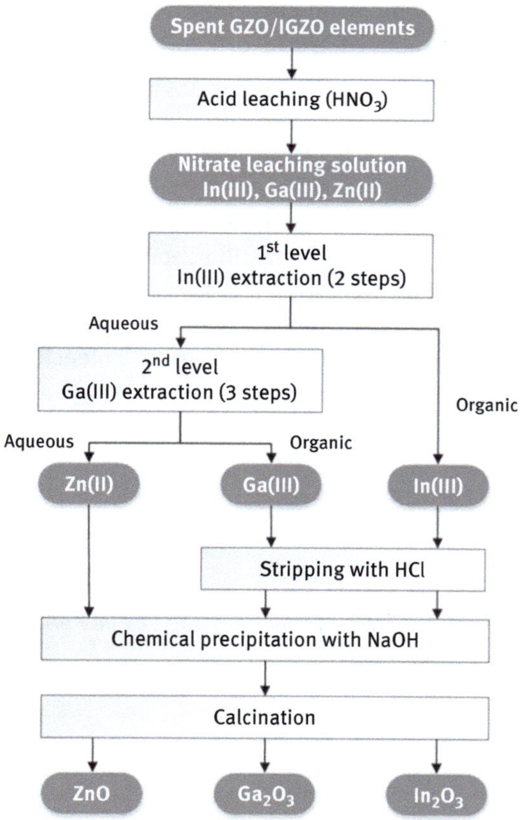

Figure 1.10: Recovery process for Zn, Ga and In separation from various LCDs and solar panels [88].

Acidic and solvating extractants EHEHPA and Cyanex 923 were used to impregnate resins which further were used to remove In(III) from H_2SO_4 (pH 1.5) or HCl (pH 2) solutions [95, 96].

Some new approaches to selective indium winning are chloride volatilization at 400°C or ultrasonic-assisted and supercritical liquid extraction [89]. Solvent extraction with aqueous biphasic systems, considered cheap, safe (reduced flammability) and nontoxic [38], was also proposed; however, In recovery was not efficient (30%). Another proposal is In electrocoagulation resulting in formation of $In(OH)_3$ precipitation at the current density 6.4 mA/cm^2, 0.003 N NaCl supporting electrolyte and 298 K [102].

However, up to now, only Sharp Corporation, a company leading in LCD production, has contributed to recover In from waste LCD panels since 2005 [22, 103]. The company developed a simple, eco-friendly and low-cost method based on anion-exchange resins. The demonstration equipment could process 240 kg of waste LCD panels/day. After dismantling, LCD panels are crushed and leached with HCl. Glass and film are removed by filtration, while leaching solution containing indium and other metal ions (impurities) is purified at anion-exchange resin. The concentrated In solution obtained after sorption of accompanying metal ions is further treated with alkaline agent to precipitate In(III) sludge (about 94% In content). The final In recovery achieved by the process was up to 90%.

1.5.3 Biometallurgy

Biometallurgy or *biohydrometallurgy* employs microorganisms for bioleaching to convert metals from solid materials into their soluble forms. Biometallurgy is considered as an alternative technology for metal recovery from very low-grade ores and concentrates, thus, it has also great potential for treatment of e-waste containing components of low content [14, 19, 62, 104, 105]. An exemplary scheme of bioleaching use for e-waste treatment is shown in Figure 1.11.

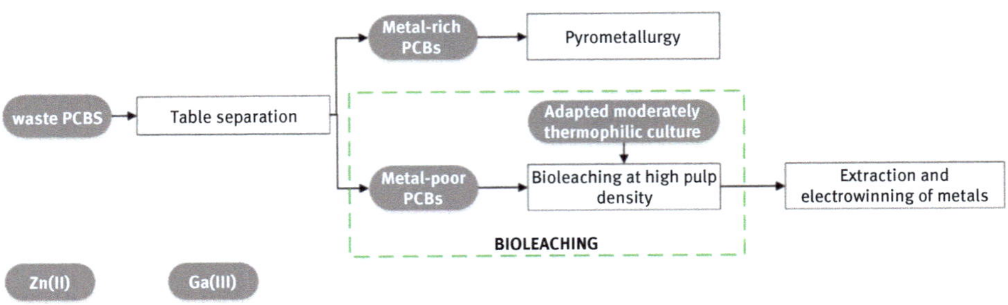

Figure 1.11: Scheme of bioleaching use for e-waste treatment [104].

General information on biometallurgy is given in *Chapter Wastes generated by automotive industry – spent automotive catalysts*, while here some examples of recent use of microorganisms for metal bioleaching from WEEE are presented. As nonmetallic components of electronic scrap (e.g. Pb) and waste PCBs are toxic for bacteria, these components should be removed prior to biometallurgical process, and only metal concentrates should be treated by bioleaching [106, 107].

Though most efforts are directed toward copper bioleaching from e-waste, also Ni, Zn, Cr or precious metal are technically feasible to be recovered by bacteria- or fungi-assisted reactions. Generally, three microbial groups are involved in this process: autotrophic bacteria, heterotrophic bacteria and heterotrophic fungi (e.g. *Acidithiobacillus ferrooxidans, Leptospirillum ferrooxidans* and *Aspergillus* sp.) [19, 55, 108–110]. Cu bioleaching is enhanced by a presence of Fe^{2+} ions in acidic solution because bio-oxidation of copper runs through indirect mechanism, involving a redox intermediate such as ferrous/ferric couple, which oxidizes the metals according to the reactions [46, 55, 106, 107, 111]:

$$4Fe^{2+} + O_2 + 4H^+ \xrightarrow{\text{bacteria}} 4Fe^{3+} + 2H_2O \tag{1.8}$$

$$2Fe^{3+} + Cu^0 \rightarrow 2Fe^{2+} + Cu^{2+} \tag{1.9}$$

Maintaining pH of a leaching solution at low level <2.5 (by acid addition) is necessary to provide suitable amount of Fe^{3+} for reaction (9), and to avoid hydrolysis of Fe^{3+} and production of jarosite precipitate $KFe_3(SO_4)_2(OH)_6$ [46, 107].

Part of copper is also leached by direct oxidation:

$$2Cu^0 + O_2 + 4H^+ \rightarrow 2Cu^{2+} + 2H_2O \tag{1.10}$$

The following optimized conditions, mixed culture of acidophilic bacteria, 10% inoculation quantity, initial pH 2.00, initial Fe(II) concentration 12 g/dm^3, 12 g/dm^3 metal concentrate dosage of 60–80 mesh particle size, 30°C temperature, were successfully applied and resulted in 96.8% of Cu leaching in 45 h, and 88.2% and 91.6%, Al and Zn leaching efficiency, respectively, in 96 h [106].

Not only *Sulfobacillus thermosulfidooxidans*, the most frequently applied bioleaching bacteria, but also fungi (*Aspergillus niger, Penicillium simplicissimum*) mobilize Cu, Sn, Al, Ni, Pb and Zn, by formation of inorganic and organic acids, to be leached at more than 90% efficiency at 30°C for 21 days [108]. For commercially interesting process, the microorganisms should be grown separately in the absence of e-scrap; therefore, a two-step process for an industrial application comprises organisms growth in the absence of electronic scrap followed by the second step of metal solubilization by the metabolites formed in the first step. Owing to such a solution, the biomass does not contain metals and can be recycled, and, on the other hand, the waste material is not contaminated by microorganisms. Moreover, higher efficiency of two-step bioleaching can be obtained by higher waste concentrations (microbial growth was inhibited in direct contact with e-scrap exceeding 10 g/dm^3), and by optimization of acid formation in the absence of e-waste. Two strains of *A. niger* were used for bioleaching of Au, Cu and Ni in the presence of gold-plated finger ICs found in computer motherboards and mobile phone PCBs. However, the efficiencies of Cu and Ni leaching after 14 days at 28°C did not exceed 30%, while Au was leached even in 87% from PCBs with a combination of *A. niger* MXPE6 + *A. niger* MX7 [112].

A mixed culture of mesophilic bacteria (*A. ferrooxidans, L. ferrooxidans, Acidithiobacillus thiooxidans*) was also applied for Cu leaching from manufacturing scrap TV circuit boards reaching Cu recovery equal even to 84%. This process was enhanced by the presence of Fe^{2+}, which was added in the form of pyrite concentrate (a source of iron and sulfur) [55]. Fe^{3+} released in reaction (8) reacts with pyrite resulting in Fe^{2+} and acid, which are essential for effective dissolution of copper in the bioleaching process. High bioleaching efficiencies achieved for PCB treatment (15 g/dm^3) in the presence of Fe^{3+} in 72 h by *A. ferrooxidans* can reach even 96.8% of Cu, 83.8% of Zn and 75.4% of Al [107]. A mixed acidophilic culture enriched from a sulfide ore mine site, containing *A. ferrooxidans, A. thiooxidans/albertensis, Acidithiobacillus caldus, L. ferrooxidans, S. thermosulfidooxidans, Sulfobacillus thermotolerans* and some members of *Alicyclobacillus genus* was proven to bioleach 99% of Cu (in 3 days) from PCB at high initial Fe^{2+} concentration (7.8 g/dm^3) and pH as low as 1.6 [105].

Bioleaching Zn, Cu, Pb, Ni, Cd and Cr from PCBs by a culture of sulfur-oxidizing bacteria (*Acidithiobacillus* sp.) and a mixed culture of biosurfactant-producing bacteria (*Bacillus subtilis* PCM 2021 and *Bacillus cereus* PCM 2019) and sulfur-oxidizing bacteria resulted in almost 50% dissolution of Zn, and over 90% of Cr, Ni and Cu were better leached in the acidic medium (48.5% and 53%, respectively), while Cr was the only metal that was more effectively bioleached in the medium containing both sulfur-oxidizing and biosurfactant-producing bacteria [109]. Effective bioleaching without biosurfactant-producing bacteria was shorter (15 days) than with these bacteria (25 days).

Metal-poor PCBs (fraction of PCBs after a shaking table separation) contain only about 8% of metals, i.e. 2.83% of Cu, 0.62% of Fe, 1.78% of Pb, 1.11% of Sn, 1.29% of Al, 0.85% of Zn and 0.35% of Ba. As metal content in this fraction is low, it is not suitable for pyrometallurgy (in the contrary to metal-rich PCBs) but can be treated by bioleaching with moderately thermophilic bacteria species, i.e. *Leptospirillum ferriphilum* and *A. caldus*. The bacteria were adapted (for 10 days) before being applied for bioleaching and after adaptation they could tolerate 80 g/dm^3 metal-poor PCBs. Recovery of Zn, Cu and Al reached almost 100%, 77% and 85% in 7 days [104].

Bioleaching of precious metals (Pd, Pt, Au) needs other microorganisms than Cu and other base metals. Therefore, separation of Au and Cu from PCBs requires bioleaching with two different types of bacteria. Bacteria such as *Bacillus megaterium, Chromobacterium violaceum, Pseudomonas aeruginosa, Pseudomonas fluorescens, Pseudomonas plecoglossicida* and fungi such as *Clitocybe* sp., *Polysporus* sp. and *Marasmius oreades* have the ability to produce cyanide applied for Au leaching [113].

For example, two-step bioleaching was proposed to increase efficiency of Au leaching from PCB. In the first step, *A. ferrooxidans* were used to completely remove copper from PCB powder because Cu presence negatively influenced Au bioleaching.

Next, *B. megaterium*, a cyanogenic bacterium, produced cyanide (in a presence of glycine – a direct precursor of cyanide) as Au leaching agent, and then gold was leached (64%) at pH 10, pulp density 2, 0.5 g/dm^3 [114]. Other strains of bacteria producing cyanide for gold leaching from e-scrap are *C. violaceum, P. aeruginosa* and *P. fluorescens* [113, 115]. The bacterium produces cyanide (at the beginning of the stationary phase), and also an enzyme, β-cyanoalanine synthase responsible for degradation and detoxification of the cyanide excess (during the late stationary and death phases) [113]. This detoxification step decreases the environmental and health risks of the bioleaching. Although Au was bioleached by cyanide for 8 days, the efficiency of gold dissolution did not exceed 30%.

Another approach was proposed by using *Desulfovibrio desulfuricans* to recover Au^{3+} as Au^0 from e-scrap leachates [53]. A demonstration bioreactor was constructed (working volume ~2 dm^3), and a three-stage bioseparation process for the recovery of Au(III), Pd(II) and Cu(II) in the presence of electrochemically-generated H_2 was proposed.

Though biometallurgy is considered to be an eco-friendly and energy saving way of metal recovery, it is still applied rather on a laboratory scale when speaking about WEEE processing. Bioleaching is limited by very slow processing and sensitivity of the microorganisms to toxic compounds or sudden changes in process conditions.

1.6 Summary

To summarize the processes presented in this chapter, some advantages and disadvantages of pyro-, hydro- and biometallurgy are given in Table 1.11.

Table 1.11: Comparison of pyrometallurgy, hydrometallurgy and biometallurgy applied for recovery of metals from WEEE.

Process	Pyrometallurgy	Hydrometallurgy	Biometallurgy
Drawbacks	– High temperature of operation	– Not applicable to complex e-scrap	– Slower and less efficient than hydrometallurgy
	– Limited selectivity of metal refining – Generation of hazardous fumes (dioxins, Hg, Pb, Cd) and emission of great volumes of gases (SO_2) – Difficulty in recovery of Al and Fe – Precious metals are obtained at the very end of the process (stay for a long time) – High energy consumption – High operation cost	– High volumes of solutions used (can be corrosive or toxic) – Causes secondary pollution with wastewaters or sludges	– Limited efficiency of metal leaching due to sensitivity of microorganisms

(*continued*)

Table 1.11: (*continued*)

Process	Pyrometallurgy	Hydrometallurgy	Biometallurgy
Advantages	– Any form of e-scrap	– Pure products and high recovery rate	– Low temperature of operation and energy consumption
	– Applicable to complex electronic scrap	– Selective leaching of various metals	– Greener than pyro- and hydrometallurgy
		– Low temperature of operation and low energy consumption	– Selective for specific metals
		– Low or no gas emission	– Low investment cost
		– No dust or low dust generation	
		– Easy working condition and simple equipment	

References

[1] Wong MH, Wu SC, Deng WJ, Yu XZ, Luo Q, Leung AOW, et al. Export of toxic chemicals – a review of the case of uncontrolled electronic-waste recycling. Environ Pollut. 2007;149:131–40

[2] New Boliden. One of the largest recycler of electronic material. Available at: http://www.boliden.com/sustainability/case-studies/largest-electronic-material-recycler-in-the-world/. Accessed September 4, 2017.

[3] Waste Electrical & Electronic Equipment (WEEE). European Commission. Available at: http://ec.europa.eu/environment/waste/weee/index_en.htm. Accessed August 22, 2017.

[4] Directive 2012/19/EU of the European Parliament and of the Council of 4 July 2012 on waste electrical and electronic equipment (WEEE) (recast). Official Journal of the European Union. L 197/38, 2012. Available at: http://eur-lex.europa.eu/legal-content/EN/TXT/?uri=CELEX:32012L0019. Accessed August 22, 2017.

[5] Directive 2011/65/EU of the European Parliament and of the Council of 8 June 2011 on the restriction of the use of certain hazardous substances in electrical and electronic equipment (recast). Official Journal of the European Union. L 174/88, 2011. Available at: http://eur-lex.europa.eu/legal-content/EN/TXT/?uri=CELEX:32011L0065. Accessed August 22, 2017.

[6] European Academies' Science Advisory Council. Priorities for critical materials for a circular economy. EASAC policy report 29, Halle, German National Academy of Sciences, 2016.

[7] Center for Sustainable Systems. Critical Materials Factsheet. Pub. No. CSS14-15. Michigan, USA: University of Michigan. Available at: https://css.snre.umich.edu 2016 . Accessed August 26, 2017.

[8] The ad hoc working group on defining critical raw materials. Annexes to the report on critical raw materials for the EU. Ref. Ares (2015) 1819558. European Commission, 2014. Available at: https://ec.europa.eu/growth/sectors/raw-materials/specific-interest/critical_en. Accessed August 26, 2017.

[9] Sverdrup H, Ragnarsdottir KV. Natural resources in a planetary perspective. Geochemical Perspect. 2014;3:129–336.

[10] European Academies' Science Advisory Council. Indicators for a circular economy. EASAC policy report 30. Halle: German National Academy of Sciences, 2016.
[11] Huang K, Guo J, Xu Z. Recycling of waste printed circuit boards: a review of current technologies and treatment status in China. J Hazard Mater. 2009;164:399–408.
[12] Kahhat R, Kim J, Xu M, Allenby B, Williams E, Zhang P. Exploring e-waste management systems in the United States. Resour Conserv Recycl. 2008;52:955–64.
[13] Sinha-Khetriwal D, Kraeuchi P, Schwaninger M. A comparison of electronic waste recycling in Switzerland and in India. Environ Impact Assessment Rev. 2005;25:492–504.
[14] Priya A, Hait S. Comparative assessment of metallurgical recovery of metals from electronic waste with special emphasis on bioleaching. Environ Sci Pollut Res. 2017;24: 6989–7008.
[15] TOXNET, Toxicology Data Network. Available at: https://toxnet.nlm.nih.gov Accessed August 26, 2017.
[16] Robinson BH E-waste: an assessment of global production and environmental impacts. Sci Total Environ. 2009;408:183–91.
[17] Tsydenova O, Bengtsson M. Chemical hazards associated with treatment of waste electrical and electronic equipment. Waste Manage. 2011;31:45–58.
[18] Garlapati VK. E-waste in India and developed countries: management, recycling, business and biotechnological initiatives. Renewable Sustainable Energy Rev. 2016;54:874–81.
[19] Kaya M. Recovery of metals and nonmetals from electronic waste by physical and chemical recycling processes. Waste Manage. 2016;57:64–90.
[20] Sun Z, Cao H, Xiao Y, Sietsma J, Jin W, Agterhuis H, et al. Toward sustainability for recovery of critical metals from electronic waste: the hydrochemistry processes. ACS Sustainable Chem Eng. 2017;5:21–40.
[21] Nnorom IC, Osibanjo O. Electronic waste (e-waste): material flows and management practices in Nigeria. Waste Manage. 2008;28:1472–79.
[22] Zhang K, Wu Y, Wang W, Li B, Zhang Y, Zuo T. Recycling indium from waste LCDs: a review. Resour Conserv Recycl. 2015;104:276–90.
[23] Mester A, Fraunholcz N, Van Schaik A, Reuter MA. Characterization of the hazardous components in end-of-life notebook display. Light Metals 2005, TMS. 2005;1213–16.
[24] Chen CF. LEDs for liquid crystal display (LCD) backlighting. Chapter 19. In: Huang JJ, Hc K, Shen SC, editors. Nitride semiconductor light-emitting diodes (LEDs): materials, technologies and applications. Cambridge: Woodhead Publishing, 2014: 566–94.
[25] Huang K, Guo J, Xu Z. Recycling of waste printed circuit boards: a review of current technologies and treatment status in China. J Hazard Mater. 2009;164:399–408.
[26] Nogrady B. Your phone is full of precious metals. BBC, 18 October 2016. Available at: http://www.bbc.com/future/story/20161017-your-old-phone-is-full-of-precious-metals. Accessed August 27, 2017.
[27] Consumer electronics to be a US$ 3 trillion market by 2020: report, 2016. Available at: http://www.futuremarketinsights.com/press-release/consumer-electronics-market. Accessed August 27, 2017.
[28] Cucchiella F, D'Adamo I, Lenny Koh SC, Rosa P. Recycling of WEEEs: an economic assessment of present and future e-waste streams. Renewable Sustainable Energy Rev. 2015;51:263–72.
[29] Hagelüken C, Corti CW. Recycling of gold from electronics: cost-effective use through 'Design for Recycling'. Gold Bulletin. 2010;43:209–20.
[30] Kozłowski J. Economic aspects and recovery of metals from electronic and electrotechnic waste. Rudy Metale (In Polish). 2008;53:555–63.
[31] Kang HY, Schoenung JM. Electronic waste recycling: a review of U.S. infrastructure and technology options. Resour Conserv Recycl. 2005;45:368–400.

[32] Li JH, Gao S, Duan HB, Liu LL. Recovery of valuable materials from waste liquid crystal display panel. Waste Manage. 2009;29:2033–39.
[33] Zhang L, Wu B, Chen Y, Xu Z. Treatment of liquid crystals and recycling indium for stripping product gained by mechanical stripping process from waste liquid crystal display panels. J Cleaner Prod. 2017;162:1472–81.
[34] Swain B, Mishra C, Hong HS, Cho SS. Beneficiation and recovery of indium from liquid-crystal-display glass by hydrometallurgy. Waste Manage. 2016;57:207–14.
[35] Rocchetti L, Amato A, Beolchini F. Recovery of indium from liquid crystal displays. J Cleaner Prod. 2016;116:299–305.
[36] Amato A, Rocchetti L, Fonti V, Ruello ML, Beolchini F. Secondary indium production from end-of-life liquid crystal displays. Phys Status Solidi C. 2016;13(10–12):979–83.
[37] Rocchetti L, Amato A, Fonti V, Ubaldini S, De Michelis I, Kopacek B, et al. Cross-current leaching of indium from end-of-life LCD panels. Waste Manage. 2015;42:180–87.
[38] Fontana D, Forte F, De Carolis R, Grosso M. Materials recovery from waste liquid crystal displays: a focus on indium. Waste Manage. 2015;45:325–33.
[39] Guan J, Wang S, Ren H, Guo Y, Yuan H, Yan H, et al. Indium recovery from waste liquid crystal displays by polyvinyl chloride waste. RSC Adv. 2015;5:102836–43.
[40] Duan H, Hou K, Li J, Zhu X. Examining the technology acceptance for dismantling of waste printed circuit boards in light of recycling and environmental concerns. J Environ Manage. 2011;92:392–99.
[41] Montero R, Guevara A, De La Torre E. Recovery of gold, silver, copper and niobium from printed circuit boards using leaching column. J Earth Sci Eng. 2012;2:590–95.
[42] Umicore, Precious Metals Refining. Available at: http://pmr.umicore.com/en/recyclables/electronic-scrap/. Accessed September 4, 2017.
[43] Park YJ, Fray DJ. Recovery of high purity precious metals from printed circuit boards. J Hazard Mater. 2009;164:1152–58.
[44] Shokri A, Pahlevani F, Cole I, Sahajwalla V. Selective thermal transformation of old computer printed circuit boards to Cu-Sn based alloy. J Environ Manage. 2017;199:7–12.
[45] Janyasuthiwong S, Ugas R, Rene ER, Alessandra C, Esposito G, Lens PNL. Effect of operational parameters on the leaching efficiency and recovery of heavy metals from computer printed circuit boards. J Chem Technol Biotechnol. 2016;91:2038–46.
[46] Chen S, Yang Y, Liu C, Dong F, Liu B. Column bioleaching copper and its kinetics of waste printed circuit boards (WPCBs) by *Acidithiobacillus ferrooxidans*. Chemosphere. 2015;141:162–68.
[47] Isildar A, Van De Vossenberg J, Rene ER, Van Hullebusch ED, Lens PNL. Two-step bioleaching of copper and gold from discarded printed circuit boards (PCB). Waste Manage. 2016;57:149–57.
[48] Mäkinen J, Bachér J, Kaartinen T, Wahlström M, Salminen J. The effect of flotation and parameters for bioleaching of printed circuit boards. Min Eng. 2015;75:26–31.
[49] Silvas FPC, Jiménez Correa MM, Caldas MPK, De Moraes VT, Espinosa DCR, Tenório JAS. Printed circuit board recycling: physical processing and copper extraction by selective leaching. Waste Manage. 2015;46:503–10.
[50] Zhang Z, Zhang FS Selective recovery of palladium from waste printed circuit boards by a novel non-acid process. J Hazard Mater. 2014;279:46–51.
[51] Alzate A, López ME, Serna C. Recovery of gold from waste electrical and electronic equipment (WEEE) using ammonium persulfate. Waste Manage 2016;57:113–20.
[52] Williams PT. Valorization of printed circuit boards from waste electrical and electronic equipment by pyrolysis. Waste Biomass Valor. 2010;1:107–20.

[53] Creamer NJ, Baxter-Plant VS, Henderson J, Potter M, Macaskie LE. Palladium and gold removal and recovery from precious metal solutions and electronic scrap leachates by *Desulfovibrio desulfuricans*. Biotechnol Lett. 2006;28:1475–84.
[54] Kamberović Ž, Korać M, Ranitović M. Hydrometallurgical process for extraction of metals from electronic waste-part II: development of the processes for the recovery of copper from printed circuit boards (PCB). Metalurgija-MJoM. 2011;17(3):139–49.
[55] Bas AD, Deveci H, Yazici EY. Bioleaching of copper from low grade scrap TV circuit boards using mesophilic bacteria. Hydrometallurgy. 2013;138:65–70.
[56] Diaz LA, Lister TE, Parkman JA, Clark GG. Comprehensive process for the recovery of value and critical materials from electronic waste. J Clean Prod. 2016;125:236–44.
[57] Savvilotidou V, Hahladakis JN, Gidarakos E. Leaching capacity of metals–metalloids and recovery of valuable materials from waste LCDs. Waste Manage. 2015;45:314–24.
[58] Yang J, Retegan T, Steenari BM, Ekberg C. Recovery of indium and yttrium from Flat Panel Display waste using solvent extraction. Sep Purif Technol. 2016;166:117–24.
[59] Ferella F, Belardi G, Marsilii A, De Michelis I, Vegliò F. Separation and recovery of glass, plastic and indium from spent LCD panels. Waste Manage. 2017;60:569–81.
[60] Ryan A, O'Donoghue L, Lewis H. Characterising components of liquid crystal displays to facilitate disassembly. J Clean Prod. 2011;19:1066–71.
[61] Cui J, Forssberg E. Mechanical recycling of waste electric and electronic equipment: a review. J Hazard Mater. 2003;B99:243–63.
[62] Cui J, Zhang L. Metallurgical recovery of metals from electronic waste: a review. J Hazard Mater. 2008;158:228–56.
[63] Hagelüken C. Improving metal returns and eco-efficiency in electronics recycling. Proceedings of the 2006 IEEE International Symposium on Electronics and the Environment, San Francisco, 2006: 218–23.
[64] Hagelüken C. Recycling of electronic scrap at Umicore's integrated metals smelter and refinery. World of Metallurgy – ERZMETALL. 2006;59(3):152–61.
[65] Aurubis Recycling Technology. Available at: https://www.aurubis.com/products/recycling/technology. Accessed September 4, 2017.
[66] Glencore Recycling. Available at: http://www.glencorerecycling.com/en/AboutUs/Pages/home.aspx. Accessed September 4, 2017.
[67] APME. Plastics recovery from waste electrical & electronic equipment in non-ferrous metal processes, 8036/GB/07/00, APME (Association of Plastics Manufacturers in Europe) Report, 2000.
[68] Willner J, Fornalczyk A. Electronic scraps as a source of precious metals. Przem Chem. 2012;91(4):517–22.
[69] Rocchetti L, Vegliò F, Kopacek B, Beolchini F. Environmental impact assessment of hydrometallurgical processes for metal recovery from WEEE residues using a portable prototype plant. Environ Sci Technol. 2013;47:1581–88.
[70] Barbieri L, Giovanardi R, Lancellotti I, Michelazzi M. A new environmentally friendly process for the recovery of gold from electronic waste. Environ Chem Lett. 2010;8:171–78.
[71] Hino T, Agawa R, Moriya Y, Nishida M, Tsugita Y, Araki T. Techniques to separate metal from waste printed circuit boards from discarded personal computers. J Mater Cycles Waste Manag. 2009;11:42–54.
[72] Sheng PP, Etsell TH. Recovery of gold from computer circuit board scrap using aqua regia. Waste Manage Res. 2007;25:380–83.
[73] Torres R, Lapidus GT. Closed circuit recovery of copper, lead and iron from electronic waste with citrate solutions. Waste Manage. 2017;60:561–68.

[74] Birloaga I, Vegliò F. Study of multi-step hydrometallurgical methods to extract the valuable content of gold, silver and copper from waste printed circuit boards. J Environ Chem Eng. 2016;4:20–29.
[75] Behnamfard A, Salarirad MM, Veglio F. Process development for recovery of copper and precious metals from waste printed circuit boards with emphasize on palladium and gold leaching and precipitation. Waste Manage. 2013;33:2354–63.
[76] Petter PMH, Veit HM, Bernardes AM. Evaluation of gold and silver leaching from printed circuit board of cellphones. Waste Manage. 2014;34:475–82.
[77] Dunn J, Wendell E, Carda DD, Storbeck TA. Chlorination process for recovering gold values from gold alloys. US Patent, US5004500, 1991.
[78] Syed S. A green technology for recovery of gold from non-metallic secondary sources. Hydrometallurgy. 2006;82:48–53.
[79] Hyk W, Kitka K. Highly efficient and selective leaching of silver from electronic scrap in the base-activated persulfate – ammonia system. Waste Manage. 2017;60:601–08.
[80] Jiang P, Harney M, Song Y, Chen B, Chen Q, Chen T, et al. Improving the end-of-life for electronic materials via sustainable recycling methods. Proc Environ Sci. 2012;16:485–90
[81] Akcil A, Erust C, Gahan CS, Ozgun M, Sahin M, Tuncuk A. Precious metal recovery from waste printed circuit boards using cyanide and non-cyanide lixiviants – a review. Waste Manage. 2015;45:258–71.
[82] Hilson G, Monhemius AJ. Alternatives to cyanide in the gold mining industry: what prospects for the future. J Clean Prod. 2006;14:1158–67.
[83] Marsden JO, House CI The chemistry of gold extraction, 2nd ed. Littleton, Colorado: Society for Mining, Metallurgy and Exploration Inc., 2006.
[84] Zhang L, Xu Z. A review of current progress of recycling technologies for metals from waste electrical and electronic equipment. J Clean Prod. 2016;127:19–36.
[85] Zhang W, Ren J, Liu S, Yuan Z. Mechanism and clean procedure to extract gold from printed circuit board. Procedia Environ Sci. 2016;31:171–77.
[86] Silveira AVM, Fuchs MS, Pinheiro DK, Tanabe EH, Bertuol DA. Recovery of indium from LCD screens of discarded cell phones. Waste Manage. 2015;45:334–42.
[87] Grimes SM, Yasri NG, Chaudhary AJ. Recovery of critical metals from dilute leach solutions – separation of indium from tin and lead. Inorg Chim Acta. 2017;461:161–66.
[88] Chen WS, Wang YC, Chiu KL. The separation and recovery of indium, gallium, and zinc from spent GZO(IGZO) targets. J Environ Chem Eng. 2017;5:381–90.
[89] Argenta AB, Reis CM, Mello GP, Dotto GL, Tanabe EH, Bertuol DA. Supercritical CO_2 extraction of indium present in liquid crystal displays from discarded cell phones using organic acids. J Supercrit Fluids. 2017;120:95–101.
[90] Kato T, Igarashi S, Ishiwatari Y, Furukawa M, Yamaguchi H. Separation and concentration of indium from a liquid crystal display via homogeneous liquid–liquid extraction. Hydrometallurgy. 2013;137:148–55.
[91] Yang J, Retegan T, Ekberg C. Indium recovery from discarded LCD panel glass by solvent extraction. Hydrometallurgy. 2013;137:68–77.
[92] Yang J, Ekberg C, Retegan T. Optimization of indium recovery and separation from LCD waste by solvent extraction with bis(2-ethylhexyl)phosphate (D2EHPA). Int J Chem Eng. 2014;2014:1–9.
[93] Virolainen S, Ibana D, Paatero E. Recovery of indium from indium-tin-oxide by solvent extraction. Hydrometallurgy. 2011;107(1–2):56–61.
[94] Li J, Gao S, Duan H, Liu L Recovery of valuable materials from waste liquid crystal display panel. Waste Manage. 2009;29:2033–39.
[95] Yuan Y, Liu J, Zhou B, Yao S, Li H, Xu W. Synthesis of coated solvent impregnated resin for the adsorption of indium (III). Hydrometallurgy. 2010;101:148–55.

[96] Wei S, Liu J, Zhang S, Chen X, Liu Q, Zhu L, et al. Stoichiometry, isotherms and kinetics of adsorption of In(III) on Cyanex 923 impregnated HZ830 resin from hydrochloric acid solutions. Hydrometallurgy. 2016;164:219–27.
[97] Jeon C, Cha JH, Choi JY. Adsorption and recovery characteristics of phosphorylated sawdust bead for indium(III) in industrial wastewater. J Ind Eng Chem. 2015;27:201–06.
[98] Paiva AP. Recovery of indium from aqueous solutions by solvent extraction. Sep Sci Technol. 2001;36(7):1395–419.
[99] Gupta B, Deep A, Malik P. Liquid-liquid extraction and recovery of indium using Cyanex 923. Anal Chim Acta. 2004;513(2):463–71.
[100] Gupta B, Mudhar N, Begum ZI, Singh I. Extraction and recovery of Ga(III) from waste material using Cyanex 923. Hydrometallurgy. 2007;87(1–2):18–26.
[101] Ahmed IM, El-Nadi YA, El-Hefny NE. Extraction of gallium(III) from hydrochloric acid by Cyanex 923 and Cyanex 925. Hydrometallurgy. 2013;131–132:24–28.
[102] Chou WL, Huang YH. Electrochemical removal of indium ions from aqueous solution using iron electrodes. J Hazard Mater. 2009;172:46–53.
[103] Tsujiguchi M. Indium recovery and recycling from an LCD panel. In panel. In: Matsumoto M, Umeda Y, Masui K, Fukushige S, editors. Design for innovative value towards a sustainable society. Dordrecht: Springer, 2012.
[104] Xia MC, Wang YP, Peng TJ, Shen L, Yu RL, Liu YD, et al. Recycling of metals from pretreated waste printed circuit boards effectively in stirred tank reactor by a moderately thermophilic culture. J Biosci Bioeng. 2017;123(6):714–21.
[105] Mäkinen J, Bachér J, Kaartinen T, Wahlström M, Salminen J. The effect of flotation and parameters for bioleaching of printed circuit boards. Min Eng. 2015;75:26–31.
[106] Zhu N, Xiang Y, Zhang T, Wu P, Dang Z, Li P, et al. Bioleaching of metal concentrates of waste printed circuit boards by mixed culture of acidophilic bacteria. J Hazard Mater. 2011;192: 614–19.
[107] Yang Y, Chen S, Li S, Chen M, Chen H, Liu B. Bioleaching waste printed circuit boards by *Acidithiobacillus ferrooxidans* and its kinetics aspect. J Biotechnol. 2014;173:24–30.
[108] Brandl H, Bosshard R, Wegmann M. Computer-munching microbes: metal leaching from electronic scrap by bacteria and fungi. Hydrometallurgy. 2001;59:319–26.
[109] Karwowska E, Andrzejewska-Morzuch D, Łebkowska M, Tabernacka A, Wojtkowska A, Telepko A, et al. Bioleaching of metals from printed circuit boards supported with surfactant-producing bacteria. J Hazard Mater. 2014;264:203–10.
[110] Ilyas S, Anwar MA, Niazi SB, Ghauri MA. Bioleaching of metals from electronic scrap by moderately thermophilic acidophilic bacteria. Hydrometallurgy. 2007;88:180–88.
[111] Xiang Y, Wu PX, Zhu NW, Zhang T, Liu W, Wu J, et al. Bioleaching of copper from waste printed circuit boards by bacterial consortium enriched from acid mine drainage. J Hazard Mater. 2010;184(1–3):812–18.
[112] Madrigal-Arias JE, Argumedo-Delira R, Alarcón A, Mendoza-López MR, García-Barradas O, Cruz-Sánchez JS, et al. Jiménez-Fernández M. Bioleaching of gold, copper and nickel from waste cellular phone PCBs and computer goldfinger motherboards by two *Aspergillus niger* strains. Brazilian J Microbiol. 2015;46:707–13.
[113] Natarajan G, Buck Tay S, Shan Yew W, Ting YP. Engineered strains enhance gold biorecovery from electronic scrap. Min Eng. 2015;75:32–37.
[114] Arshadi M, Mousavi SM. Enhancement of simultaneous gold and copper extraction from computer printed circuit boards using *Bacillus megaterium*. Bioresour Technol. 2015;175: 315–24.
[115] Natarajan G, Ting YP. Gold biorecovery from e-waste: an improved strategy through spent medium leaching with pH modification. Chemosphere. 2015;136:232–38.

Martyna Rzelewska and Magdalena Regel-Rosocka

2 Wastes generated by automotive industry – Spent automotive catalysts

Abstract: *Rhodium, ruthenium, palladium,* and *platinum* are classified as platinum group metals (PGM). A demand for PGM has increased in recent years. Their natural sources are limited, therefore it is important, and both from economical and environmental point of view, to develop effective process to recover *PGM* from waste/secondary sources, such as spent *automotive catalysts.* Pyrometallurgical methods have always been used for separation of PGM from various materials. However, recently, an increasing interest in hydrometallurgical techniques for the removal of precious metals from secondary sources has been noted. Among them, *liquid-liquid extraction* by contacting two liquid phases: aqueous solution of metal ions and organic solution of extractant is considered an efficient technique to separate valuable metal ions from solutions after leaching from spent catalysts.

Keywords: automotive catalyst, metal recovery, platinum group metals (PGMs)

2.1 Availability of platinum group metals (PGM)

The largest PGM mines are located in the South Africa (The Bushveld Complex), Russia (The Urals deposits), North America (Canada) and Zimbabwe (The Great Dyke) [1]. Figure 2.1 presents world mine production of PGM in the period of 2009–2013.

Natural ores contain small amounts of PGM, i. e. less than 10 g/t (0.001 %) [3]. Thus, the technological process of platinum ore processing is very energy-consuming and also causes environmentally damaging activities. For example, obtaining 1 kg of platinum requires extraction of about 150 tons of ore from a depth of 1,000 m. In comparison to the manufacturing process, the content of precious metals in waste is quite high, because from 2 tons of spent catalytic converters (automotive catalysts) about 1 kg of platinum can be obtained [4]. Despite of the increasing demand for PGM, their prices have been decreasing since 2010. However, in 2017 all the prices have increased, particularly for *Pd* and *Rh*. Figure 2.2 presents prices of four PGM for the last 8 years.

This article has previously been published in the journal *Physical Sciences Reviews*. Please cite as: Rzelewska, M., Regel-Rosocka, M. Wastes generated by automotive industry – spent automotive catalysts. *Physical Sciences Reviews* [Online] **2018**, *3*. DOI: 10.1515/psr-2018-0021

https://doi.org/10.1515/9783110547061-002

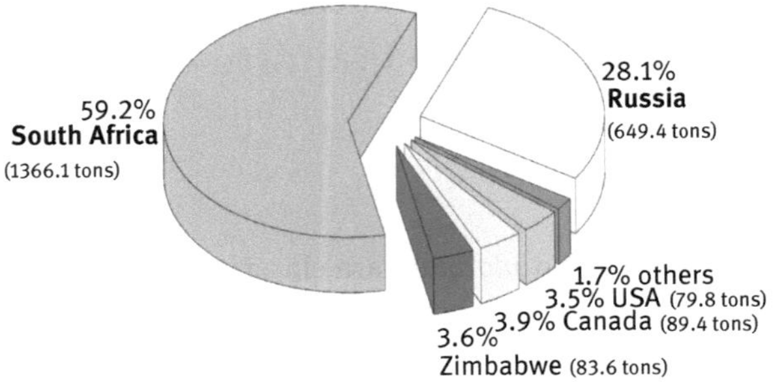

Figure 2.1: World mine production of PGM in 2009–2013 [2].

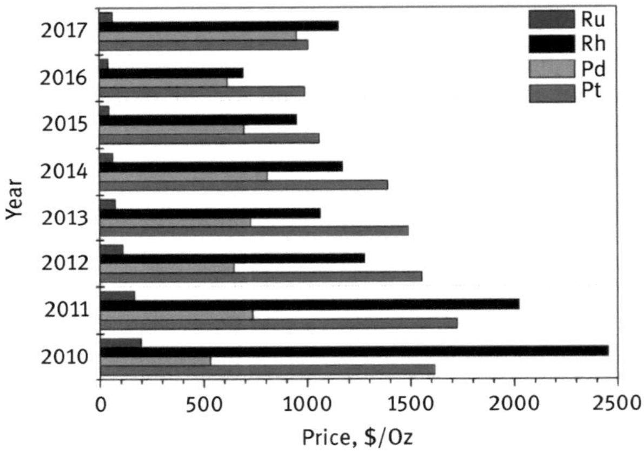

Figure 2.2: Prices of platinum group metals from 2010 to September 2017, Johnson Matthey, September 2017 (1 troy ounce (oz) = 31.1 g) [1].

2.2 Applications of PGM

The recovery from secondary sources and reuse of PGM is not only economically viable, but also environmentally friendly. PGM are very important for the industry due to their specific properties, such as low chemical reactivity or high temperature resistance. Compared to non-precious metals, PGM are less reactive and more resistant to various chemicals and high temperatures [5].

About 60 % of the yearly world production of platinum metals, especially *Pt*, Pd and Rh, are used to produce automobile catalytic converters, called also automotive/automobile catalysts [6]. The cars (98 %) currently produced have automobile catalysts containing PGM. However, they are used not only for automotive catalyst manufacturing; the main areas of applications of PGM are presented in Figure 2.3.

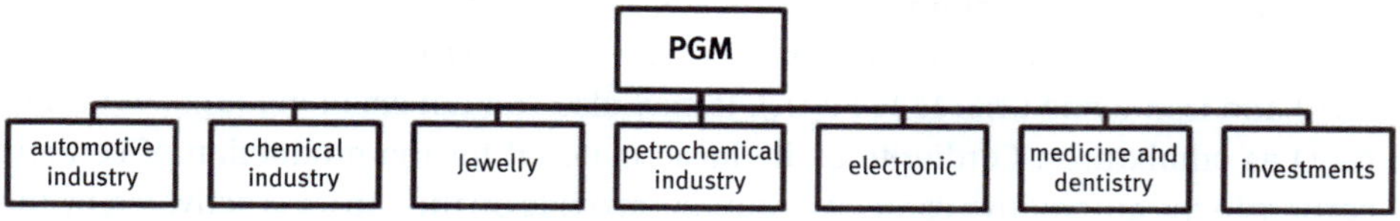

Figure 2.3: Main areas of applications of platinum group metals.

PGM are also used in chemical and petrochemical industry (catalysts for many chemical processes), automotive industry (fuel cells), and jewelry. They have wide application as the investments and for production of hard disks, electronic components (mainly palladium) and as dental restorations [5–9]. These areas of application are also main *secondary sources* of PGM. Both the recovery of noble metals from waste solutions as well as preconcentration techniques to facilitate their determination at very low levels are important issues to be developed.

2.3 PGM in automotive catalysts

The main element of *automotive catalyst* is a ceramic or metallic carrier which is covered with an intermediate layer of Al_2O_3 (0.01 mm thick) and an active layer – comprising noble metals, mainly platinum, palladium and rhodium, the content of which varies from 10^{-5} to 10^{-1} wt% [10]. The ceramic carrier has a honeycomb structure and is made of cordierite, while the metallic carrier is a spiral or a rolled film made of an acid-resistant steel [7]. Figure 2.4 illustrates an automotive catalyst before and after grinding.

Figure 2.4: An automotive catalyst before and after grinding.

Cordierite, magnesium aluminosilicate, is a mineral composed of magnesium oxide (3%–35 % MgO), aluminum oxide (32%–35 % Al_2O_3), silicon oxide (45%–51 % SiO_2) and iron oxide (2%–12 % Fe_2O_3). It may also contain small amounts of CaO or Na_2O as admixtures. Cordierite is the basic material for the production of ceramic carriers in the automotive catalysts. Aluminum oxide on the carrier is activated by the addition of CeO_2, NiO, BaO, CaO and La_2O_3. The role of cerium oxide, for example, is increasing the catalytic efficiency of the converter (oxygen storage ability) and stabilization of PGM dispersions [11].

In the 1970s, car manufacturers in the United States proposed the use of first production catalytic converter containing Pt/Pd/Rh to reduce emissions of automotive exhaust. In Western Europe catalysts were voluntarily implemented later, in the 1980s [12, 13]. Further very rigorous regulations concerning the exhaust gas emissions from vehicles forced car manufacturers to modernize and improve performance of the existing catalytic converters. Since the 1970s, the global demand for PGM for production of automobile catalysts has grown rapidly and is still at a high level, as almost each car currently produced has a catalyst containing Pt, Pd, and Rh.

The catalytic converters neutralize more than 90 % of the exhaust gases, particularly, they catalyze the reduction of NO_x and hydrocarbons (up to 80 %) and the oxidation of carbon monoxide (up to 90 %) to N_2, CO_2 and H_2O [3, 5, 13]. For comparison, one car sold in 1960 produced more exhaust gases than 100 cars produced in 2014 with catalytic converters containing PGM. As the production of cars is still increasing, the demand for PGM is growing [4, 6, 14].

There are many different techniques of metal recovery from solid materials. Regardless of the source of PGM (ores or waste materials containing PGM), the techniques of recovery of PGM are divided into pyrometallurgical, hydrometallurgical, and biometallurgical. Selection of the proper metallurgical technique depends primarily on the type of metal and its ore (low or high grade, oxide or sulfide, and polymetallic) or waste materials. An important factor to be considered is also the amount and availability of the metal and the cost of energy needed to process the ore or waste.

2.4 Recovery of PGM from spent automotive catalysts

2.4.1 Pyrometallurgy

Pyrometallurgy, i. e. thermal treatment (defined in *Chapter Electronic waste*), is the main technique used to recover precious metals from waste materials, i. e. secondary sources such as spent automotive catalysts, on the industrial scale.

Umicore (Hoboken, Belgium) is one of the largest producers of precious metals recycled from waste from the electronics industry, spent automobile catalysts and spent chemical catalysts (processing about 200 different types of materials) [13], with annual capacity of over 50 tones PGM (25 tones of each Pt and Pd, 5 tones of Rh) [13,

15]. The company carries out a unique process of metals' recovery based on copper, lead and nickel as collector metals, focusing particularly on recovery of precious metals. Among various industrial wastes and by-products from the non-ferrous metal industry (listed in **Chapter Electronic waste**) used in the process as a feed material are also spent automotive catalysts.

The crucial apparatus of the process is a new smelter, which serves as a gateway and upgrading step for e-scrap fractions [13]. Also in this process, similarly to Noranda and Boliden (discussed in **Chapter Electronic waste)**, plastics or organic residues contained in the feed materials play role of a reducing agent and a fuel substitute. Mixture of 6 % of plastics and 1 % of coke as a reducing agent, instead of 4.5 % coke, yielded in similar process efficiency (metal recovery rates and operating stability), and resulted in saving energy sources and utilization of polymeric wastes.

Over 95 % of the feed is turned into useful products, and the final slag is used for construction purposes. The feed material must be a mixture of e-scrap and other materials to have appropriate chemical and physical properties. Also, to avoid contamination of the environment with Cd and Hg (remaining, non-recyclable part of the feed), these metals are extracted and safely disposed of.

Umicore recycling process operates in two technological streams (Figure 2.5):

a. Precious Metals Operations (PMO) focused on efficient refining of an extended range of complex and valuable raw materials, containing precious metals,
b. Base Metals Operations (BMO) focused on processing of by-products from the PMO at the lowest possible cost.

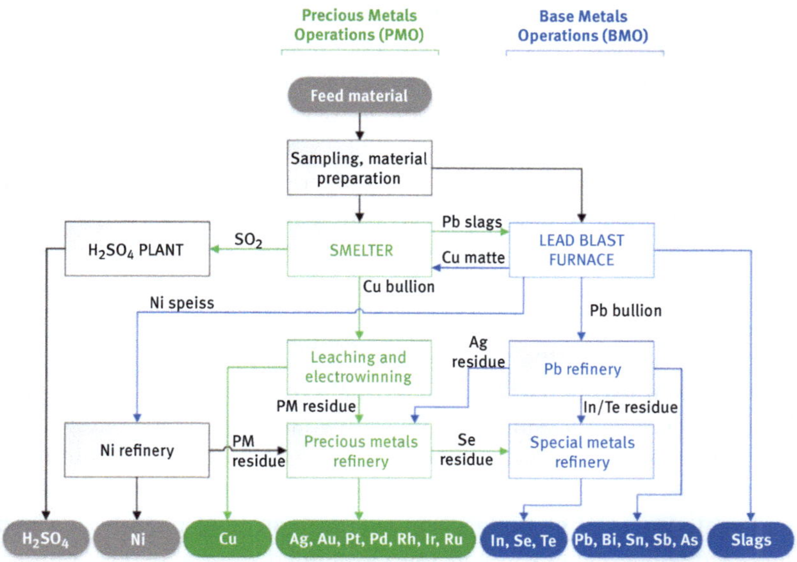

Figure 2.5: Umicore process flowsheet [13, 15, 16].

The first step of the PMO – smelting in IsaSmelt furnace (equipped with submerged lance combustion technology) – results in separation of precious metals in a copper bullion from mostly all other metals. Other metals are concentrated in a lead slag which is further treated at the BMO. The next step is leaching out the copper in the leaching and electrowinning plant to collect the precious metals in a residue that is further refined at the precious metals refinery. It should be emphasized that the raw materials can enter the flowsheet at the most optimal process step, determined by their physical aspect, their analytical fingerprint and their value [13]. As Umicore indicates:

> The precious metals refinery combines classical methods (cupellation) with unique in-house developed processes (silver refinery), to enable Hoboken to treat all possible variations and ratios of precious metals (silver and gold) and platinum group metals (platinum, palladium, rhodium, iridium and ruthenium)

[13].

BMO processing consists of three main steps: the blast furnace, the lead refinery and the special metals plant. Impure lead bullion, nickel speiss, copper matte and depleted slag are products of reduction of the lead slag (from the smelter) in the blast furnace. The impure lead bullion, containing non-precious metals, is processed in the lead refinery. After further refinement of the special metals residues and sidestreams of PMO, the final products are high quality indium, selenium, tellurium and antimony. The copper matte is returned to the smelter, and nickel speiss is leached to obtain nickel sulfate, while the precious metals remain in the residue and are refined at PMO [13].

The Umicore's plant works continuously so it needs uninterrupted supply of waste material and a lot of energy, therefore Umicore process is profitable only on a large (industrial) scale. An alternative route to pyrometallurgical recovery of precious metals from secondary sources is *hydrometallurgy*. As the advantages of hydrometallurgy the following features are emphasized: the possibility to process waste materials on a small scale, operating temperatures less than 100 °C and the absence of ash after the process (see also **Chapter Electronic waste**).

2.4.2 Hydrometallurgy

2.4.2.1 Extraction of PGM from model solutions

The processes of separation and purification of PGM are difficult and complicated due to, on the one hand, their resistance to chemical attack and, on the other hand, the ability to form many chemical compounds of PGM in chloride solutions. For these reasons it is very important to select effective and economical way to recover these metals from the secondary sources. At first, PGM must be leached from the solid automotive catalysts, and then separated from aqueous solution (*leachate*). Liquid-liquid extraction, i. e. contacting two liquid phases: an aqueous solution of PGM ions and an organic solution of an extractant (responsible for selective and efficient transport of metal ions from the aqueous phase) is often proposed as a hydrometallurgical technique for selective recovery of PGM ions from various, also diluted, solutions.

First of all, some examples of an extraction treatment of model PGM aqueous solutions are given in Table 2.1. The feed solutions of precious metal ions contain mineral acids, HCl, HNO_3 or H_2SO_4, and various concentrations of Pd(II), Pt(IV), Rh(III), Ru(III), or even Ag(I) and Au(III) ranging from 5×10^{-5} to 0.1 M.

Triisobutylphosphine sulfide (*Cyanex 471x*), a commercial extractant was used for effective Pd(II) recovery from nitric acid medium. Organic compounds containing S (Cyanex 471x, ILs containing a disulfide group) and N atoms (trialkylamine (*Alamine 336*) or methyltrioctylammonium chloride (*Aliquat 336*)) are known to have affinity to extract PGM, specially, Pd(II) and Pt(IV). Moreover, some ionic liquids (IL), not only synthesized in a laboratory pyridinium, pyrrolidinium and piperidinium *bis*(trifluoromethylsulfonyl)imides, betainium *bis*(trifluoromethanesulfonyl) imide but also commercial trihexyl(tetradecyl)phosphonium chloride (*Cyphos IL 101*) or bis(2,4,4-trimethylpentyl)phosphinate (*Cyphos IL 104*) were proposed as new extractants for Pd(II) or Ru(III) from model solutions.

2.4.2.2 Leaching of metal ions from solid materials

A *leaching* bases on a mass transfer from a solid to a liquid phase through eluting the desirable ingredients by dissolving these ingredients in a leaching solution, called leaching agent. Two types of dissolution are possible, i. e. chemical and physical. During the chemical dissolution, the soluble component reacts with the solvent. In the physical dissolution, there is no reaction between the ingredient and the solvent. In an ideal situation, as a result of leaching two fractions could be distinguished: one the worthless material that is free from valuable components and is dumped, and the other – the metal ion-pregnant solution (reach in separated component), called also pregnant leaching solution or leachate – that is advanced for further processing. The leaching is a preliminary step in obtaining metal ions from ores or other solid materials [34, 35].

Before the leaching, a material should be crushed to a powder, to increase the surface of mass exchange and reduce the contact time of both phases. The leaching does not require high temperatures, in most cases is carried out at ambient temperature or somewhat higher temperatures, which is one of the main advantages of hydrometallurgy. The dissolution of the solid material is performed in batch, concurrent, or countercurrent modes at slightly below atmospheric, atmospheric, or above atmospheric pressures. The required metal ions are extracted from pregnant leach solutions by various separation techniques [36].

The selection of an adequate leaching method depends primarily on the type, size and composition of the ore or waste. A great advantage of hydrometallurgy is the

Table 2.1: Examples of extraction systems for PGM recovery from model aqueous solutions.

Metal ions	Model solution/stripping phase	Organic phase	Comments	Ref.
Pd(II)	7×10^{-2} M Pd(II) in 2 M HNO_3 (0.5–5 M HNO_3) Stripping: 1 M stabilized thiosulfate solution	2×10^{-2}-0.12 M **Cyanex 471x** (triisobutylphosphine sulfide) in kerosene and 7 % octanol as a modifier	$E_{Pd(II)}$ decreases with increasing HNO_3 concentration up to 2 M then becomes constant (co-extraction of HNO_3) from 92% to 80 %, stripping 75 %	[17]
Pt(IV), Pd(II), Rh(III)	5×10^{-4}-0.1 M Pt(IV), Pd(II), Rh(III) in HCl, H_2SO_4 or HNO_3 (0.1–6 M)	0.1 M **Cyanex 923** (trialkylphosphine oxide) in toluene	$E_{Pt(IV)}$ from HCl: 70%–100 %, from H_2SO_4 or HNO_3: max. 55 %, $E_{Pd(II)}$ from HNO_3: max. 70 %, $E_{Rh(III)}$ from HNO_3: max. 35 %	[18]
Pd(II), Pt(IV)	(1) Model feed: 150 mg/dm^3 Pd(II), 550 mg/dm^3 Pt(IV), 500 mg/dm^3 Mn(II), 1000 mg/dm^3 Ni(II), 1500 mg/dm^3 Fe(III), 100 mg/dm^3 Cr(III) in 3 M HCl	(1) 0.0018–0.183 M **TBP** (tributyl phosphate) in kerosene	(1) Selective and quantitative Pd(II) extraction even at the lowest extractant concentration. 99.9 % of Pd(II) separation from raffinate in 2 stages at 3.75 A/O phase ratio. 99.8 % Pd(II) stripping from loaded organic phase at 5 O/A with 0.5 M $(NH_2)_2CS$ in 0.1 M HCl. Pd(II) enrichment factor of 18.8 times in relation to the feed.	[19]
	(2) Raffinate Pd-Fe free containing Pt(IV), Cr(III), Ni(II), Mn(II) at pH 3.2	(2) 0.002–0.22 M **Aliquat 336** in kerosene	(2) 99.7 % Pt(IV) extraction from Pd-Fe free raffinate at 0.011 M Aliquat 336 in two stages at 3A/O phase ratio 99.9 % Pt(IV) stripping in two stages with 0.5 M $(NH_2)_2CS$ in HCl at 4 O/A phase ratio Pt(IV) enrichment factor of 12 times in relation to the feed	
Pt(IV), Rh(III), Ir(III)	100 ppm Pt(IV), Rh(III) and Ir(III) in 6 M HCl Stripping of Pt(IV): 0.01 M HCl	2 M **TBP** in kerosene	$E_{Pt(IV)}$ ~ 80 %, Rh(III) and Ir(III) remain in a raffinate,	[20]

(continued)

Table 2.1: (*continued*)

Metal ions	Model solution/stripping phase	Organic phase	Comments	Ref.
Ru(III)	0.1 g/dm^3 Ru(III) in 0.1–8 M HCl with 0.5–4.5 M NaCl	0.02–0.6 M **Aliquat 336**, 0.021–0.062 M **Alamine 336** (trialkylamine) in kerosene, 20 % TBP as modifier	$E_{Ru(III)}$: with Aliquat 336: 80 %, with Alamine 336: 72 %, $E_{Ru(III)}$ with Aliquat 336 in the presence of NaCl is much higher than with Alamine 336	[21]
Pt(IV)/Rh(III)	5 × 10^{-4} M Pt(IV), Rh(III) in HCl Efficient stripping: 0.75 M $(NH_2)_2CS$ (thiourea) in 5 M HCl	**Aliquat 336** (methyltrioctylammonium chloride) in kerosene **Aliquat 336** mixed with acidic organophosphorous extractants (*bis*(2-ethylhexylphosphoric acid (**D2EHPA**), 2-thylhexylphosphonic acid mono-2-ethyl hexyl ester (**PC88A**), *bis*(2,4,4-trimethylpentyl) phosphinic acid (**Cyanex 272**)) in kerosene	$S_{Pt(IV)/Rh(III)}$ = 279.2 from 1 M HCl with Aliquat 336 $S_{Pt(IV)/Rh(III)}$ = 612.3 from 3.0 M HCl with 0.01 M Aliquat 336 100 % recycling and reuse capacity of Aliquat 336 in 10 cycles of extraction-stripping	[22]
Ag(I), Pd(II), Au(III)	100–200 mg/kg of metal ions in highly purified water, pH 2.8 Au(III) in multi-metal solution (Li(I), K(I), Ca(II), Ni(II), Zn(II), Au(III)), pH 3.4	Pyridinium, pyrrolidinium and piperidinium based **ionic liquids** with bis (trifluoromethylsulfonyl)imide anion	$E_{Pd(II)}$ 90%–100 % with all ILs (containing such functional groups as disulfide, alkenyl, and nitrile)	[23]
Pd(II), Ru(III), Rh(III)	5 × 10^{-3} M Pd(II) or Ru(III) or Rh(III) in HCl Stripping of: –Pd(II) with 0.5 M ammonia solution, –Ru(III) with 0.1 M thiourea in 0.5 M HCl	Ionic liquids: trihexyl(tetradecyl) phosphonium chloride (**Cyphos IL 101**) or bis(2,4,4-trimethylpentyl) phosphinate (**Cyphos IL 104**) in toluene	$E_{Pd(II)}$ with IL 101: 45%–97 % depending on HCl concentration $E_{Ru(III)}$ with IL 101 and IL 104 ~ 55 $E_{Rh(III)}$ with IL 101 and IL 104 25%–30 Reuse of IL: 5 cycles of extraction-stripping	[24–30]
Pd(II), Rh(III), Ru(III)	Pd(II), Rh(III), Ru(III) in 0.1–10 M HNO_3	**Ionic liquid**: betainium bis(trifluoromethanesulfonyl)imide	$E_{Pd(II)}$ ~100 $E_{Rh(III)}$ max 65 % $E_{Ru(III)}$ 45 %	[31]

(*continued*)

Table 2.1: (*continued*)

Metal ions	Model solution/stripping phase	Organic phase	Comments	Ref.
Pd(II), Pt(IV)	5×10^{-5} M Pd(II) or 5×10^{-5} M Pd(II) and 5×10^{-5} M Pt(IV) in 3 M HCl Stripping: 28 % ammonia solution	Four sulfide-containing monoamides (S-MA): N-methyl-N-octyl-4-thiapentanamide (**MO4**), N-methyl-N-octyl-3-thiapentanamide (**MO3**), N-methyl-N-octyl- phenyl-3-thiapentanamide (**MOPh**) N,N-dioctyl-3-thiapentanamide (**DO3**) in chloroform	After 10–50 min ~ 100 The order of the extractabilities for Pd(II): **MO4 > DO3 ≈ MO3 > MOPh**	[32]
Rh(III), Pd(II), Pt(IV)	1×10^{-3} M Ru(III) in 0.5–10 M HCl/10 M HCl and 10 M HNO_3	N-Methyl-N-n-octyl-chloroacetamide (**MonoAA**) in chloroform or N-methyl-N-octylamine (**TrisAA**) and trimethylamine in chloroform	TrisAA – $E_{Rh(III)}$ = 95 % (from 0.5 to 2 M HCl), MonoAA – $E_{Rh(III)}$ = 75 % (from low concentration HCl)	[33]

E – efficiency of extraction, %, $S_{Pt(IV)/Rh(III)}$ – separation factor of Pt(IV)/Rh(III), A/O – aqueous/organic phase ratio.

possibility of selective separation and efficient recovery of metals from low-grade raw or waste materials, which cannot be realized by other techniques [13].

Two general techniques of leaching are distinguised, and selection of the adequate depends on the size of the raw material [11, 12, 35]:

- **percolation** techniques – leaching of components from stationary ore body,
- **agitation** techniques – leaching of shredded material in stirred tanks.

A division of the leaching techniques into percolation and agitation is presented in Figure 2.6.

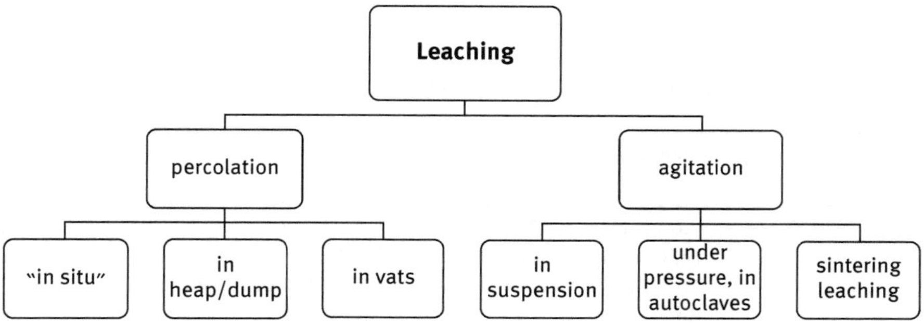

Figure 2.6: Various techniques of leaching [34, 35].

The selection of leaching agent (lixiviant) depends on the composition of the material and properties of the separated metals. In the case of metal leaching from secondary raw materials, agressive lixiviants at high concentration are used. The most frequently as the leaching solutions are used the following [12, 34, 35]:

- mineral acids: H_2SO_4, HCl, HNO_3, and aqua regia (mixture of 3:1 vol. HCl:HNO_3),
- hydroxides: NaOH, NH_4OH,
- chlorides: NH_4Cl, $CuCl_2$, and $FeCl_3$,
- carbonates: Na_2CO_3, $(NH_4)_2CO_3$, and
- water.

As the leaching is the first stage in the preparation of a material for further processing, the leaching solution should be selective for separated metal ions, easy to regenerate, non-volatile, non-toxic, non-flammable, cheap, easily accessible, safe to use and store, and should be well soluble in water. Due to the type of leaching agent, chlorine, ammoniacal, alkaline, cyanide or acid (H_2SO_4, HCl) leaching are distinguished [11, 12, 35]. The leaching can also be carried out using microorganisms, such as bacteria, fungi, algae, and yeast [37–39].

2.4.2.3 Leaching of PGM from spent automobile catalysts

Searching for an effective method of PGM recycling from spent automotive catalysts is nowadays an important and extensively investigated issue due to the high demand for Pd, Pt and Rh. Processing of waste materials to recycle PGM reduces waste, as well as the exploitation of increasingly poorer ores, however, makes sense only if the cost of PGM recovery is significantly lower than the profit from their sale. Thus, the profit depends not only on the type of metal and its price but mainly on the complexity of its treatment to obtain final saleable product. Spent automotive converters are considered to be PGM reach secondary raw materials, however, they also comprise of many base metals. Different compositions of spent automobile catalysts, containing PGM are shown in Table 2.2.

Table 2.2: The content of platinum group and base metals in different spent automobile catalysts.

Spent automobile catalysts	Composition	Ref.
1	3.68 % MgO, 0.43 % CaO, 13.94 % Al_2O_3, 12.54 % SiO_2, 20.0 % Fe, 0.01 % Cu, 0.04 % Ni, 0.16 % **Pt**, 0.12 % **Pd**, 0.042 % **Rh**	[40]
2	20 % Al, 1,3 % Ce, 0.030 % Zr, 0.113 % **Pt**, 0.023 % **Pd**, 0.014 % **Rh**	[41]
3	1.1 % Cr, 3.5 % Fe, 0.4 % **Pt**, 65.4 % O, 29.3 % Al, 0.3 % Si	[42]
4	13.70 % **Pt**, 1.30 % **Rh**, 16.90 % Fe, 2.46 % Ni, 1.15 % Ca, 1.35 % Mg, 0.15 % Cu, 0.17 % Zn, 0.17 % Pb, 0.62 % As, 0.84 % Re, 8.25 % Si	[43]
5	20.3 % Al, 5.9 % Mg, 1.2 % Ba, 1.5 % Zr, 6.2 % Ce, 2300 ppm **Pt**, 530 ppm **Rh**	[44]
6	0.035 % **Pt**, 0.013 % **Pd**, 0.008 % **Rh**, 44.3 % Al, 38.3 % Si, 14.8 % Mg, 1.6 % Fe, 0.56 % Ce, 0.51 % Ni	[45]

The composition of various automobile catalysts differs significantly and depends on the following parameters [12]:

- type of active component, with or without an admixture (platinum with palladium and/or rhodium as an admixture provides high efficiency and stability of the catalyst),
- type of the carrier,
- presence of promoters (e. g. alkali metal oxides or lanthanide oxides, responsible for the proper management of oxygen in the catalyst), and
- application of the catalyst.

Generally, the automotive catalytic converters contain 300–1000 µg/g of Pt, 200–800 µm/g of Pd, and 50–100 mg/kg of Rh [45].

For the first time issues concerning the recycling of PGM from spent automotive catalysts were discussed at conferences in the United States in the 1990s [46, 47]. A first paper about recycling of PGM was published in 1988 [36]. In this article six different ways of platinum, palladium and rhodium recovery from spent automobile catalysts were discussed and compared. As a result of the discussion, leaching from the spent catalyst with concentrated hydrochloric acid was proposed as the optimum and profitable technique.

As spent automobile catalysts contain not only PGM (Pt, Pd, and Rh), but also non-precious (base) metals, such as Cu, Fe, Mg, Al, Ca, Zn and Bi, it is necessary to dissolve them and further separate only the precious metals from ions of other metals existing in the leachate. The metal ions are leached from the spent catalysts usually with concentrated hydrochloric acid, but also with sulfuric acid, and then separated in various extraction and stripping systems [48–55]. PGM generally are very resistant to HCl dissolution, and to form platinum, rhodium and palladium chloro-complexes in chloride solutions a high reduction potential > 0.74 V is required therefore addition of an oxidizing agent is necessary [56]. Thus, to dissolve chemically resistant PGM the *chloride leaching* by HCl involves addition of *chloride precursors* (e. g. $AlCl_3$, $CuCl_2$) or *oxidants* such as HNO_3 (*aqua regia*), H_2O_2, $NaClO_3$ [45, 57, 58].

Some attempts to leach PGM with cyanides at elevated temperature (160 °C) and pressure (0.6 MPa) were made but cyanidation of PGM is not a technique commonly used for leaching of these metals from the automotive catalysts. Recovery of PGM from the pregnant cyanide solutions could be realized by ion exchange, adsorption onto activated carbon, thermal hydrolysis, or zinc cementation and electrowinning [59].

As alumina washcoat is easily dissolved by acids, many PGM recovery methods tend to minimize dissolution of the alumina catalyst support. In an ideal situation PGM should be dissolved from the spent automotive catalyst leaving the alumina support unattacked. In fact, γ-alumina dissolved in HCl under oxidizing conditions has a negative impact on the economy of the PGM recovery process, because of high consumption of the acids and other reagents. To significantly minimize this dissolution, a thermal pretreatment (heating in 1,200 °C) of the spent catalysts is proposed to convert γ-alumina to α-alumina – a compound which is only slightly soluble in both acidic and alkaline aqueous solutions [60].

The concept of minimizing Al dissolution is also realized by leaching with mixture of HCl-$AlCl_3$-HNO_3 that extracts PGM with good efficiency, and does not dissolve the entire alumina washcoat [60]. A Canadian company Duometal Inc. was probably one of the first to successfully apply chloride leaching with high temperature mixture of $AlCl_3$, HCl and HNO_3. As rhodium dissolution is always lower than platinum one, it is important to maintain high temperature (80–90 °C) and $AlCl_3$/HCl ratio to increase rhodium recovery. The solid after leaching was additionally washed with aqueous solution of $AlCl_3$, H_2SO_4 and water to recover the unleached PGM. To separate Pt(IV) and Pd(II) from the leachate, sorption with Emberen 345 resin at pH 3 was carried out and finally the precious metals were precipitated by cementation with Al powder. The final recovery of platinum and palladium from catalysts reached 97%–98 % [45].

Another proposal to recover PGM efficiently from spent automotive catalysts, and to reduce quantity of the reagents used includes prereduction of the catalyst samples (particle size < 0.59 mm) with formic acid (HCOOH) [58]. At HCl concentration less than 4 M leaching of PGM from prereduced samples was significantly enhanced. On the contrary, at higher HCl concentration an effect of prereduction on PGM leaching was negligible. The optimal conditions of prereduction were established as follows: 15 vol% HCOOH, 10 % pulp density, 60 °C, 1 h. After this pretreatment, leaching with 2 M HCl at 90 °C for 2 h resulted in more than 80 % Pt(IV), 85 % Pd(II), and 60 % Rh (III) in the leachant. Moreover, addition of an oxidant 1.5 M $NaClO_3$ enhanced leaching efficiency of PGM up to 95 %.

There are many proposals of leaching/liquid-liquid extraction systems for processing of PGM from secondary sources. PGM are selectively extracted from leachates by different types of extractants, e. g. hydroxyoximes, dialkyl sulfides, alkyl derivatives of 8-hydroxyquinoline, hydrophobic amines, esters of pyridine-carboxylic acids, trialkylphosphine oxides and organophosphoric acids [61], some of them are presented in Table 2.1. As concentration of HCl in the leach solutions is reported to range from 1 up to 12 M HCl [38, 56], PGM will be present in chloride solutions as various chloro-complexes. Thus, these chloro-complexes can be extracted to the organic phase according to different reactions, e. g. by anion exchange, addition or chelation.

Table 2.3 presents some proposals of the leaching/liquid-liquid extraction systems for hydrometallurgical recovery of PGM from automotive catalysts. For the presented systems of the leaching/extraction the most effective group of PGM extractants from the acidic leachates are amines. It can be explained by formation of anionic chloro-complexes in the acid solutions that are exchanged for anions paired with protonated nitrogen of the amine. The extraction efficiency of Pt(IV) from chloride solutions with trioctylamine (*TOA*) or Alamine 336 equals to more than 90 % [64, 65]. Also, amide DMDPHTDMA extracts quantitatively Pt(IV) and simultaneously almost does not transfer Al(III) and Fe(III) to the organic phase.

A proposal of flowsheet of HCl-$AlCl_3$-HNO_3 leaching of PGM from spent automotive catalysts and liquid-liquid extraction of PGM from the leachate, including also chloride removal and absorption of gases evolved during the leaching step is presented in Figure 2.7. One of the advantages of the treatment proposed is a selective extraction of PGM, and leaving the γ-alumina washcoat as much unattacked as possible [60].

Gemini Industries Inc. (USA) developed a hydrometallurgical technology for recycling of soluble alumina substrate catalysts, particularly petroleum refinery and petrochemical catalysts containing Pt, Re, Ir and Pd [66]. However, it is suggested that also platinum and palladium automotive catalysts can be treated in

Table 2.3: Proposals of leaching/liquid-liquid extraction systems for hydrometallurgical recovery of PGM from the automotive catalysts.

Materials	Leaching solutions and conditions	Concentration of metal ions in solutions after leaching	Organic phase	Extraction efficiency	Ref.
Automotive catalyst containing Pt, Pd, Rh and Cu, Fe, Mg, Ca, Zn, Bi	6 M HCl	0.292 mg/dm^3 Rh(III), 0.0017 mg/dm^3 Pt(IV) and 0.009 mg/dm^3 Pd(II) in 6 M HCl	**TBP** in petroleum ether, kerosene or hexane	Separation of Pt(IV) and Pd(II) from Rh(III) after 1 min, $E_{Pt(IV)}$ ~94 %, $E_{Pd(II)} < 50$ %	[50]
Spent automotive catalyst crushed into 4 x 2 x 2 cm pieces	1.66 M HCl-0.29 M $AlCl_3$-3.5 M HNO_3 (mixture) temperature: 95 °C time: 2.5 h	1070 ppm Pt(IV), 50 ppm Rh(III) Recovery of leaching: 85 % Rh(III), 95 % Pt(IV), Pd(II) was not detected in the catalyst	7-(4-ethyl-1-methyloctyl)-8-hydroxyquinoline (**Kelex 100**) or alkyl substituted 8-hydroxyquinoline (**LIX 26**)	No data	[60]
15 g of material from a spent catalyst	100 cm^3 of concentrated HCl, 50 cm^3 of a 10 % $NaClO_3$ solution (adding dropwise), temperature: 70 °C	520 mg/dm^3 of Pt(IV), 5.5 mg/dm^3 Pd(II), 1.3 mg/dm^3 Rh(III), 204 mg/dm^3 Fe(III), 8000 mg/dm^3 Al(III), 2.0 mg/dm^3 Cu(II), 3.4 mg/dm^3 Ni(II), 0.2 mg/dm^3 Co(II), Mn(II) <0.5 mg/dm^3 in 5 M HCl	0.1 M N,N'-dimethyl-N,N'-diphenyltetradecylmalonamide (**DMDPHTDMA**) in 1,2-dichloroethane	$E_{Pt(IV)} = 100$ %, $E_{Pd(II)} = 40$ %, $E_{Al(III)} = 5$ %, $E_{Fe(III)} = 9$ %, Rh(III), Cu(II), Ni(II) remain in a raffinate	[62]
Spent automotive catalyst	8 M HCl, room temperature, time: 30 min	Final solution 625 mg/dm^3 Pt(IV) and 104 mg/dm^3 Rh(III)	2–24 % trioctylamine (**TOA**) in toluene	$E_{Pt(IV)} = 83–95$ %, $E_{Rh(III)} < 20$ %, Stripping (8 M HNO_3): $S_{Pt(IV)} = 82\%–99$ %, $S_{Rh(III)} < 1$%	[64]

(continued)

Table 2.3: (*continued*)

Materials	Leaching solutions and conditions	Concentration of metal ions in solutions after leaching	Organic phase	Extraction efficiency	Ref.
Spent automotive catalyst	Oxidative roasting in air at 500 °C for 2 h, chloride leach solutions: 15 % HCl and 0.33 % $AlCl_3$, temperature: 90 °C time: 90 min	19.4 mg/dm^3 Pt(IV), 10.8 mg/dm^3 Pd(II), 3.3 mg/dm^3 Rh(III) in HCl (pH = 1.72)	5 % **Alamine 336** in kerosene	$E_{Pt(IV)}$ and $E_{Pd(II)}$ above 99 %, Rh(III) remains in the raffinate	[65]
Spent automotive catalyst	50 cm^3 of concentrated HCl and HNO_3 (9:1), temperature: 100 °C, time: 2 h	1.84 g/kg Pt, 0.94 g/kg Fe, 0.26 g/kg Zn, 0.06 g/kg Cu, 0.03 g/kg Mn, 0.92 g/kg Pb(II), 0.06 g/kg Ni(II), 0.03 g/kg Cr(III), 0.02 g/kg Ti(IV), 0.08 g/kg of Al(III)	**Cyanex 921** (trioctylphospine oxide) in kerosene	$E_{Pt(IV)}$, $E_{Zn(II)}$ and $E_{Fe(III)}$ more than 95 %, $E_{Cu(II)}$ = 50 %, Stripping (1 g/dm^3 phenanthroline or 1 g/dm^3 citric acid): $S_{Pt(IV)}$ = 64 %, $S_{Fe(III)}$ = 23 %, $S_{Pt(IV)}$ = 98 %, $S_{Fe(III)}$ = 50 %	[63]

E – efficiency of extraction, %, S – efficiency of stripping, %

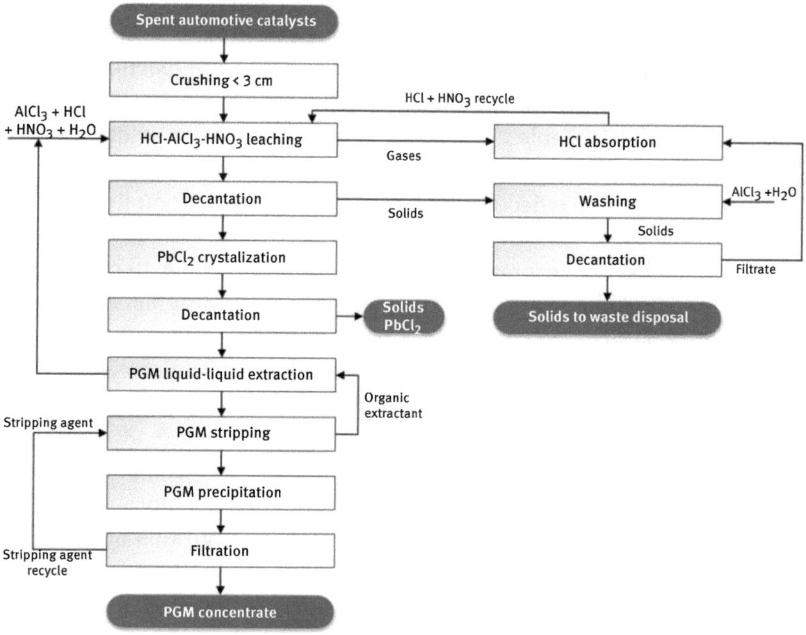

Figure 2.7: Flowsheet of HCl-$AlCl_3$-HNO_3 leaching and extraction of PGM from spent automotive catalysts [60].

Gemini process based on the dissolution of an inert carrier (Al_2O_3) in sulfuric acid [45]. Precious metals are collected as a leach cake that is then treated with aqua regia.

In contrast to pyrometalurgical processes, the company recovers more Pt and Pd (99.99 % purity, 90 % recovery from waste material) and significantly more Re, and it operates at low temperatures, and does not produce waste slag

2.4.2.4 Adsorption and ion exchange

The *adsorption* process means an anion-exchange process which takes place only on the surface of an adsorbent. Thus, the loading or the exchange capacity of adsorbents is relatively smaller in comparison with ion-exchange resins. The characteristics of surface and the pores of adsorbent are important parameters that influence the capacity and decide on their applications. Ion exchange technique is realized by passing a metal ion-containing solution through bed of solid *ion-exchange* resin (IX). The metal ions are exchanged with the ions of the resin carrying accessible cations, anions or amphoteric ions. Afterwards, the pregnant/loaded resin is washed with another aqueous solution to elute the adsorbed metal ions and to regenerate the resin. This is a technique useful for concentration, purification and separation of chemically similar metallic elements present in aqueous solutions [35]. Ion-exchange resins interact electrostatically with the ions leading to stoichiometric capture of the sorbed ion. The exchange is influenced by relative ionic sizes, valence and concentrations of the ions displaced. These stoichiometric ion-exchange reactions distinguish ion-exchange resins from adsorbents like an activated charcoal.

Though some examples in Table 2.4 refer to model solutions or solutions from leaching of other materials than automotive catalysts, they all are acidic chloride or chloride/nitrate solutions which reflect conditions used for spent automotive catalyst dissolution.

Strongly basic anion exchangers react with anionic chloro-complexes of Pd(II) by interaction with protonated amine groups on the resins, e. g. Lewatit MonoPlus SR-7 [73]. These resins are characterized by good kinetic properties, also, they can be regenerated with ammonia solution and reused repeatedly without decreasing their adsorption capacity. High affinity of Purolite S-984 for noble metal ions was revealed in both single and tertiary component solutions at agitation speed 180 rpm and temperature 40 °C [74]. In the case of weakly basic anion exchangers, e. g. Purolite S-984, A-830 [74, 75], the anion exchange mechanism (of anionic chloro-complexes of precious metals) takes place with the polyamine functional groups of the resin protonated in the concentrated HCl. However, in weakly acidic solutions a chelating mechanism takes place (inner-sphere complexation with Pd(II), Pt(IV) or Au(III)) because the functional groups of the resin are deprotonated. Thus, Purolite S-984 is not selective when it is used as a sorbent of metal ions that form stable anionic chloro-complexes (tertiary component Pd(II)-Pt(IV)-Au(III) solutions).The sorption process with weakly basic anion exchangers can be controlled by various steps (simultaneously or one of the steps) [76]: (1) transport of Pd(II) in the bulk solution (can be neglected when stirring is vigorous), (2) Pd(II) diffusion across the so-called liquid film surrounding resin beads, (3) Pd(II) diffusion in the liquid contained in the pores of the resin particle and along the pore walls – intraparticle diffusion, and (4) sorption of Pd(II) chloro-complexes on the resin and its desorption from the resin surface.

Polymeric sorbent (MP-102) coordinates Pd(II) and Ni(II) by the phosphine oxide groups, preferably in a bidentate mode (chelation), while Cu(II) adsorption proceeds through both amine and phosphine oxide groups. High selectivity of Pd(II) adsorption results from higher energy gain due to coordination of Pd(II) than Cu(II) and Ni(II).

Maximum monolayer adsorption capacity (MMAC) of triazine-hexamine (TAPEHA) polymer is very high (almost 520 mg Pd(II)/g). Pd(II) adsorption onto TAPEHA particles is a spontaneous, physisorptive, and exothermic process that proceeds according to ligand exchange mechanism. Similarly to previously described commercial Purolite, Amberlite or Lewatit Mono Plus resins, in low pH the dialkylamino groups of the TAPEHA are protonated and positively charged, and attract anionic chloro-complexes of Pd(II). Additionally, 1,3,5-triazine compound is not only a cross-linking agent but also the presence of triazine N atoms (–NH–CH_2–NH–) positively affects Pd(II)

Table 2.4: Adsorbents and ion-exchangers used for removal of Pd(II) or Pt (IV) from various solutions.

Sorbent/matrix	Modification/sorbent parameters	Conditions of sorption/desorption	Ref.
Lewatit MonoPlus SR-7/crosslinked polystyrene	Unmodified Strongly basic IX, **MMAC**: 197.0 mg Pd(II)/g MBS: 0.57–0.67 mm	**Model feed solutions**: – Pd(II) in chloride and chloride–nitrate(V) solutions, molar ratios of $HCl:HNO_3$ = 1:9, 1:4, 1:1, 4:1, 9:1 – 10 % HCl, 5 % $AlCl_3$, 100 mg/dm^3 Pd(II) – effect of Al(III) presence **Eluting solutions**: HCl, HNO_3, NH_4OH, five cycles of sorp-tion/desorption	[73]
Lewatit TP 214/crosslinked polystyrene	Unmodified Chelating IX, with thiourea groups, **MMAC**: 33.22 mg Pt(IV)/g, **MBS**: 0.55 mm	**Model feed solutions**: Pd(II) in chloride, pH 1.5 **No elution studies**	[67]
Amberlite IRA-478RF/polyacry-late	Unmodified Strongly basic IX, gel, with the tertiary amine and quaternary ammonium functional groups	**Model feed solutions**: – Pd(II) in chloride and chloride–nitrate(V) solutions (HCl: HNO_3) – Pd(II), Pt(IV), Au(III) mixture (molar ratio 1:1:1) in 0.1 M HCl **Eluting solutions**: HCl, HNO_3, H_2SO_4, NaOH, NH_4OH, $(NH_2)_2CS$ (thiourea), three cycles of sorption/desorption	[74]
Amberlite XAD-1180/polystyrene-divinylbenzene	IX modified with 2-mercaptobenzothiazole **MMAC**: 50 mg Pd(II)/g	**Model feed solutions**: – Pd(II), pH 4 – Leachate of Pd(II) from spent catalyst **Eluting solution**: 2 % $(NH_2)_2CS$ (quatitative Pd(II) deso-rption), three cycles of sorption/desorption	[79]

(*continued*)

Table 2.4: (*continued*)

Sorbent/matrix	Modification/sorbent parameters	Conditions of sorption/desorption	Ref.
Amberlite XAD-7/polystyrene-divinylbenzene	IX modified with Cyphos IL 101 (tetradecyl(tri-hexyl)phosphonium chloride **MMAC**: 71 mg Pd(II)/g, 74.6 mg Pt(IV)/g	**Model feed solutions**: Pd(II) or Pt(IV) in HCl solutions **Eluting solution:** – for Pd(II): HNO_3 and $(NH_2)_2CS$ in HCl, five cycles of sorption/desorption – for Pt(IV): HNO_3, five cycles of sorption/desorption	[68, 69]
Dowex MSA-1, Dowex MSA-2/styrene-divinylbenzene	Unmodified IX, working anion exchange capacities: 0.0066 and 0.0050 g/cm^3, for Dowex MSA-1, Dowex MSA-2	**Model feed solutions:** Pd(II) in chloride and chloride–nitrate(V) solutions ($HCl:HNO_3$) **No elution studies**	[70]
Purolite S-984/polyacrylate	Unmodified Weakly basic IX, chelating, functional groups of the polyamine type **MMAC**: 504.3 mg Pd(II)/g	**Model feed solutions:** – Pd(II) in chloride and chloride–nitrate(V) solutions ($HCl:HNO_3$) – Pd(II), Pt(IV), Au(III) mixture (molar ratio 1:1:1) in 0.1 M HCl **Eluting solutions**: HCl, HNO_3, H_2SO_4, NaOH, NH_4OH, $(NH_2)_2CS$, three cycles of sorption/desorption	[74]
Purolite A-830/polyacrylate cross-linked with divinylbenzene	Unmodified Weakly basic IX, polyamine, complex amine groups **MMAC**: 356.88 mg Pd(II)/g, **MBS**: 0.3 – 1.2 mm	**Model feed solutions:** – Pd(II) in chloride and chloride–nitrate(V) solutions ($HCl:HNO_3$) – 10 % HCl, 5 % $AlCl_3$, 100 mg/dm^3 Pd(II) – effect of Al(III) presence **Eluting solutions**: HCl, HNO_3, NH_4OH, five cycles of sorption/desorption	[75]

(*continued*)

Table 2.4: (*continued*)

Sorbent/matrix	Modification/sorbent parameters	Conditions of sorption/desorption	Ref.
Varion ADAM/polyacrylate	Unmodified Weakly basic IX, **MMAC**: 121.48 mg Pd(II)/g, **MBS**: 0.315–1.250 mm	**Model feed solutions:** Pd(II) in HCl-NaCl solutions **No elution studies**	[76]
Modified polyacrylate	Modified with thiourea and glutaraldehyde Chelating **MMAC:** 10.26 mg Pt(IV)/g, 14.92 mg Pd(II)/g	**NiS fire-leaching solution**: Pt(IV), Pd(II), Rh(III), Ru(III), Au (III), Cu(II), Ni(II), Zn(II), Sn(II) in aqua regia, optimal pH 1.50 **Eluting solution**: HCl, $(NH_2)_2CS$, or a mixture of $(NH_2)_2CS$ and HCl, five cycles of sorption/desorption	[71]
MP-102	Modified with aminophosphine oxide groups **MMAC:** 1 mmol Pd(II)/g at pH 2	**Model feed solution**: Pd, Ni and Cu in 0.01–1 M HCl – up to 96 % selectivity for Pd in presence of Cu and Ni at pH 1 **Eluting solution:** 1 M $(NH_2)_2CS$ with 0.01 M HNO_3, 99 % of Pd(II) elution	[72]
S,N-containing sorbent	Hetero chain polymer, thio-methylated polyamines, **MMAC**: 1.4 g Pt(IV)/g, 3.0 g Pd(II)/g, 0.6 g Rh(III)/g	**Real leaching solution – aqua regia and leaching with HCl and H_2O_2:** 1.5×10^{-4} –5×10^{4}wt% Pt(IV), 0.170–0.189 wt% Pd(II), 0.018–0.021 wt% Rh(III)	[10]
1,3,5-Triazine-pentaethylenehexa-mine (**TAPEHA**)	Unmodified **MMAC**: 517.2 mg Pd(II)/g **BET**: 10.981 m^2/g, **TPV**: 2.781 cm^3/g	**Model feed solution**: 50 mg/dm^3 Pd(II), optimum pH 2 (HCl or HNO_3) with different Cl^- concentrations **No elution studies**	[77]

(*continued*)

Table 2.4: (*continued*)

Sorbent/matrix	Modification/sorbent parameters	Conditions of sorption/desorption	Ref.
Alumina nanopowder	Functionalized with polyethyleneimine, **MMAC**: 97.7 mg Pd(II)/g at pH 6 **BET**: 155 m^2/g	**Model feed solution**: Pd(II) in chloride solution **Eluting solution**: 0.2 M $(NH_2)_2CS$, in three cycles decrease in Pd(II) adsorption from 91.4% to 90.8 %	[80]
Activated carbon	Unmodified **MMAC**: 35.71 mg Pd(II)/g, 45.45 mg Pt(IV)/g, **BET**: 922.33 m^2/g **TPV**: 0.569 cm^3/g	**Model feed solution**: 100 mg/dm^3 Pd(II) or Pt(IV), optimum pH 2 **No elution studies**	[11-3]
Activated carbon (bituminous coal)	Activated by steam at 750 °C in a furnace, chemically treated with tris(hydroxymethyl)aminomethane and thiophosphoric acid (commercial extractant MSP-8)	**Model feed solution**: Pd(II), Pt(II), Ni(II), Cr(III) in 0.1 M HCl **No elution studies**	[78]

MMAC – maximum monolayer adsorption capacity, MBS – mean bead size, BET – specific surface area, TPV – total pore volume adsorption by chelating of Pd species. TAPEHA resin was shown to be stable, acid-resistant and strong when contacted with 2 M HNO_3 and 0.1 M NaCl for 36 h [77]. However, there are no information on desorption and reuse of the regenerated TAPEHA resin.

Addition of S-containing groups (e. g. Lewatit TP 214 or carbon activated with thiophosphoric acid) at a surface of sorbents increases the selectivity of removal of the precious metals from mixtures of metal ions due to covalent interaction between Pd(II) or Pt(IV) anionic chloro-complexes and sulfur atoms [78].

Amberlite XAD 1180 was used to recover Pd(II) from leachate (mixture of 10 % HCl and 5 % H_2O_2) after treatment of spent catalyst (5 % palladium on activated carbon) [79]. pH of the solution before adsorption was adjusted to the optimum value 4, and 90 % of Pd(II) was adsorbed onto Amberlite modified with mercaptobenzothiazole.

Palladium is efficiently desorbed mostly with thiourea ($(NH_2)_2CS$) solutions because Pd(II) forms stable complexes with thiourea and disturbs the equilibrium between the resin surface and the adsorbate. Another way to reuse the resin (amine-functionalized alumina) with adsorbed Pd(II) is a proposal of its application as a catalyst for the catalytic reduction of 4-nitrophenol and Cr(VI), and also hydrogen generation from ammonia borane. Pd-containing resins demonstrated an excellent catalytic activity and reusability toward energy and environmental applications [80].

2.4.2.5 Other hydrometallurgical proposals

Among a great number of hydrometallurgical approaches to recovery and separation of PGM from model and real solution, some other than the previously described are noteworthy. Not only classical liquid-liquid extraction is proposed for efficient extraction of Pd(II) or Pt(IV) but also a developing technique for separation of multi-metal complicated systems called “three-liquid-phase extraction” (TLPS). TLPS bases on unique separation selectivity from three coexisting liquid phases with different physicochemical properties, for example 1-butyl-3-methylimidazolium hexaflurophosphate (S201), 1-butyl-3-methylimidazolium hexafluorophosphate ($[C_4mim][PF_6]$) and the aqueous hydrochloric acid solutions containing Pd(II), Pt(IV) and Rh(III) [81]. An efficient separation of Pd(II)/Pt(IV)/Rh(III) is obtained by partition of metal ions among three different phases of S201-H_2O-$[C_4mim][PF_6]$ system. It means that most of Pd(II) is extracted into the 50 vol% S201/nonane organic top phase, while most of Pt(IV) is transported into IL bottom phase $[C_4mim][PF_6]$, and finally, Rh(III) remains in the aqueous middle phase without being extracted. Some difficulties to control the phase forming behavior are emphasized as a drawback of TLPS, limiting its potential application.

Another concept is represented by selective precipitation of Pt from chloride solutions containing base metals, using S-bearing liquids such as sodium thiosulfate solution ($Na_2S_2O_3$). Pt (99.5 %) was precipitated as sulfide from 4 M HCl solution.

Separation factors of Pt(IV) from such base metals as Fe(III), Co(II), and Cr(III), are as follows: $\beta_{Pt/Fe} = 112 > \beta_{Pt/Co} = 9 > \beta_{Pt/Cr} = 5$ [82].

Johnson Matthey Company proposes joining pyro- and hydrometallurgy in one process for PGM refining from various waste materials, also from spent automotive converters. The process involves three key steps [83]:

– smelting – the material is melted in temperatures over 1,200 °C for 12 h in large revertory furnaces to separate non-metallic components,
– chemical leaching of a silver bullion (from smelting) by concentrated HCl to leach Ag(I), Pt(IV) and Pd(II) and to separate out Au (gold impurities are sent for further refining), and
– chemical separation – multistage process including dissolution, series of solvent extractions, evaporation, precipitation and filtration to separate five PGM (Pt, Pd, Rh, Ru, and Ir) in a form of a PGM sponge.

2.4.3 Biometallurgy

Biometallurgy or biohydrometallurgy is used as an alternative to pyrometallurgy or hydrometallurgy to recover metals from very low-grade materials. As it is highlighted in **Chapter Electronic waste** biometallurgy is considered an eco-friendly, cost and energy saving process. This process is carried out at temperature and pressure of the environment, without complicated equipment. Biometallurgy covers, generally, bioleaching, biosorption and biodeposition [84, 85].

2.4.3.1 Bioleaching

Bioleaching utilizes natural leaching ability of various microorganisms (bacteria or fungi) to convert metals into their soluble forms through direct or indirect mechanism [86, 87]. Direct leaching (Figure 2.8) means that microorganism cells contact by adhesion directly with the surface of a material to bio-oxidize it. The electrons are acquired directly from the reduced materials. Moreover, the bacteria produce various enzymes, aminoacids or organic acids which enhance metal leaching. In indirect

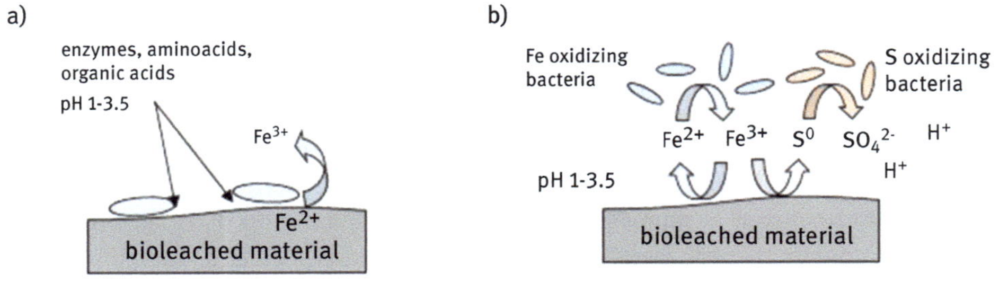

Figure 2.8: Mechanism of a) direct and b) indirect bioleaching.

mechanism the Fe oxidizing bacteria regenerate the oxidizing agent (Fe^{3+}), while S oxidizing bacteria produce sulfuric acid (Figure 2.8). Such a typical bioleaching with acidiphilic bacteria was described in **Chapter Electronic waste**.

Various microorganisms that play an important role in the recovery of metals, particularly the precious ones, from ores and industrial waste (electronic scrap, spent catalysts) are listed in Table 2.5.

Biometallurgy is applied mainly for sulfide ores or requires another source of sulfur for the microorganisms to produce sulfuric acid which provides acidic living conditions for the microbes (pH < 3.5), see Figure 2.8 [91].

Table 2.5: Biometallurgical systems for recovery of precious and base metals from spent automotive catalysts.

Microorganism type	Strains	Metals recovered	Ref.
Bacteria	*Streptomyces erythraeus, Spirulina platensis, Pyrobaculum islandicum, Pyrococcus furiosus, Archaeoglobus fulgidus, Ferroglobus placidus, Thermotoga maritima, Pyrobaculum aerophilum, Shewanella algae strain BRY, Geobacter metallireducens, Geobacter sulfurreducens, Desulfuromonas palmitatis, Geovibrio ferrireducens, Geothrix fermentans, Desulfovibrio vulgaris, Chromobacterium violaceum, Pseudomonas fluorescens, Pseudomonas plecoglossicida, Streptomyces erythreus, Desulfovibrio desulfuricans, Desulfovibrio fructosivorans, Rhodopseudomonas palustris, Escherichia coli*	Au, Ag, Pt, Pd	[85, 88, 93, 94, 97, 98, 100, 101, 115–117, 119, 120, 122, 124]
Bacteria	*Acidithiobacillus ferrooxidans, Acidithiobacillus thiooxidans*	Fe, Cr	[84, 92]
Yeast	*Saccharomyces cerevisiae*	Ag, Pt	[89, 93, 99, 102, 103]
Algae	*Chlorella vulgaris, Fucus vesiculosus, Dealginated Seaweed Waste, Sargassum fluitans, Sargassum natanss, Ascophyllum nodosum*	Au, Pt, Pd	[89, 90, 93]

As PGM require strongly oxidizing conditions to be dissolved, it is rather not possible to bioleach them with microorganisms. Thus, an opposite approach to metal bioleaching from spent automotive catalysts assumes the use of *Acidithiobacillus* strains to leach base metals (Fe, Cr and Al) from metallic support, leaving PGM not leached and, thus, separating them from the non-precious metals [92]. Fourteen-day-long contact with bacteria led to conclusion that the leaching of Fe and Cr proceeded as biological oxidation while Al leaching did not depend on the presence of the microorganisms and was attributed to chemical oxidation.

Recovery of precious metals, particularly PGM, from spent automotive converters is not carried out by bioleaching with *Acidithiobacillus* bacteria because these metals require strongly acidic or oxidizing conditions. The most frequently used for PGM treatment are the strains of *Desulfovibrio* but their role is not leaching but biosorption.

2.4.3.2 Biosorption

Biosorption is an analogous technique to the sorption, and covers accumulation and concentration of metal ions from solutions by biomaterials, i. e. either microorganisms (algae, fungi, bacteria actinomycetes, and yeast) or biopolymers and biowaste materials [93].

Biosorption is considered as a physiochemical, very complex metabolism dependent process, and is proposed to be an alternative to hydrometallurgy. It makes possible recover metal ions from very low concentrated leachates, which is not always possible when classical hydrometallurgical operations are employed [37]. The efficiency of metal ions recovery is influenced by physical and/or chemical sorption onto the cell wall of microorganisms, it means that a chemical structure of the microorganisms significantly affects the sorption properties of the biomass. Specific effects such as complexation, coordination, chelation or ionic interactions between metal ions and ligands are responsible for the mechanism of biosorption and depend on specific properties of the microorganisms, e. g. alive, dead or derived microbial cells can be used as sorbents. Alternatively, biosorption can be related to formation of various ligand groups as a result of cell metabolism [94–96]. Additionally, further removal of adsorbed metal ions depends on metabolism of the biomass used, and includes precipitation of metals in the form of sulfides or phosphates, sequestration by metal-binding proteins, peptides or siderophores, transport and internal compartmentalization [97].

PGM can be biosorbed from leachates after chemical dissolution of spent automotive catalytic converters. Pt and Pd ions are proposed to be biosorbed onto biomass (bacteria or yeasts) and biopolymer (*chitosan*) derivatives [97–99]. The mechanism of Pd or Pt recovery by the *biomass* is assigned to two steps: biosorption and bioreduction [100].

Desulfovibrio desulfuricans (NCIMB 8307), *Desulfovibrio vulgaris* (NCIMB 8303) and *Desulfovibrio fructosivorans* (DSM 3604) grown anaerobically for 2 days were used as biosorbents of Pd(II) and Pt(IV) at pH 2 for 4 days [97]. Adsorption of 19 and 9 % of Pd and Pt, respectively, in respect to dry biomass cells of *Desulfovibrio desulfuricans* were the highest results obtained in this system. Such microbial biopolymers (present in the cell wall) as proteins, nucleic acids and polysaccharides are able to bind metal ions, and their interactions depend on charge of the ionisable functional groups which are available at the surface of the biomass. For example, Gram-negative bacteria such as *Desulfovibrio*, comprises a polymer of two sugar derivatives, N-acetylglucosamine and N-acetylmuramic acid which provide carboxyl and amine groups. Additionally, an outer membrane structure of these bacteria includes lipids, lipopolysaccharides and proteins, and extracellular polymeric substances such as sugar residues [97]. Thus, Pd(II) adsorption is attributed partly to ion-exchange of metal anionic complexes for chlorides electrostatically attracted to the protonated groups present on the surface of the biomass. Pt(IV) biosorption is explained by complexation or chelation on sulfur groups of the deprotonated proteins (pH range 0–2).

Selective biosorption and recovery of ruthenium from industrial effluents with purple non sulfur bacteria *Rhodopseudomonas palustris* strains was demonstrated for the first time [101] The bacteria adsorbed about 40 mg of Ru(III) per 1 g of dry biomass. Biosorption tests were carried out on undiluted real effluents of Ru-plating process (acidic or basic) at 65 °C. After biosorption the biomass was centrifuged, lyophilized, and mineralized to recover the sorbed metal, and resulted in 42%–72 % of Ru recovered from acidic or basic wastewater, respectively.

Platinum was biosorbed by the yeast *Saccharomyces cerevisiae*, which is a cheap and abundant source of biomass [99, 102]. A suitable sorbent was produced by immobilization of polyethyleneimine and glutaraldehyde on *S. cerevisiae*. High platinum uptake (150–170 mg/g) at pH less than two was attributed to chemical sorption mechanism. The biosorption is rapid and near 70 % of Pt(IV) from a 50 mg/dm^3 solution is removed in 5 min. The problem with an effective desorption of Pt(IV) must be pointed out as a drawback of the proposed method. Additionally, it is sensitive to the composition of the wastewater, resulting in limited sorption efficiency due to high inorganic ion content and complex speciation of Pt(IV) in real solutions. Thus, biosorption with the yeast requires additional pretreatment steps. Pt sorbed at yeast-based biomass (immobilized on polyvinyl alcohol cryogels) was applied directly as an electro-catalytic anode in a fuel cell (FC) to generate electrical energy from renewable sources (e. g. glucose and ethanol) [103].

Recently, great attention is given to application of biopolymers and biowaste materials as biosorbents of precious metals, among others Pd(II) and Pt(IV) [93]. As, very often, biosorbents are not good adsorbents themselves, they are functionalized (by physical and chemical modifications) to increase stability and sorption capacity of the biological materials. Such chelating agents (containing N and S groups) as thiourea [104], thiocarbamonyl [105], and L-lysine [106] are proposed to modify and functionalize crosslinked chitosan for adsorption of the metal ions. Sorption properties are also considerably increased by incorporation of polyethylenimine (PEI) or ethylenediamine [104, 107–109]. They introduce new reactive groups: primary, secondary and tertiary amine groups with higher reactivity than, for example, carboxylic groups of alginate and algal biomass. Some examples of biopolymer and biowaste materials applied for sorption of PGM from chloride or chloride/nitrate solutions are presented in Table 2.6.

Chitosan is a biopolymer that is frequently applied for biosorption of metal ions from aqueous solutions because it includes amino and hydroxyl groups that can react as chelation sites. It is a N-deacetylated derivative of chitin, a cationic polysaccharide composed of β-D-glucosamine and acetyl-β-D-glucosamine residues with a 1,4 linkage [110]. Depending on conditions of adsorption different mechanisms of metal ion capture are proposed, i. e. chelation or electrostatic attraction. Also, algal biomass posess amino and carboxyl groups present in proteins, carbohydrate, phosphorous and sulfonic groups on other constituents of cell walls that interact with precious metal ions. To increase its sorption capacity it is chemically modified/fuctionalized [108].

An acidic extractant *bis*(2,4,4-trimethylpentyl)dithiophosphinic acid (Cyanex 301) was immobilized into alginate matrix and was used for Pd(II) sorption [111] or reactive precipitation metal reaction with the sulfur functional group of Cyanex 301 [112]. In the letter case, alginate was applied rather for stabilization of emulsion of the extractant in the solution, and afterwards a precipitate of Pd(II) and the

Table 2.6: Biopolymer and biowaste materials for PGM sorption.

Biopolymer or biowaste	Modification/ functionalization	Parameters of biosorbent	Conditions of sorption	Ref.
Magnetic crosslinking chitosan nanoparticles	Magnetization with Fe_3O_4 Modified with ethylenediamine	**MMAC**: 171 mg Pt(IV)/g, 138 mg Pd(II)/g	**Model feed solution**: Pd(II) or Pt(IV) in chloride, mixture of Pd(II) and Pt(IV) in chlorides, pH 2 **Desorption (five cycles)**: – highest selectivity of Pd(II) (82.2 %) desorption with 5 M NH_4OH, – effective of both Pt(IV) and Pd(II) with mixture of 0.4 M HNO_3 and 1.0 M $(NH_2)_2CS$	[96]
Magnetic cellulose (nanoactive adsorbent)	Magnetization (Fe_2O_3 nanoparticles) and functionalization with thiol and amine	**MMAC**: 40.48 mg Pt(IV)/g	**Optimum conditions**: pH 2, initial concentration of Pt(IV) 25 mg/dm^3, 45 °C, adsorbent dose: 0.04 g **Desorption**: 0.5 M H_2SO_4 in 1 M $(NH_2)_2CS$, > 90 % of initial capacity in 4 cycles of regeneration	[114]

(*continued*)

Table 2.6: (*continued*)

Biopolymer or biowaste	Modification/ functionalization	Parameters of biosorbent	Conditions of sorption	Ref.
Alginate matrix	Cyanex 301 (acidic extractant) immobilized at alginate	**MMAC:** 150 mg Pd(II)/g **Size of beds:** 450 ± 30 mm	**Model feed solution:** – Pd(II) in 1 M HCl – Pd(II), Pt(IV), Ni(II), Zn(II), Cu(II) in HCl **Desorption:** 0.2 M $(NH_2)_2CS$, Pd (II) desorption efficiency 57 %	[111, 112]
Algal biomass beads *(Laminaria digitata)*	Glutaraldehyde-crosslinked polyethyleneimine	**MMAC:** 135.68 mg Pd(II)/g, 62.54 mg Pt(IV)/g in pH 2.5	**Model feed solution:** – Pd(II) or Pt(IV), pH 2.5 – Pd(II) and Pt(IV), pH 2.5	[108, 109]
Waste biomass *Corynebacterium glutamicum*	Functionalized with polyethylenimine	**MMAC:** 176.8 mg Pd(II)/g	**Model feed solution:** Pd(II) in HCl **Desorption:** acidified $(NH_2)_2CS$ in 1.0 M HCl	[106]
Active carbon	Modified with chitosan	**MMAC:** 43.48 mg Pd(II)/g, 52.63 mg Pt(IV)/g, **BET:** 362.30 m^2/g **TPV:** 0.230 cm^3/g	**Model feed solution:** 100 mg/dm^3 Pd(II) or Pt(IV), optimum pH 2 **No elution studies**	[113]
Pistachio nut shell	Carbonized at 1000 °C under an argon atmosphere	**MMAC:** 38.31 mg Pt(IV)/g		[71]

extractant were isolated (e. g. aqueous biphasic systems – ABS). This approach can be compared to other biphasic systems developed over the decades. The binding capacity reached even 350 mg Pd(II)/g in this system. Quantitative stripping of Pd (II) from the precipitated biopolymer was not achieved.

The extractant immobilized into alginate (bead shape) has great affinity with Pd (II), and selectively attracts Pd(II) in the presence of Pt(IV) [111]. Generally, when Pd(II) was in excess against Pt(IV) the selectivity coefficient was greater than 500. The presence of base metal ions (Cu(II), Ni(II) and Zn(II)) decreased the adsorption of Pd(II).

A positive effect of modification of active carbon surface with chitosan was shown, resulting in higher MMAC of Pt(IV) and Pd(II) on modified carbon in comparison to the unmodified one [113].

Application of magnetic cellulose or chitosan was proposed to produce nano-active adsorbent for Pt(IV) or Pd(II) removal from model solutions [104, 114]. Magnetic polymers have some advantages that make them prospective for recovery of metal ions by adsorption, i. e. they are chemically stable, magnetic susceptible and therefore easily to be collected and precipitated in a magnetic field. The MMAC of grafted magnetic cellulose with thiol/amine was equal to 40.48 mg of Pt(IV) per one g of adsorbent. It should be emphasized that regeneration of the adsorbent with 0.5 M H_2SO_4 in 1 M thiourea was successful and the capacity in four adsorption cycles was stable and maintained over 90 % of the first cycle capacity [114]. Also, magnetic chitosan nanoparticles were found to be stable and have good resuability for over five cycles [104].

In some cases desorption of PGM from sorbent is difficult that means that sorbent recycling is impossible. However, biosorbents, e. g. algal-based, are relatively cheap, therefore they can be destroyed by thermal degradation (incineration) to recover the precious metals [108, 112].

2.4.3.3 Biodeposition

PGM are proposed to be at first chemically leached and then microorganisms are used to separate Pd from pregnant leaching solution. A series of works of Macaskie is an interesting example of such a PGM treating, called electrobiotechnology. 80 % of metal ions are recovered by microwave-assisted leaching using 50 % diluted aqua regia. The dissolution of acid mixture is necessary for the leachate to be more biocompatible. *Desulfovibrio desulfuricans, Escherichia coli* or *Rhodobacter sphaeroides* were used for hydrogenase-mediated reduction of soluble PGM to cell-bound insoluble metals (e. g. Pd(II)→Pd(0)) [85, 115–118]. The bacteria were treated in a special way, because free cells did not reduce PGM – they were immobilized on a solid alloy which formed a biomass-coated Pd-Ag alloy electrode. Then, the biomass prepared in such a way was used in a flow-through electrobioreactor with electrochemically generated hydrogen for recovery of Pd, Pt and Rh from aqua regia leachates (pH 2.5) of spent automotive catalysts. The efficiency of PGM recovery amounted up to 90 % (15 min of flow residence time). However, in the previous research Yong et al. [119] observed that uptake of Pt(IV) and Rh(III) by *Desulfovibrio desulfuricans* from real leachate is negligible in the presence of Pd(II) (pH 7, in the presence of formate as electron donor). It was concluded that biosorption is effective at pH < 3.5 even for leachates rich in Cl^- but biodeposition, as it is inhibited by chloride excess, would be appropriate for nitrate solutions.

Application of the active biofilm layer in the flow-through electrobioreactor is considered as a prospective recovery technology and way of manufacturing of bioinorganic catalyst for the reductive dehalogenation of chlorophenol and PCB species [120] or flame retardant materials [121]. Such a biocatalyst is proposed also for reduction of Cr(VI) to Cr(III) in formate solution (formic acid-NaOH, pH 3.0) [122, 123]. The Pd coated biomass was shown to remove, after 3 days of continuous operation of such a bioreactor, Pd and Pt in 90 %, and Rh in 70 % as a black precipitation [85]. Reduction of PGM ions to elemental precious metals is attributed to enzymatic reduction by the sulfate-reducing bacteria in the presence of hydrogen as the electron donor. In other words, such sulfate-reducing bacteria have a broad metal-reducing activity coupled to hydrogenase or/and cytochromes. However, the enzyme responsible for Pd(II) reduction has not yet been identified.

The mechanism of Pd(0) deposition assumes an initial biosorption of Pd(II) from model solutions and leachates from spent automotive catalysts onto biomass ligands, e. g. amine groups, at pH 2–3 – as it was described previously in ***Biosorption*** paragraph, followed by hydrogenase-mediated metal reduction to form clusters of Pd(0) which remain associated with the biomass [119]. Even 100 % of Pd(II) reduction was achieved in 5 or 30 min anaerobically at pH 7 or 3, respectively, and Pd(0) was adsorbed at the surface of the biomass, which was confirmed by energy dispersive X-ray microanalysis (EDAX) and X-ray diffraction analysis (XRD). Pd(II) biosorption was selective in relation to Pt(IV) and Rh(III). Biosorption of Pd(II) from leachates (15 %) is inhibited by high chloride (aqua regia) content.

To sum up, the bioconversion of palladium and platinum-containing spent automotive catalysts leads to formation of a new type of catalysts (biocatalysts) for treatment of environmental contaminants or clean electricity generation [117, 124]. A proposal of an "integrated biorefinery" to convert agri-food wastes and precious metal wastes into power [117] is shown in Figure 2.9.

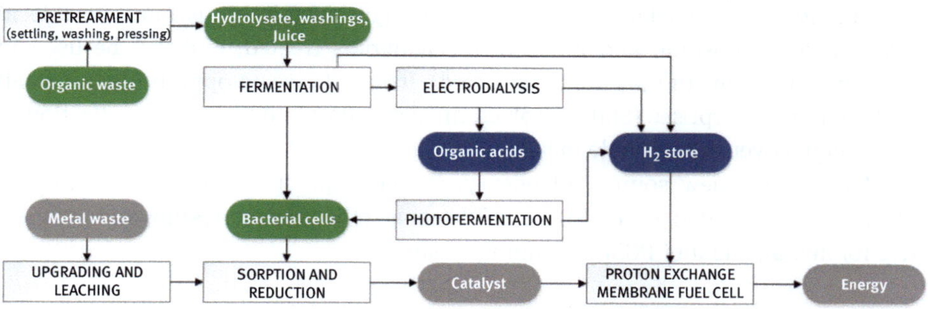

Figure 2.9: The proposal of "integrated biorefinery" for power production from agri-food and precious metal wastes [117].

In the final step of integrated biorefinery proton exchange membrane FCs are proposed for generation of pure energy. The role of an anode and cathode catalyst in FC, composed of precious metals (mainly Pt), is splitting of H_2 into $2H^+$ and $2e^-$. The electrons generate the current, while the protons in a reaction with the oxygen at the cathode form water. Biomass (yeast or bacteria) covered with Pt(0) or Pd(0) is an attractive material to produce active FC biocatalysts (e. g. Bio-Pd anode) and to generate energy (68 mW of power per 16 cm^2 anode [117]) when FC is fed with bioproduced hydrogen [103, 117]. Such a solution involving utilization of waste biomass and metal waste into biocatalyst formation to produce clean energy is considered inexpensive and eco-friendly.

2.5 Summary

A demand for PGM (palladium, platinum, rhodium, and ruthenium) has increased in recent years. Their natural sources are limited, therefore it is important, and both from economical and environmental point of view, to develop effective processes to recover PGM from secondary sources, particularly from spent automotive catalysts.

Searching for an effective method of PGM recycling from spent automotive catalysts is nowadays an important and extensively investigated issue. The processing of secondary materials to recycle PGM reduces waste, as well as the exploitation of increasingly poorer ores, however, makes sense only if the cost of PGM recovery is significantly lower than the profit from their sale. Thus, the profit depends not only on the type of metal and its price but mainy on the complexity of its treatement to obtain final saleable product. Spent automotive converters are considered to be PGM reach secondary raw materials, however, they also comprise of many base metals.

Up to now, PGM are recovered from spent automotive catalysts with pyrometallurgical processes. However, as these operations are high energy consuming and generate emission of harmful gasses, they could be in the future replaced by hydrometallurgy and/or biometallurgy. Though none of the processes is drawback-free, hydrometallurgy is considered more eco-friendly than pyrometallurgy, and less time consuming than biometallurgy. Additionally, hydrometallurgy is more flexible in relation to composition and grade of the materials treated, and can be used in small mobile plants (if they are constructed). It would give an opportunity for small and medium enterprises (SME) involved in spent automotive catalysts collection to treat and recover the valuable metals.

Looking for new sources of energy, the concept of membrane FCs built of biocatalysts deposited with Pd(0) or Pt(0) seems to be very interesting and prospective for management of PGM-containing waste.

Funding: This work was supported by the 03/32/DSPB/0810 grant.

References

[1] Johnson Matthey base prices of platinum group metals. Available at: www.platinum.matthey.com. Accessed: September 10, 2017.

[2] Brown TJ, Wrighton CE, Raycraft ER, et al. World mineral production 2009–13. Keyworth: Nottingham British Geological Survey, 2015.

[3] Crundwell FK, Moats MS, Ramachandran V, Robinson TG, Davenport WG. Extractive metallurgy of nickel, cobalt and platinum-group metals. Oxford: Elsevier, 2011.

[4] Saternus M, Fornalczyk A. An used auto catalytic converters as a source of PGM. Rudy Metale. 2009;54(2):59–67.

[5] Godlewska-Żyłkiewicz B, Pyrzyńska K. Platynowce – zastosowanie i metody oznaczania. (PGM – application and methods of determination). Warszawa: Wydawnictwo MALAMUT, 2012 (in Polish).

[6] PGM Applications, International Platinum Group Metals Association. Available at: http://ipa-news.com/index/pgm-applications/?PHPSESSID=9d78cf16444ff2af103bb7b091dfdacd. Accessed August 1, 2017.
[7] Wołowicz A. Use of palladium and its compounds with particular consideration of catalysis. Przem Chem. 2013;92(7):1237–45 (in Polish).
[8] Froehlich P, Lorenz T, Martin G, Brett B, Bertau B. Valuable metals-recovery processes, current trends and recycling strategies. Angew Chem Int Ed. 2017;56:2544–80.
[9] Cieszyńska A, Regel-Rosocka M, Wiśniewski M. Extractive methods for recovery and separation of noble metals. Przem Chem. 2011;90(8):1579–85 (in Polish).
[10] Eskina VV, Dalnova OA, Filatova DG, Baranovskaya VB, Karpov YA. Separation and preconcentration of platinum-group metals from spent autocatalysts solutions using a hetero-polymeric S,N-containing sorbent and determination by high-resolution continuum source graphite furnace atomic absorption spectrometry. Talanta. 2016;159:103–10.
[11] Serry MA, Abd El-Raof FM, Ahmed SE. Composition and properties of traditional cordierite-mullite ceramics for thermo-mechanical applications. Int Ceramic Rev. 2011;39–43.
[12] Hoffmann JE. Recovering platinum-group metals from auto catalysts. JOM. 1988;40:40–4.
[13] Umicore Precious Metals Refining. Available at: http://pmr.umicore.com/en/metals-products/precious-metals/. Accessed August 23, 2017.
[14] Fornalczyk A, Saternus M. Removal of platinum group metals from the used auto catalytic converter. Metalurgija. 2009;48(2):133–6.
[15] Hagelüken C. Recycling of electronic scrap at Umicore's integrated metals smelter and refinery. World Metall – Erzmet. 2006;59(3):152–61.
[16] Willner J, Fornalczyk A. Electronic scraps as a source of precious metals. Przem Chem. 2012;91(4):517–22 (in Polish).
[17] Ahmed IM, Nayl AA, Daoud JA. Extraction of palladium from nitrate solutions by CYANEX 471X. Int J Miner Process Processing. 2011;101:89–93.
[18] Gupta B, Singh I. Extraction and separation of platinum, palladium and rhodium using Cyanex 923 and their recovery from real samples. Hydrometallurgy. 2013;134–135:11–8.
[19] Lee JY, Raju B, Nagaphani Kumar B, Rajesh Kumar J, Park HK, Ramachandra Reddy B. Solvent extraction separation and recovery of palladium and platinum from chloride leach liquors of spent automobile catalyst. Sep Purif Technol. 2010;73:213–8.
[20] Nguyen TH, Sonu CH, Lee MS. Separation of Pt(IV), Pd(II), Rh(III) and Ir(IV) from concentrated hydrochloric acid solutions by solvent extraction. Hydrometallurgy. 2016;164:71–7.
[21] Panigrahi S, Dash T, Nathsarma KC, Sarangi K. Extraction of ruthenium using both tertiary and quaternary amine from chloride media. Sep Sci Technol. 2014;49:545–52.
[22] Lee JY, Kumar JR, Kim JS, Kim DJ, Yoon HS. Extraction and separation of Pt(IV)/Rh(III) from acidic chloride solutions using Aliquat 336. J Ind Eng Chem. 2009;15:359–64.
[23] Lee JM. Extraction of noble metal ions from aqueous solutions by ionic liquid. Fluid Phase Equilib. 2012;319:30–6.
[24] Rzelewska M, Janiszewska M, Regel-Rosocka M. Application of quaternary phosphonium salts as extractants of Ru(III) and Rh(III) from model aqueous solutions. Chemik. 2016;70:375–7.
[25] Rzelewska M, Wiśniewski M, Regel-Rosocka M. Effect of composition and ageing of chloride solutions on extraction of Rh(III) and Ru(III) with phosphonium ionic liquids Cyphos IL 101 and IL 104. Sep Sci Technol. 2017. DOI: 10.1080/01496395.2017.1295996.
[26] Rzelewska M, Baczyńska M, Wiśniewski M, Regel-Rosocka M. Phosphonium ionic liquids as extractants for recovery of ruthenium(III) from acidic aqueous solutions. Chem Pap. 2016;71:1065–72.
[27] Regel-Rosocka M, Rzelewska M, Baczyńska M, Janus M, Wiśniewski M. Removal of palladium(II) from aqueous chloride solutions with Cyphos phosphonium ionic liquids as metal ion carriers

from liquid-liquid extraction and transport across polymer inclusion membranes. Physicochem Probl Miner Process. 2015;51:621–31.

[28] Cieszynska A, Wisniewski M. Extraction of palladium(II) from chloride solutions with Cyphos®IL 101/toluene mixtures as novel extractant. Sep Purif Technol. 2010;73:202–7.

[29] Cieszynska A, Wisniewski M. Selective extraction of palladium(II) from hydrochloric acid solutions with phosphonium extractants. Sep Purif Technol. 2011;80:385–9.

[30] Cieszynska A, Wisniewski M. Extractive recovery of palladium(II) from hydrochloric acid solutions with Cyphos IL 104. Hydrometallurgy. 2012;113:79–85.

[31] Sasaki K, Takao K, Suzuki T, Mori T, Arai T, Ikeda Y. Extraction of Pd(II), Rh(III) and Ru(III) from HNO3 aqueous solutions to betainium bis(trifluoromthanesulfonyl)imide ionic liquid. Dalton Trans. 2014;43:5648–51.

[32] Narita H, Morisaku K, Tamura K, et al. Extraction properties of palladium(II) in HCl solution with sulfide-containing monoamide compounds. Ind Eng Chem Res. 2014;53:3636–40.

[33] Narita H, Morisaku K, Tanaka M. Highly efficient extraction of rhodium(III) from hydrochloric acid solution with amide-containing tertiary amine compounds. Solvent Extr Ion Exch. 2015;33(4):407–17.

[34] Chmielewski T. Ługowanie metali z rud, koncentratów, półproduktów i odpadów. (Leaching of metals from ores, concentrates, intermediates and waste). Physicochem Probl Miner Process. 1996;30:217–31 (in Polish).

[35] Gupta CK. Chemical metallurgy: principles and practice. Weinheim: Wiley-Vch Verlag GmbH & Co. KGaA, 2003.

[36] Kucharski M. Recykling metali nieżelaznych. (Recycling of non-ferrous metals). Kraków: Wydawnictwa AGH, 2010 (in Polish).

[37] Brandl H, Bosshard R, Wegmann M. Computer-munching microbes: metal leaching from electronic scrap by bacteria and fungi. Hydrometallurgy. 2011;59:319–26.

[38] Szubert A, Grotowski A. Nowoczesne metody przeróbki rud miedzionośnych. (Modern methods of copper ore processing) Rudy Metale. 2008;53(8):467–72 (in Polish).

[39] Jadhav UU, Hocheng H. A review of recovery of metals from industrial waste. J Achievements Mater Manuf Eng. 2012;54(2):159–67.

[40] Kirichenko AS, Seregin AN, Volkov AI. Developing a technology for recycling automotive exhaust-gas catalysts. Metallurgist. 2014;58(3-4):250–5.

[41] Suoranta T, Zugazua O, Niemela M, Peramaki P. Recovery of palladium, platinum, rhodium and ruthenium from catalyst materials using microwave-assisted leaching and cloud point extraction. Hydrometallurgy. 2015;154:56–62.

[42] Fornalczyk A, Kraszewski M, Willner J, et al. Dissolution of metal supported spent auto catalysts in acids. Arch Metall Mater. 2016;61(1):233–6.

[43] Barakat MA, Mahmoud MHH. Recovery of platinum from spent catalyst. Hydrometallurgy. 2004;72:179–84.

[44] Kim CH, Woo SI, Jeon SH. Recovery of platinum-group metals from recycled automotive catalytic converters by carbochlorination. Ind Eng Chem Res. 2000;39:1185–92.

[45] Kasuya R, Miki T, Morikawa H, Tai Y. Dissolution of platinum in catalyst materials using hydrochloric acid: a new method based on the use of complex oxides. Miner Eng. 2016;87:25–31.

[46] Hennion FJ. Availability of precious metals from spent catalysts. Platinum group metals – an in depth view of the industry. Williamsburg, VA, USA: IPMI Seminar; 1983.

[47] Meng X, Han KN Recovery of platinum and palladium from spent automobile catalytic converters by leaching with solutions containing halogen salts, ammonium and oxidants. Proceedings of the 1995 TMS Fall Symposium, Point Clear, AL, USA, TMS, 1995.

[48] Dong H, Zhao J, Chen J, Wu Y, Li B. Recovery of platinum metals from spent catalysts: a review. Int J Miner Process. 2015;145:108–13.

[49] Poṡpiech B. The hydrometallurgical technology for recovery of precious and non-ferrous metals from spent catalysts. Przem Chem. 2012;91(10):2008–10 (in Polish).
[50] Dragulovic S, Ljubomirovic Z, Stanojevic Simsic Z, et al. Recovery of rhodium from secondary raw materials for usage in electronic devices. Optoelectron Adv Mat. 2011;5(12):1370–5.
[51] Sun P, Lee M. Recovery of platinum from chloride leaching solution of spent catalysts by solvent extraction. Mater Trans. 2013;54(1):74–80.
[52] Sun P, Lee M. Recovery of platinum from spent petroleum catalysts by a hydrometallurgical method. Korean J Met Mater. 2013;51(12):873–81.
[53] Jha MK, Lee J, Kim M, Jeong J, Kim BS, Kumar V. Hydrometallurgical recovery/recycling of platinum by the leaching of spent catalysts: a review. Hydrometallurgy. 2013;133:23–32.
[54] Poṡpiech B. Studies on platinum recovery from solutions after leaching of spent catalysts by solvent extraction. Physicochem Problem Miner Process. 2012;48(1):239–46.
[55] Fornalczyk A. Industrial catalysts as a source of valuable metals. J Achievements Mater Manuf Eng. 2012;55(2):864–9.
[56] Jimenez De Aberasturi D, Pinedo R, Ruiz De Larramendi I, Ji RDL, Rojo T. Recovery by hydrometallurgical extraction of the platinum-group metals from car catalythic converters. Miner Eng. 2011;24:505–13.
[57] Nogueira CA, Paiva AP, Oliveira PC, Costa MC. Rosa da Costa AM. Oxidative leaching process with cupric ion in hydrochloric acid mediafor recovery of Pd and Rh from spent catalytic converters. J Hazard Mater. 2014;278:82–90.
[58] Trinh HB, Lee JC, Srivastava RR, Kim S, Ilyas S. Eco-threat minimization in HCl leaching of PGMs from spent automobile catalysts by formic acid prereduction. ACS Sustainable Chem Eng. 2017;5:7302–9.
[59] Schoeman E, Bradshaw SM, Akdogan G, Snyders CA, Eksteen JJ. The extraction of platinum and palladium from a synthetic cyanide heap leach solution with strong base anion exchange resins. Int J Miner Process. 2017;162:27–35.
[60] Boliński L Platinum and rhodium recovery from scrapped automotive catalyst by oxidative acid chloride leaching. MSc thesis. Department of Mining and Metallurgical Engineering, McGill University, Montreal, Canada, 1991.
[61] Regel-Rosocka M, Wisniewski M, Borowiak-Resterna A, Cieszynska A, Sastre AM. Selective extraction of palladium(II) from hydrochloric acid solutions with pyridinecarboxamides and ACORGA®CLX50. Sep Purif Technol. 2007;53:337–41.
[62] Malik P, Paiva AP. A novel solvent extraction route for the mutual separation of platinum, palladium and rhodium in hydrochloric acid media. Solvent Extr Ion Exch. 2011;28:49–72.
[63] Pietrelli L, Fontana D. Automotive spent catalysts treatment and platinum recovery. Int J Environ Waste Manage. 2013;11(2):222–31.
[64] Jaree A, Khunphakdee N. Separation of concentrated platinum(IV) and rhodium(III) on acidic chloride solution via liquid-liquid extraction using tri-octylamine. J Ind Eng Chem. 2011;17:243–7.
[65] Yin CY, Nikoloski AN, Wang MW. Microfluidic solvent extraction of platinum and palladium from a chloride leach solution using Alamine 336. Miner Eng. 2013;45:18–21.
[66] Gemini Industries Incorporated. Available at: http://www.gemini-catalyst.com/ourProcess.htm. Accessed September 20, 2017.
[67] Morcali MH, Zeytuncu B, Aktas S, Yucel O, Gulluoglu AN. Platinum adsorption from chloride media using carbonized biomass and commercial sorbent. Min Metall Proc. 2013;30:129–36.
[68] Navarro R, Garcia E, Saucedo I, Guibal E. Platinum(IV) recovery from HCl solutions using Amberlite XAD-7 impregnated with a tetraalkyl phosphonium ionic liquid. Sep Sci Technol. 2012;47:2199–210.

[69] Navarro R, Saucedo I, Gonzalez C, Guibal E. Amberlite XAD-7 impregnated with Cyphos IL-101 (tetraalkylphosphonium ionic liquid) for Pd(II) recovery from HCl solutions. Chem Eng J. 2012;185:226–35.
[70] Wołowicz A, Hubicki Z. Sorption of palladium(II) complexes onto the styrene-divinylbenzene anion exchange resins. Chem Eng J. 2009;152:72–9.
[71] Morcali MH, Zeytuncu B. Investigation of adsorption parameters for platinum and palladium onto a modified polyacrylonitrile-based sorbent. Int J Miner Proc. 2015;137:52–8.
[72] Ricoux Q, Bocokić V, Méricq JP, Bouyer D, Zutphen S, Faur C. Selective recovery of palladium using an innovative functional polymer containing phosphine oxide. Chem Eng J. 2015;264:772–9.
[73] Wołowicz A, Hubicki Z. Ion exchange recovery of palladium(II) from acidic solutions using monodisperse Lewatit SR-7. Ind Eng Chem Res. 2012;51:16688–96.
[74] Wołowicz A, Hubicki Z. Polyacrylate ion exchangers in sorption of noble and base metal ions from single and tertiary component solutions. Solvent Extr Ion Exch. 2014;32:189–205.
[75] Wołowicz A, Hubicki Z. Applicability of new acrylic, weakly basic anion exchanger Purolite A-830 of very high capacity in removal of palladium(II) chlorocomplexes. Ind Eng Chem Res. 2012;51:7223–30.
[76] Wołowicz A, Hubicki Z. Investigation of macroporous weakly basic anion exchangers applicability in palladium(II) removal from acidic solutions – batch and column studies. Chem Eng J. 2011;174:510–21.
[77] Sayın M, Can M, Imamoglu M, Arslan M. 1,3,5-Triazine-pentaethylenehexamine polymer for the adsorption of palladium(II) from chloride-containing solutions. React Funct Polym. 2015;88: 31–8.
[78] Kasaini H, Everson RC, Bruinsma OSL. Selective adsorption of platinum from mixed solutions containing base metals using chemically modified activated carbons. Sep Sci Technol. 2005;40:507–23.
[79] Sharma S, Rajesh N. Augmenting the adsorption of palladium from spent catalyst using a thiazole ligand tethered on an amine functionalized polymeric resin. Chem Eng J. 2016;283:999–1008.
[80] Nagarjuna R, Sharma S, Rajesh N, Ganesan R. Effective adsorption of precious metal palladium over polyethyleneimine-functionalized alumina nanopowder and its reusability as a catalyst for energy and environmental applications. ACS Omega. 2017;2:4494–504.
[81] Zhang C, Huang K, Yu P, Liu H. Ionic liquid based three-liquid-phase partitioning and one-step separation of Pt(IV), Pd(II) and Rh(III). Sep Purif Technol. 2013;108:166–73.
[82] Siame J, Kasaini H Selective precipitation of Pt and base metals in liquid-liquid chloride systems. International Conference and Environmental Engineering (ICCEE' 2013), Johannesburg, South Africa, 2013, 88–94.
[83] Johnson Matthey. The process of PGM refining. Available at: http://www.jmrefining.com/the-process. Accessed October 10, 2017.
[84] Zhang L, Xu Z. A review of current progress of recycling technologies for metals from waste electrical and electronic equipment. J Clean Prod. 2016;127:19–36.
[85] Yong P, Rowson NA, Farr JPG, Harris LR, Macaskie LE. A novel electrobiotechnology for the recovery of precious metals from spent automotive catalysts. Environ Technol. 2003;24:289–97.
[86] Priya A, Hait S. Comparative assessment of metallurgical recovery of metals from electronic waste with special emphasis on bioleaching. Environ Sci Pollut Res. 2017;24:6989–7008.
[87] Tributsch H. Direct versus indirect bioleaching. Hydrometallurgy. 2001;59:177–85.
[88] Murray AJ, Zhu J, Wood J, Macaskie LE. A novel biorefinery: Biorecovery of precious metals from spent automotive catalyst leachates into new catalysts effective in metal reduction and in the hydrogenation of 2-pentyne. Min Eng. 2017;113:102–8.

[89] Godlewska-Żyłkiewicz B. Biosorption of platinum and palladium for their separation/preconcentration prior to graphite furnace atomic absorption spectrometric determination. Spectrochim Acta B. 2003;58:1531–40.

[90] Dziwulska U, Bajguz A, Godlewska-Żyłkiewicz B. The use of algae *Chlorella vulgaris* immobilized on Cellex-T support for separation/preconcentration of trace amounts of platinum and palladium before GFAAS determination. Anal Lett. 2004;37:2189–203.

[91] Akcil A, Veglio F, Ferella F, Okudan MO, Tuncuk A. A review of metal recovery from spent petroleum catalysts and ash. Waste Manage. 2015;45:420–33.

[92] Willner J, Kaduková J, Fornalczyk A, Mrážiková A, Marcinčáková R, Velgosová O. Possibilities of metals extracton from spent metallic automotive catalytic converters by using biometallurgical method. Arch Metall Mater. 2015;60:1877–90.

[93] Das N. Recovery of precious metals through biosorption – A review. Hydrometallurgy. 2010;103:180–9.

[94] De Vargas I, Macaskie LE, Guibal E. Biosorption of palladium and platinum by sulfate-reducing bacteria. J Chem Technol Biotechnol. 2002;79:49–56.

[95] Gadd GM. Bioremedial potential of microbial mechanisms of metal mobilisation and immobilisation. Curr Opin Biotechnol. 2000;11:271–9.

[96] White C, Wilkinson SC, Gadd GM. The role of microorganisms in biosorption of toxic metals and radionuclides. Int Biodeterior Biodegrad. 1995;35:17–40.

[97] Kashefi K, Tor JM, Nevin KP, Lovley DR. Reductive precipitation of gold by dissimlatory Fe(III)-reducing *Bacteria* and *Archaea*. Appl Environ Microbiol. 2001;67:3275–9.

[98] Chassary P, Vincent T, Marcano JS, Macaskie LE, Guibal E. Palladium and platinum recovery from bicomponent mixtures using chitosan derivatives. Hydrometallurgy. 2005;76:131–47.

[99] Mack CL, Wilhelmi B, Duncan JR, Burgess JE. A kinetic study of the recovery of platinum ions from an artificial aqueous solution by immobilized *Saccharomyces cerevisiae* biomass. Miner Eng. 2008;21:31–7.

[100] Creamer NJ, Baxter-Plant VS, Henderson J, Potter M, Macaskie LE. Palladium and gold removal and recovery from precious metal solutions and electronic scrap leachates by *Desulfovibrio desulfuricans*. Biotechnol Lett. 2006;28:1475–84.

[101] Colica G, Caparrotta S, De Philippis R. Selective biosorption and recovery of ruthenium from industrial effluents with *Rhodopseudomonas palustris* strains. Appl Microbiol Biotechnol. 2012;95:381–7.

[102] Mack CL, Wilhelmi B, Duncan JR, Burgess JE. Biosorptive recovery of platinum from platinum group metal refining wastewaters by immobilised *Saccharomyces cerevisiae*. Water Sci Technol. 2011;63(1):149–55.

[103] Dimitriadis S, Nomikou N, McHale AP. Pt-based electro-catalytic materials derived from biosorption processes and their exploitation in fuel cell technology. Biotechnol Lett. 2007;29:545–51.

[104] Zhou L, Xu J, Liang X, Liu Z. Adsorption of platinum(IV) and palladium(II) from aqueous solution by magnetic cross-linking chitosan nanoparticles modified with ethylenediamine. J Hazard Mater. 2010;182:518–24.

[105] Baba Y, Noma H, Nakayama R, Matsushita Y. Preparation of chitosan derivatives containing methylthiocarbamoyl and phenylthiocarbamoyl groups and their selective adsorption of copper(II) over iron(III). Anal Sci. 2002;18:359–61.

[106] Fujiwara K, Ramesh A, Maki T, Hasegawa H, Ueda K. Adsorption of platinum(IV), palladium(II) and gold(III) from aqueous solutions onto L-lysine modified crosslinked chitosan resin. J Hazard Mater. 2007;146:39–50.

[107] Won SW, Park J, Mao J, Yun YS. Utilization of PEI modified *corynebacterium glutamicum* biomass for the recovery of Pd(II) in hydrochloric solution. Bioresour Technol. 2011;102:3888–93.

[108] Wang S, Vincent T, Roux JC, Faur C, Guibal E. Pd(II) and Pt(IV) sorption using alginate and algal-based beads. Chem Eng J. 2017;313:567–79.
[109] Wang S, Vincent T, Roux JC, Faur C, Guibal E. Innovative conditioning of algal-based sorbents: Macro-porous discs for palladium sorption. Chem Eng J. 2017;325:521–32.
[110] Guibal E. Interactions of metal ions with chitosan-based sorbents: a review. Sep Purif Technol. 2004;38:43–74.
[111] Guibal E, Vincent T. Palladium recovery from dilute effluents using biopolymer-immobilized extractant. Sep Sci Technol. 2006;41:2533–53.
[112] Vincent T, Guibal E, Chiarizia R. Palladium recovery by reactive precipitation using a Cyanex 301-based stable emulsion. Sep Sci Technol. 2007;42:3517–36.
[113] Sharififard H, Soleimani M, Zokaee Ashtiani F. Evaluation of activated carbon and bio-polymer modified activated carbon performance for palladium and platinum removal. J Taiwan Inst Chem Eng. 2012;43:696–703.
[114] Anbia M, Rahimi F. Adsorption of platinum(IV) from an aqueous solution with magnetic cellulose functionalized with thiol and amine as a nano-active adsorbent. J Appl Polym Sci. 2017;134:45361.
[115] Redwood MD, Deplanche K, Yong P, Baxter-Plant VS, Macaskie LE Biomass-supported palladium catalysts on *Desulfovibrio* and *Rhodobacter*. In: Harrison STL, Rawlings DE, Petersen J, editors. Proceedings of 16th International Biohydrometallurgy Symposium, Cape Town, South Africa, 2005: 335–42.
[116] Lloyd JR, Yong P, Macaskie LE. Enzymatic recovery of elemental palladium by using sulfate-reducing bacteria. Appl Environ Microbiol. 1998;64:4607–9.
[117] Yong P, Mikheenko IP, Deplanche K, Redwood MD, Macaskie LE. Biorefining of precious metals from wastes: an answer to manufacturing of cheap nanocatalysts for fuel cells and power generation via an integrated biorefinery? Biotechnol Lett. 2010;32:1821–8.
[118] Humphries AC, Mikheenko IP, Macaskie LE. Chromate reduction by immobilized palladized sulfate-reducing bacteria. Biotechnol Bioeng. 2006;94(1):81–90.
[119] Yong P, Rowson NA, Farr JPG, Harris IR, Macaskie LE. Bioaccumulation of palladium by *Desulfovibrio desulfuricans*. J Chem Technol Biotechnol. 2002;77:593–601.
[120] Baxter-Plant VS, Mikheenko IP, Macaskie LE. Sulphate-reducing bacteria, palladium and the reductive dehalogenation of chlorinated aromatic compounds. Biodegradation. 2003;14:83–90.
[121] Deplanche K, Snape TJ, Hazrati S, Harrad S, Macaskie LE. Versatility of a new bioinorganic catalyst: Palladized cells of *Desulfovibrio desulfuricans* and application to dehalogenation of flame retardant materials. Environ Technol. 2009;30:681–92.
[122] Mabbett AN, Sanyahumbi D, Yong P, Macaskie LE. Biorecovered precious metals from industrial wastes: single-step conversion of a mixed metal liquid waste to a bioinorganic catalyst with environmental application. Environ Sci Technol. 2006;40:1015–21.
[123] Yong P, Liu W, Zhang Z, Beauregard D, Johns ML, Macaskie LE. One step bioconversion of waste precious metals into *Serratia* biofilm-immobilized catalyst for Cr(VI) reduction. Biotechnol Lett. 2015;37:2181–91.
[124] Macaskie LE, Mikheenko IP, Yong P, et al. Today's wastes, tomorrow's materials for environmental protection. Hydrometallurgy. 2010;104:483–7.

Maciej Staszak

3 Rubber industry

Abstract: Following chapter presents short introductory description of rubber and rubber industry. The main problem of rubber industry is the way of the usage of spent tires. Furthermore very important group of problems arise considering the metal and nonmetal additives which are significant component of the vulcanized rubber. The key attention is dedicated to typical ways of rubber usage in utilization and recovery of metals from spent rubber materials concentrating specifically on used tires processing. The method of recovery of rare metals from rubber tires was described. The rubber debris finds widest use in the field of waste metal solutions processing. The environmental pollution caused by metals poses serious threat to humans. Several applications of the use of waste rubber debris to remove metals from environmental waters were described. Moreover, the agriculture usage of waste tire rubber debris is described, presenting systems where the rubber material can be useful as a soil replacement.

Keywords: spent tires, recovery systems, rubber-based adsorbents

3.1 Introduction

Natural and synthetic rubbers are two main products of chemical rubber industry. While natural rubber is produced based on caoutchouc, sometimes called India rubber, the synthetic one is the product of chemical synthesis from petroleum by-products. Natural rubber, which is obtained from natural rubber plant trees, is mainly composed of polymer of isoprene, mainly poly-cis-isoprene, as presented in Figure 3.1. Isoprene is naturally found in many species of trees like oaks, poplars, or eucalyptus.

Natural rubber comes from latex which is stable aqueous emulsion of polymer microparticles. The latex itself is a substance which main function is to protect the trees from herbivores. There are over 12,000 of trees species produce latex but most of the species latex is not suitable for industrial use [1].

3.2 Rubber vulcanization

The natural and synthetic rubber in its uncured, primary state has not any useful industrial parameters. Its viscous and sticky character excludes it from wider use. The rubber easily deforms and is inelastic and breakable in low temperatures. The chemical

This article has previously been published in the journal *Physical Sciences Reviews*. Please cite as: Staszak, M. Rubber industry. *Physical Sciences Reviews* [Online] **2018**, *3* (4). DOI: 10.1515/psr-2018-0022

https://doi.org/10.1515/9783110547061-003

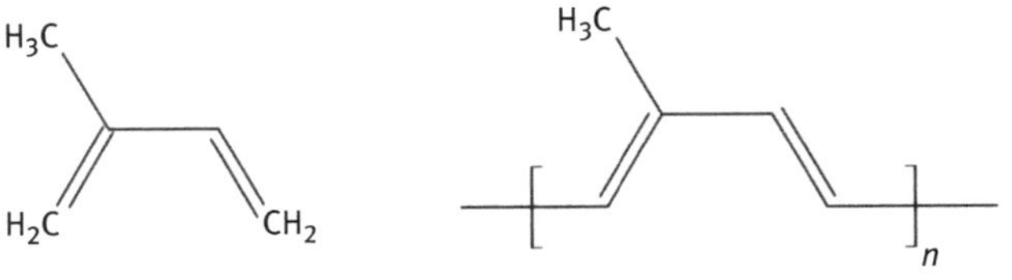

Figure 3.1: Isoprene and polyisoprene.

process of converting natural rubber into much more useful materials with beneficial mechanic properties is the process of vulcanization. The most popular method of vulcanization is the process conducted in high temperatures from about 140°C to 180°C and under high pressure. Prior to the process, the additives like sulfur are added and mixed with the rubber. The sulfur itself does not guarantee high quality of the product. The accelerators like zinc oxide (ZnO) or stearic acid must also be added to provide high efficiency of the process of vulcanization. From chemical point of view, these substances are catalysts but they play also additional important roles, for example ZnO increases the heat conductivity which, for example, is important and required property for tires durability.

The industrially widest method of vulcanization is the sulfur vulcanization. The addition of sulfur causes formation of disulfide bonds in the polymer structure which connects polymer chains and forms cyclic structures (Figure 3.2). The polymer structure is then changed and stiffened by the disulfide bridges which are the main reason of new valuable rubber properties.

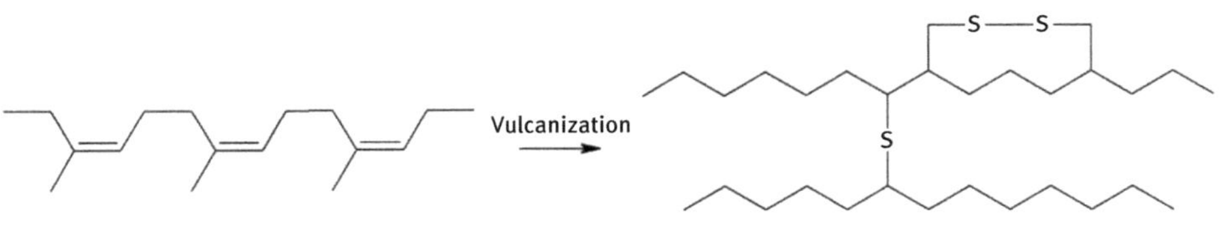

Figure 3.2: Generic scheme of sulfur vulcanization [2].

In present vulcanization processes [3–6], natural rubber is usually formed with the presence of accelerators in amounts from 0.5% to 1%, together with various concentrations of carbon black which may reach up to 45% for tire production, low amounts of phenols and aromatic amines for antioxidation reasons, silica carbide (nanometer-sized reinforcing filler), and from 5% to 8% of sulfur. Such types of vulcanized natural rubber are generally recognized as natural rubber composites or natural rubber nanocomposites depending on the dimension scale.

A complex structure of a rubber can be presented as a heterogeneous network of macromolecules [7]. The network is considered to be an ensemble of zones with numerous potential energies for separation of active chains from temporary nodes and attachment of hanging chains to the network connections. The network is cross-linked by means of chemical and physical bonds, and by mixes and aggregates of filler. The vulcanized rubber structure [8], its mechanical properties, and thermal stability, formed with additives based on sulfur and accelerators, depend on two essential attributes. One is the amount of the cross-links being present in the rubber composition, and the second important attribute is the sulfide bond type, e. g. mono-, di-, or polysulfide bridges in the rubber structure.

The rate of vulcanization might be increased by raising the process temperature. Nevertheless, at higher temperature, the quality of sulfur cross-links is lower on the expense of the physical properties. Such situation occurs due to the sulfur bonds and rubber chains dissociation [9–11]. Therefore, the vulcanization rate can be regulated by supplementary amounts of sulfur and accelerator in place of higher temperature. The higher amount of sulfur addition leads to growth in cross-link density and the amount of polysulfide bridging, which consequently leads to a decrease in the stability on aging. Higher amounts of accelerators cause that the efficiency of sulfur vulcanization having mono- and disulfide bridges is improved and results in higher stability considering aging. Unfortunately, the latter is the reason of the decrease in dynamic properties of the rubber. Consequently, it is important to maintain ideal vulcanization conditions which are an optimization task between productivity and the quality of the final products, e. g. automobile tires [12, 13].

The elemental composition of most widely used radial tires contains three main elements [14]: carbon (73%), iron (13%), and hydrogen (6%) – which results in total 92% of tire weight. The remaining 8% of the tire is composed from about 20 other chemical elements in smaller amounts, including the nonmetallic elements like abovementioned sulfur, as well as nitrogen and oxygen. The most typical heavy metals are copper, zinc, tin, and cobalt. Every substance which composes the rubber material has a particular objective and provides different ways to improve physical properties, simplify production, and enhance the final product performance. The quantitative composition of these substances will depend on the type of tire and its producer. One of the typical examples is enhancing the mechanical properties of tires when brass is used as a coating for the steel cord threads. This helps during the process of wire drawing and improves adhesion to the belt compound. Bronze plating keeps a firm and protective coating on tire's bead wire for manufacturing and guarantees appropriate adhering to vulcanized rubber.

3.2.1 Rubber waste

One of the most significant sources of rubber waste is the tire management and utilization. For example, the UK produces around 4.3×10^8 kg of waste tires a year [15, 16], while Japan produces 2.1×10^8 kg of tire wear debris [17]. The economics of international financial flows of tire industry evidently shows the magnitude of the

Table 3.1: Exported values of new rubber pneumatic tires, in US dollars [18].

Exporters	Exported value in 2014	Exported value in 2015	Exported value in 2016
China	16,446,634	13,842,691	12,893,129
Germany	6,295,369	5,420,488	5,490,394
Japan	6,496,494	5,387,441	4,784,154
United States of America	5,622,782	5,096,080	4,479,954
Thailand	3,505,808	3,401,750	3,557,764
Republic of Korea	4,002,360	3,371,560	3,413,127
World summary	**83,899,621**	**72,788,289**	**69,595,539**

rubber industrial potential, see short excerpt from the International Trade Centre database in Table 3.1 (only six first countries and world summary are presented).

As the largest part among waste polymers in world, waste rubber tires decompose slowly due to its cross-linked structure and existence of stabilizers and other additives. Therefore, traditionally they are treated by incineration or landfilling [19]. Waste tires are utilized as an additive in road pavements [20], rubber roofs, drainage systems, and floor mats. Merging of waste rubber tire with polymeric substances became an important subject in recent years [21, 22]. Numerous studies described in the literature showed the effective application of waste rubber tire as an adsorbent for several types of pollutant, for example dyes [23], pesticides [24], phenols [25], and various metal ions [26–30].

At present, the significant increase in waste tires imposes serious threats to human health and environment [31, 32]. Numerous approaches were studied to convert waste tire rubbers into highly value-added products, such as fuel [33–35], adsorbent (activated carbon) [36–39], or filler [40–43]. The fabrication of adsorbents [44] from activated carbon for wastewater remediation has been of interest in the field of tire recycling. Typical preparation procedures for activated carbon production are performed by the pyrolysis of waste tire rubber at high temperature (500–900°C), followed by chemical activation with HNO_3, H_2O_2, or KOH [45–53].

The rubber tires typically contain reinforcing fillers like carbon black, used to strengthen the rubber and support abrasion resistance, reinforcing steel fibers or textile, usually in the form of a cord, used to provide the supporting strength or tensile component in tires [54]. They contain also extenders like petroleum oils, used to control viscosity, decrease internal friction during processing, and improve low temperature elasticity in the vulcanized product. Tires are constructed with two types of design: nonbelted and steel-belted. The nonbelted design uses textile material as stiffening cord, while most popular steel-belted radial type tires utilize steel cord.

The carbon black metal composition measured by inductively coupled plasma technique is presented in Table 3.2 [55]. The results indicate that carbon black is the potential source of various metals, but their concentrations are usually too small to recover, mainly from the economical point of view.

Table 3.2: Metal and ash content in typical carbon black from tire rubber.

Metal	ppm	Metals	ppm
Aluminum	205	Molybdenum	6
Antimony	<1.0	Nickel	56
Arsenic	<1.0	Phosphorus	210
Barium	16	Potassium	301
Beryllium	<1.0	Selenium	<1.0
Bismuth	<1.0	Silicon	1,960
Boron	16	Silver	<1.0
Cadmium	<1.0	Sodium	330
Calcium	9,333	Strontium	<1.0
Chromium	40	Thallium	<1.0
Cobalt	<1.0	Tin	2
Copper	44	Titanium	962
Iron	1,600	Vanadium	26
Lead	69	Zinc	45.866
Magnesium	533	Ash, wt%	8.12

The rubber automotive tire usage is a source of heavy metals which reach the atmosphere in significant amounts [56]. Substantial quantities of manganese, iron, cobalt, nickel, copper, zinc, cadmium, and lead are present in the tire tread which is spread especially in the case of studded tires. However, ZnO and zinc sulfide are present in the tire rubber and tire tread dust in the highest amounts [57].

Waste tire treatment is typically performed by pyrolysis that is a thermal processing at high temperatures. Such solutions are often proposed in the industry. Exemplary, in the Finnish Metso company, one of the technology targets is recycling where the pyrolysis process is one of the elements of the resources recovery [58]. The tire pyrolysis system involves an indirectly fired rotary oven, steel recovery system, char management, crushing and pelletizing circuit, oil condensing, and gas cleaning system. The schematic of tire recovery system used in industry is presented in Figure 3.3.

Very interesting analysis of pyrolysis of tires was presented in work [59]. Not only the main metals form tires (zinc (Zn) and iron (Fe)) but others such as barium (Ba), bromine (Br), calcium (Ca), chlorine (Cl), chromium (Cr), copper (Cu), potassium (K), manganese (Mn), phosphorus (P), lead (Pb), rubidium (Rb), sulfur (S), silicon (Si), strontium (Sr), titanium (Ti), and yttrium (Y) were considered by authors. The behavior of these elements during pyrolysis and the effect of temperature were studied. With the increasing pyrolysis temperature from 500°C to 650°C, tires showed increases in the contents of ash, fixed carbon, C, S, and the yield of gas. The amount of metals after pyrolysis process is different in comparison to the raw material. There are no relationship between temperature and the metal contents, as it is seen in Table 3.3. The results showed that the contents of the metals are various

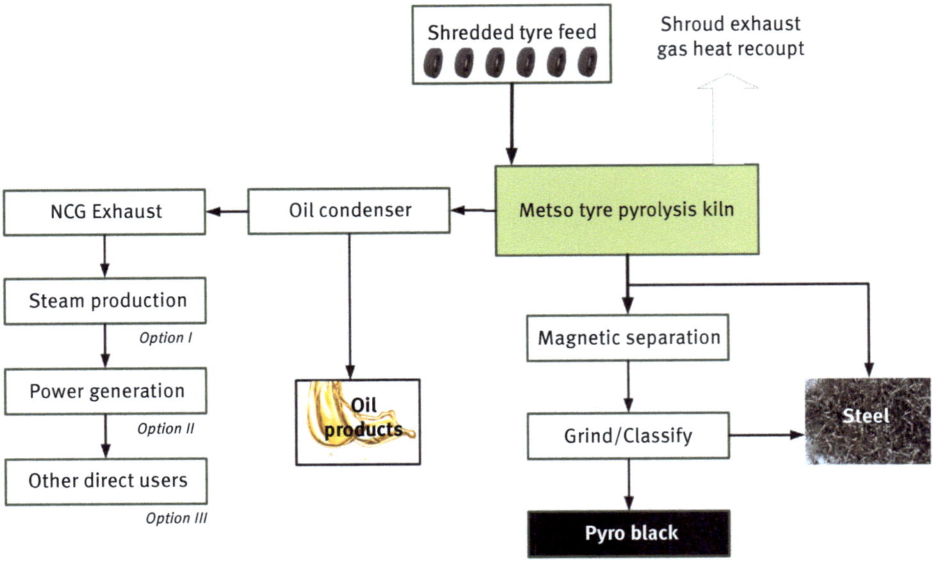

Figure 3.3: Used tires recovery system.

Table 3.3: Elements composition before and after pyrolysis of tires, in mg/kg.

Pyrolysis temp., °C	Ba	Br	Ca	Cl	Cr	Cu	Fe	K	Mn
Raw material – tire	11.8	5.4	787.0	626.0	2.9	13.2	828.0	794.0	10.2
500	11.7	13.2	1,940.0	794.0	5.3	29.3	1,800.0	1,010.0	23.4
600	11.0	11.7	1,410.0	775.0	5.4	25.5	1,420.0	920.0	15.3
650	11.6	9.3	1,330.0	762.0	5.3	25.2	1,370.0	1,040.0	12.1
Pyrolysis temp., °C	P	Pb	Rb	S	Si	Sr	Ti	Y	Zn
Raw material – tire	373.0	20.1	4.2	5,250.0	9,060.0	4.5	29.4	2.3	16,700.0
500	884.0	19.3	4.9	10,800.0	9,520.0	7.1	42.6	1.7	31,200.0
600	709.0	27.3	5.2	9,780.0	8,300.0	6.8	41.1	1.5	23,800.0
650	686.0	30.6	5.4	9,650.0	8,000.0	9.6	41.7	2.8	24,500.0

and depend on the process conditions. This should be taken into consideration in the further processing of this type of materials.

Besides the abovementioned pyrometallurgical processes, the hydrometallurgical ones, based mainly on leaching and extraction, are widely used. Hydrometallurgical processes are energy efficient and flexible considering the metal contents in wastes for the recovery of important metals, compared with pyrometallurgical ones. Many investigations have been conducted on the leaching behavior and the recovery systems of heavy metals from industrial wastes like heavy oil ashes [60], coal ashes [61], waste catalyst [62], sewage sludge, and residue from waste water treatment plant [63] or spent batteries [64].

3.3 Additives containing metals

3.3.1 Zinc processing

ZnO is a chemical compound which is extensively used in the rubber industry due to its specific and useful properties which allow it to work as activator during the process of sulfur vulcanization. The largest single market for ZnO is the tire industry, utilizing more than half of the total worldwide requirement of 1.2×10^6 tons [65]. In general, ZnO is applied in rubber mixtures in amounts from 3% to 8%. Heavy-duty tires contain high amounts of ZnO for better heat conductivity and strengthening. This is due to the negative effect of heat increase which is critical for centrifugal forces arising in tires at higher speeds compared with their solid rubber counterparts.

The additional advantages of the ZnO use are its biochemical activity preserving plantation latex from decomposition, improved dielectric strength minimizing the corona electrical discharge risk, stabilization to light and UV aging effect, heat stabilization which impedes the devulcanization, reinforcement which depends on the ZnO particle size and its reactivity with the rubber, improved rubber to metal bonding, and improved tack retention over long time of storage. The activator ZnO is an additional significant pollution source [66], tire debris holds substantial quantities of zinc which is potential source of contamination of soils from tire wear. Tire wear debris has been known for long time to be the source of Zn to the environment [67–72].

The study of recovery of zinc from tire ashes [73] using acid and alkaline solutions from both viewpoints of environmental protection and resource utilization is described by following procedure. The uniformly milled rubber samples were filtered out using a screen using 0.5 mm screen. The metal concentrations in the ashes were measured by dissolving the ash in aqua regia (mixture of nitric acid and hydrochloric acid) at an ambient temperature for 24 h, and determining the metal contents in the resulting solution by inductively coupled plasma atomic emission spectrometry (ICP-AES). After such procedure, the tire ash composition was as follows (in wt%): Zn (2.1), Fe (6.8), Al (0.05), Mg (0.05), Co (0.02), Cu (0.04), Ti (0.02), and Mn (0.05). Leaching experiments were performed by submerging the dried ash in leachant. The leaching temperature was set at 296 and 333 K. After filtering out the solid remainders, the concentrations of metals in leach solutions were measured by ICP-AES after dilution, and the percent leaching was estimated from a mass balance. Solvent extraction was used for metal recovery from leach liquor. Equal volumes of the aqueous feed and the organic solution were shaken for 1 h, and small portion of the organic phase was taken out to be mixed with stripping solution. The metal content in both aqueous phases after the forward and back extractions was determined, and the percent extraction was calculated from a mass balance. Authors state that no pH adjustment was performed to the aqueous feeds. The typical industrial extractants were proposed in this work. Extractants used were hydroxyoximes: LIX54 (α-acetyl-m-dodecyl acetophenone) and LIX84 (anti-2-hydroxy-5-nonyl acetophenone oxime); acidic organophosphorus extractant: D2EHPA (bis(2-ethylhexyl) phosphoric acid); and basic extractants: Aliquat336 (tricaprylmethyl ammonium chloride), Alamine336 (tricaprylamine), and TOA (tri-n-octylamine). The extractants were diluted with kerosene to the concentration of 100 g/L. Moreover, recovery of the metals from ammonia leach liquor was performed by precipitation, during which the precipitate of metal was obtained by keeping the solution at 313 K to evaporate ammonia. The yield was computed from a difference in the metal concentration in the aqueous feed before and after the precipitation. The results presented a considerable prospect of selective zinc recovery from the ashes with hydrometallurgical treatments. The pulverized tire ash is a promising source for zinc due to its high metal content. Alkaline leaching caused effective separation of zinc from the undesired metals, though the percent leaching was lower than that in the acid leaching. As downstream processing, solvent extraction was used to the acidic leaching liquors. The acidic extractant was found to be suitable for removing the coexisting iron, while the amine extractants were suitable for extracting zinc. Utilizing well-known TOA as extractant, 67% of zinc recovery was attained without any treatment to the leaching solution.

The solvent extraction is potential method for the recovery of zinc using Cyanex®272 from spent automobile tire ash after prior leaching in hydrochloric acid [74]. An extraction yield of 93.42% zinc by 0.032 M Cyanex®272 was obtained from 10 g/L spent automobile tire ash leach liquor at 25°C and a stirring time of 25 min. The stripping effectiveness of about 92% of zinc from the organic phase was achieved.

3.3.2 Arsenic processing

Arsenic pollution has become widespread, particularly in developing countries [75], where a substantial fraction of the residents depends on groundwater. The World Health Organization provisional guideline of 0.01 mg/dm^3 has been approved as the drinking water standard [76]. The arsenite As^{3+} is more acutely toxic and mobile than arsenate As^{5+}. Arsenic is moved and transferred by natural weather and geochemical processes, biological activity, and other anthropogenic behaviors. The arsenic pollution treatment technology must conform maximized adsorption capacity of As for a particular sorbent material, wide accessibility, and economical effectiveness. The activated carbons manufactured from coconut shells, wood char, lignin, petroleum coke, bone-char, carbon black, rice hulls, sugar, peach pits, and fertilizer waste [77, 78] can be expected as good selection. Graphitic carbon nitride can be used as the sorbent material which can be synthesized through a low-temperature solvolysis of sodium cyanide in formamide [79].

Huge amount of used tires are a potential danger to the environment. Standard pyrolytic-tire char (PyrC) can be used as starting point to obtain some novel adsorbent materials for arsen treatment. The low-cost protocol [80] containing two stages of the process, involving acid-cleaning of the novel material $PyrC_{350}$® surface and calcination at temperature of 350°C, permits production of pyrolytic char which is optimal for As^{3+} adsorption from aqueous solutions at environmentally relevant pH values 4–8.5. The $PyrC_{350}$® optimized material is manufactured through a detailed selection of the material considering As^{3+} uptake. $PyrC_{350}$® reaches As^{3+} uptake capacity of 31 mg/g at pH 7 that is about 400% improved versus raw pyrolytic tire carbon. The char powder used was obtained from a commercial pyrolysis of used rubber tires at 450°C, in oxygen-free atmosphere under vacuum for 4 h [81]. In the procedure, an initial step is the acid treatment of PyrC which is suspended in 2 M HNO_3 for 2 h to remove metal traces remaining from pyrolysis from the surface. Next, the material is separated by centrifugation with the decantation of supernatant. The solid material is washed with deionized water to obtain pH of 7.0 ± 0.1. Finally, the solid is gathered by centrifugation and dried, obtaining acidic pyrolytic material $PyrC_{acid}$. Subsequently, the material is calcined to obtain $PyrC_{350}$®, at temperature from 350°C to 500°C under nitrogen atmosphere, due to following reasons: to remove surface carboxylates which inhibit As^{3+} adsorption [82, 83], to avoid oxidation of the carbon matrix and the graphite [84, 85], to optimize the formation of the ZnS and Fe_3C phases. The substance is then heated with a temperature increase rate of 2°C/s under nitrogen atmosphere at 1 bar pressure.

The proposed chemical equilibrium reactions describing adsorption of arsenic read [80]:

Adsorption of As^{3+} onto PyrC:

$$\equiv C_xOH_2 + H_3AsO_3 \overset{K_{eq}}{\longleftrightarrow} C_xOH_2[H_3AsO_3], \quad K_{eq} = 8.1 \pm 0.2 \tag{3.1}$$

Adsorption of As^{3+} onto $PyrC_{acid}$:

$$\equiv COOH + H_3AsO_3 \overset{K_{eq}}{\longleftrightarrow} COOH[H_3AsO_3], \quad K_{eq} = 0.3 \pm 0.2 \tag{3.2}$$

$$\equiv C_xOH_2 + H_3AsO_3 \overset{K_{eq}}{\longleftrightarrow} C_xOH_2[H_3AsO_3], \quad K_{eq} = 8.1 \pm 0.2 \tag{3.3}$$

Carbon-matrix sites [84] are marked as C_xOH_2.

Adsorption of As^{3+} onto $PyrC_{350}$®:

$$\equiv COOH + H_3AsO_3 \overset{K_{eq}}{\longleftrightarrow} COOH[H_3AsO_3], \quad K_{eq} = 0.3 \pm 0.2 \tag{3.4}$$

$$\equiv Fe_3C + H_3AsO_3 \overset{K_{eq}}{\longleftrightarrow} Fe_3[H_3AsO_3], \quad K_{eq} = 9.5 \pm 0.2 \tag{3.5}$$

$$\equiv C_xOH_2 + H_3AsO_3 \overset{K_{eq}}{\longleftrightarrow} C_xOH_2[H_3AsO_3], \quad K_{eq} = 12.5 \pm 0.2 \tag{3.6}$$

$$\equiv ZnSH_2 + H_3AsO_3 \overset{K_{eq}}{\longleftrightarrow} ZnSH_2[H_3AsO_3], \quad K_{eq} = 4.2 \pm 0.2 \tag{3.7}$$

According to powder X-ray diffraction measurements, two types of specific sites exist: the ZnS and Fe_3C. Reaction (7) illustrates the binding properties of ZnS [86, 87] and reaction (5) the binding properties of Fe_3C [88, 89].

The key advantage of novel adsorbent is its low price and high effectiveness. According to United States Environmental Protection Agency (US-E.P.A.) for arsenic adsorbent design and costing tool [90], activated alumina (AA-400 G), iron-impregnated activated alumina (AA-FS50), granular ferric oxide (Bayoxide E-33, ferric hydroxide, titanium dioxide (Metsorb G), iron-modified zeolite (Z-33 Revision B), and sulfur-modified iron version III can be used for drinking-water treatment. Their cost is \$2.2/kg, \$2.8/kg, \$12.1/kg, \$6.6/kg, \$13.2/kg, \$0.5/kg, and \$8.8/kg, respectively. Because no production plant exists for the novel material, the estimated cost of $PyrC_{350}$® is \$2.5/kg. Together with its reported effectiveness (retained 99%, 88%, and 92% of As^{3+} uptake capacity for three cycles respectively), it places it as very promising adsorbent material for arsenic removal from rubber material.

3.3.3 Mercury processing

Rubber from tire wastes can be used to produce carbon adsorbents for mercury in aqueous solution. The proposed initial material is reduced in size from residual rubber obtained from tire wastes [28]. Next rubber was heated at 400°C or 900°C for 2 h in nitrogen atmosphere at flow rate of 225 ml/min. The heating increase rate was 10°C 1/min. The acid chemical treatments of residual rubber were performed by the acid solution of H_2SO_4, HNO_3, or a mixture of these acids at different H_2SO_4/HNO_3 ratios. The system was given 24 h to react. In the other case, rubber was first heated at 400°C for 2 h in the same nitrogen conditions, and then processed with a aqueous mixtures of H_2SO_4 and HNO_3, and also vice versa. For the testing purposes, mercuric chloride was used in the preparation of the adsorptive solutions. The heat treatments of residual rubber result in the release of volatile elements and in the mass loss. The tire pyrolysis studies [19, 91] found that the quantity of char obtained exceeds the amount of carbon black present in the original tires. Therefore, the final product of pyrolysis can be considered as a combination of carbon black and char created by tire rubber degradation.

The mercury adsorption process is faster for the preparation scheme performed by heat treatment of residual rubber at 400°C or 900°C and using the H_2SO_4, HNO_3, or 1:3 H_2SO_4/HNO_3 mixtures in the chemical processing of residual rubber. However, the process is slower when using the H_2SO_4/HNO_3 solutions containing a high amount of H_2SO_4 and when residual rubber is in turn heated and processed chemically, or vice versa. To estimate the amount of mercury adsorbed the Freundlich eq. (3.8) can be used, which reads [92]:

$$X_e = K_f C_e^{(1/n)} \tag{3.8}$$

where X_e is the concentration of mercury adsorbed mg/g at equilibrium. K_f expressed in mg/g is adsorption capacity of the adsorbent, and n is the variation of the adsorption with concentration. These coefficients are two constants which encompass all factors influencing the adsorption process. The value of K_f reaches about 108.9 mg/g for HNO_3-treated residual rubber at 900°C. For adsorbents prepared from different sources, values of K_f of 50.11 [93], 28.8 [94], 4.52 [95], 42.6 [96], and 6.18 [97] have been reported earlier.

The chemical composition of the adsorbents does not seem to be an important factor taking into consideration their adsorption capacity. On the other side, the micropore content in the products obtained from residual rubber appears to play a more significant role on the strength of adsorption of mercury, despite the low development of the microporosity in the adsorbents. In fact, the adsorption process of mercury is more efficient if the adsorbent is a nonporous solid, its mesopore volume is high, or its pore size distribution in the macropore region is wide.

The waste tire rubber can be a very effective way to remove mercury from the environment, e. g. from a contaminated soil [98]. In the soil, which was processed by the rubber, the mercury concentration in the rainwater leachate was 1.2 ng/mL, which is considerably lower than that for the unprocessed soil characterized by value of 84 ng/mL. Additionally, the rubber processing inhibited the development of metallic mercury Hg^0 in the mercury-contaminated soil. The tire rubber is found to be successful stabilizer for mercury under acidic to neutral pH conditions. It is important to mention that the sulfur groups present in the rubber may attribute to the reduction of mercury leaching. Mercury appears to be adsorbed to the sulfur-active sites through the surface complexes formation.

The study [99] of adsorption of mercury at the rubber pieces also allowed estimating the effectiveness of this process in the experiments conducted at laboratory scale. Vulcanized rubber pieces were used as a model adsorbent to demonstrate the influence of sulfur, zinc, and carbon black on mercury adsorption in water. The expected and confirmed fact was that pieces with smaller size generated higher rate of adsorption. The mechanism of mercury adsorption is closely related with sulfur amount which was used as a vulcanizing agent in rubber fabrication. Because mercury adsorption capacity is associated with the cross-linking density in the rubber pieces, it appears that mercury reacted with the sulfur in the cross-linking network. Consequently, the investigated adsorption process is related to the specific reaction and may be regarded as chemisorption. The influence of ZnO was found not to affect directly the adsorption but indirectly by acting as the vulcanization activator. With the higher the amount of carbon black loading, the lower was the adsorption rate due to the reduced mobility of rubber structures. This is caused by the interaction of carbon black particles with rubber molecules, thereby impeding the diffusion of mercury into the rubber.

3.3.4 Nickel, lead, and cadmium processing

Human exposure to nickel ions at elevated amounts is related to important and dangerous health consequences like dermatitis, nausea, coughing, chronic bronchitis, gastrointestinal distress, reduced lung function, and lung cancer [100, 101] due to its toxicity, persistency, and bioaccumulation in the food chain [102].

The potential of scrap tire as adsorbent for removal of nickel ions from water can be performed by particular preparation of activated carbon [45]. Lead and nickel treatment by rubber waste-prepared adsorbent is an alternative use of waste rubber tire usage in environmental processing [48]. These metal pollutants are a potential hazard to the environment because of their toxicity, persistence, and bioaccumulation in the food chain [102]. The method of adsorbent preparation is as follows. The waste ground tire crumbs after initial washing and drying at 100°C were carbonized to 500°C for 5 h. Subsequently, the processed rubber material was treated with the hydrogen peroxide (H_2O_2) solution for 24 h at 60°C to oxidize adhering organic impurities. The material was cleaned with deionized water three times to eliminate H_2O_2 and dried at 110°C for 2 h in vacuum. The dried material processed was heated up to 900°C for 2 h. Subsequently, treatment with 1 M HCl solution was performed to remove the ash content and then washed with deionized water. After cleaning process, material was prepared by drying at 100°C for 24 h. The experiments performed using fixed bed column, under optimized conditions of column flow rates, increased metal ion concentration, and higher adsorbent mass and showed the metal removal capacity of 98.43% under optimized conditions.

Another metal of interest is cadmium which is highly toxic especially in wastewater and surface water. The cadmium removal technique proposed is as follows [103]. The waste tire was carefully washed with a detergent and rinsed with deionized water to guarantee that there are no detergents or soil particles staying on the surface of the rubber material. Afterwards, the rubber material was dried at 100°C. Small portions of tires were inserted into tubular furnace to obtain a powder with different particle sizes, under a continuous purging with nitrogen. The pyrolysis process was applied to obtain the raw carbon powder at the temperature of 900°C. The powder was then been milled to obtain the particles of the same size (100–250 μm). The entire pyrolysis procedure was a standard method without optimization, to obtain the raw carbon powder which is additionally activated using different conditions. There are several possibilities of chemical activation agents such as KOH, NaOH, HCl, and H_2SO_4 [104]. The H_2O_2 is investigated due to its environmental friendly properties. The chemical activation technique can be optimized by altering factors that can influence the manufacture and efficiency of the activated carbon. Variables such as temperature and time were fixed at 200°C and 2 h, respectively. Activating acidic agents and numerous ratios of raw carbon to activating agent is significant in gaining desirable properties of the adsorbent that are appropriate in adsorbing inorganic pollutants. The most significant characteristic of the optimization was the selection of the acidic activating agents which were applied to impregnate the raw carbon with, which is phosphoric acid and H_2O_2. The temperature at which the activation was executed was set at 200°C. The ratios between the raw carbon and the activation agents were 1:5 and 1:10 but the raw carbon was fixed at 0.3 g. Subsequently, the activation process, the activated carbon was left to cool down for a few minutes and it was mixed with 10% HCl to remove the inorganic pollutions. The excess phosphoric acid in the adsorbent was removed with hot water and the H_2O_2 with cold water. The activated carbon adsorbent exhibited high efficiency in the removal of cadmium and lead from wastewater. The maximum adsorption capacities of cadmium and lead were determined to be 201 and 196 mg/g, respectively.

3.3.5 Copper processing

The adsorption capacity of tire rubber toward Cu^{2+} ions in three forms untreated, chemically activated, and physically activated was analyzed [105]. The adsorption capacity of physically activated rubber was slightly higher than the chemically activated which, consecutively, was higher than the capacity of the untreated material. The influence of activation temperature and time was studied in the physical activation process. Up to 97% of the Cu^2 ions were adsorbed from aqueous solution when untreated, chemically or physically activated tire rubber was employed as an adsorbent with initial metal concentrations of 20 mg/ml and 40 ppm, respectively. The higher the adsorbent concentration, the greater is the elimination of metal ions from the aqueous solution. Additionally, the higher the Cu^{2+} ion concentration in the presence of a constant amount of an adsorbent caused intensified metal ion loading per unit weight of the adsorbent. The adsorption process was improved with the increasing of pH value for all three adsorbents tested.

Chemical activation of tire rubber was performed by impregnation with 15 wt% $ZnCl_2$, 1:3 ratio of rubber to $ZnCl_2$. The suspension prepared was stirred at 85°C for 5 h maintaining the volume constant by addition of water. Subsequently, the material was filtered and dried at 105°C. The impregnated material was carbonized by heating in a fluidized bed under nitrogen flow at 650°C, 850°C, and 1,000°C for 15 min. The resulting activated carbon was washed with 1 M HCl followed by rinsing several times with distilled water at 90°C to entirely eliminate any chloride. The carbon was subsequently dried at 105°C.

Physical activation of tire rubber was performed by direct carbonization using nitrogen followed by carbon dioxide at various temperatures: 650°C, 850°C, and 1,000°C for 15 min to enable the porosity development analysis. The carbonization unit was a fluidized bed reactor inserted inside the heating element of a vertical furnace. The experiment was conducted in two stages; initial carbonization of the sample under an inert atmosphere using nitrogen followed by activation with carbon dioxide.

The other method of copper processing with waste tire rubber is to perform an incineration of the rubber material [106]. The material was obtained by processing of waste tire rubber in order to prepare the desired ash. Therefore, it was first washed with detergent solution and afterwards with dilute HCl to remove soil remains. The clean and dry chunks were burnt at 500°C in a muffle furnace for 4 h. The cooled ash was then washed with a very dilute acidic solution, such as

0.0010 mol/L of HCl to eliminate the residues of salts of metals, such as Na, K, and Ca. Afterwards, the mixture was filtered out and the adsorbent was washed with 100 mL of double distilled water and dried at 105°C for 2 h before use. The results obtained show that waste tire rubber ash can be used as a suitable adsorbent to remove copper from wastewaters with a good effectiveness and at low price. Numerous factors were studied and maximum adsorption was found to be in the pH range 4.5–6.0 within 120 min contact. The adsorption was intensified with increasing contact time and initial metal concentration; the maximum detected capacity of the material was 34.3 mg/g of adsorbent. The conclusion stated that waste tire rubber ash has a potential to be an efficient adsorbent for the removal of Cu^{2+} from wastewater.

3.3.5.1 Rare earth elements (REE)

The possibility of recovery of REE from Nd–Fe–B permanent magnets of electronic waste by an oxidation–reduction process using waste tire rubber-derived carbon is reported in the literature [107]. The two-step oxidation–reduction approach, using waste tire rubber-derived carbon as a reducing agent, is proposed as an alternative to hydrometallurgical processes [108–110], magnetic and leaching procedures from waste magnets [111], hydrometallurgical treatment of magnetic sludge using leaching and precipitation techniques [112], separation of rare earth chlorides from iron alloys [113], or recovery of rare earth metals via chemical vapor transport technique based on the differences in thermal stability of the vapor compounds [114]. Nd–Fe–B magnets were obtained from electronic waste devices (hard drives). The magnets were heated up to 300°C under argon stream (1 L/min) for demagnetization and the elimination of organic compounds prior to grinding, using a ring mill. The magnet powder was prepared by heating in the furnace in 1,000°C. Waste tire rubber scraps were independently processed in a high temperature tubular furnace at 1,550°C for 10 min under argon stream. Carbon derived from waste rubber was used as a reducing agent in the stoichiometric ratio for the reduction of the iron, cobalt, and nickel oxides, present in the magnet powder. The established mixture was then heated up to 1,450°C in a tubular furnace under argon stream for 90 min. Successively, the material was kept in the cold zone in an argon atmosphere for 10 min. The rear earth oxides recovered using waste tire rubber-derived carbon revealed a high level of hardness, making them potential material for abrasion and polishing applications.

3.3.5.2 Vanadium processing

Removal of metals from recycled tire oils by liquid–liquid extraction using superheated water appears to be environmentally friendly approach [115]. Superheated liquid extraction is a method in which a solvent at high temperature and pressure without reaching the critical point is used [116, 117]. The utilization of water for the elimination of contaminants from solid materials establishes a promising option [118, 119]. Aqueous solution of nitric acid can be used for the extraction of metals from plant material [120] and the addition of sodium dodecyl sulfate (typical anionic surfactant) as a micelle formation surfactant can be effectively utilized for the extraction of polycyclic aromatic hydrocarbons from soils [121]. Such a method of pressurized liquid–liquid extraction using modified superheated water as extractant, for removing metals and inorganic nonmetal species from used industrial oils [122, 123], showed to be efficient, inexpensive, environmentally friendly, and straightforwardly manageable by industry. It is an interesting alternative to commonly used methods.

The method proposed for vanadium processing is as follows. The polluted oil and the extractant were pumped in a continuous mode at the pressure required to keep the liquid state at the preset temperature. The oil and extractant streams entered mixed and the sections of the immiscible phases were directed to the extraction coil in the oven at the working temperature. The sections of oil and extractant circulated through the extraction coil at high pressure and temperature, allowing the metals to be transported from the oil into the water phase. Subsequently, the oil/water phase segments were cooled in a water bath at 25°C. Both phases were separated after collection in a vessel. Effectiveness higher than 90% was obtained in the optimum working environment, at the temperature of 175°C, with the use of oil/extractant flow rates at 5:10 ratios together with 20% HNO_3 (1 M) and KCl (10^{-3} M) and diaminoethane-tetraacetic acid (EDTA) as extractant.

3.3.6 Agricultural use focused on metal content

In some cases, the waste tire rubber can be utilized as a source of zinc in the agricultural systems [124]. The authors proposed following method. The zinc extraction from used tire was performed by using acidic solutions. Waste tire rubber and its ash were bathed using various ratios of HNO_3, H_2SO_4, and water. The solutions obtained were analyzed for metals zinc, lead, cadmium, and iron. The extracted mixture with the highest zinc amount was then applied as a source of zinc for hydroponics (the method of growing plants without soil) -grown tomatoes. The highest retrieval of zinc from rubber and its ash was obtained using HNO_3 at concentration of 4 mol/dm^3. A mixture of nutrients containing two levels of zinc (equal to and twice of full Johnson solution composition [125]) provided by the leachate of tire and ash were examined. In addition, another full Johnson nutrient solution with $ZnSO_4$ as source of zinc was prepared as the control sample. Tomato plantlets grown in the nutrient solutions provided with the leachates of waste tire and its ash yielded comparable shoot dry matter yields, though providing zinc by the tire leachate at the double Johnson solution concentration resulted in a decrease in shoot and root dry matter yield in comparison to the other prepared samples. Lower plant growth in the double Johnson solution treatment seems to be joined with very low pH (2.3) of the leachate since the zinc concentration in the leaves is not high enough to imply that zinc phytotoxicity took place. No significant difference in other heavy metals was found between model sample with $ZnSO_4$ and the tire leachate solutions tested. The double Johnson solution treatment considerably enlarged the tomato root and shoot lead concentration, but these levels stay lower than soil grown tomatoes, and very small amounts of lead moved into tomato fruits. Concentration of cadmium in the tomato tissues was lower than the atomic absorption spectrophotometry detection limit. Consequently, the health risk from cadmium content is not problematic for plants grown in the hydroponics nutrition solution provided by the ground tire rubber leachate.

The rubber from used tires is a potential substitute for soil. The plant cultivation in laboratory scale [126] shows that metal amounts in the plants tested (chrysanthemum) decreased for potassium, calcium, magnesium, and copper but increased for zinc. There was no accumulation of other heavy metals like cadmium, chromium, nickel, lead, or sodium in the tissue due to rubber application. Though, plant development decreases and the possibility of zinc toxicity may restrain the effectiveness of ground tires as an alternative for conventional organic modifications. The soilless media usefulness is studied with the use of Petunia "Stereo Red" and Impatiens "Impulse Violet" which were grown in a media that contained 0, 5, 10, and 20 volume percent of ground tire rubber [127]. The media was tested for pH equal 5.0 and 6.5. Shoot weights dropped with increasing rubber content, due to growing serious zinc toxicity. Lower shoot weights at pH 6.5 than at pH 5.0 were attributed to manganese deficiency that was not connected with the presence of rubber. Extrapolation from the experimental data indicated that Petunia "Stereo Red" would have been damaged by as little as a 2% inclusion of tire rubber. The resulting high risk of zinc toxicity pointed that ground tire rubber is not suitable material be included in soilless potting media.

Another way of incorporation of crumb rubber into artificial medium for plant growth is to use it in green roofs [128]. Such a roof is partly or totally wrapped with flora and a growing medium. The study performed an analysis of the amount of zinc that impacts the microbial ecosystem. The method of leaching with bacteria or fungi is known as a bioleaching and it is described widely in the literature. The bacteria *Salmonella enterica* was used to perform the biotests. The study showed that crumb rubber contains materials that can leach into rainwater and are inhibitory to *S. enterica* serovar *Typhimurium*. The crumb rubber inhibitory material leached out over an extended period of time, which can be minimized by pretreatment with an acidic solution. In the light of the results of the study, crumb rubber can be a practical green roof soil additive on a large scale. The study showed that *S. enterica* serovar *Typhimurium* can colonize crumb rubber and that crumb rubber extract may provide nutrients that are usable by this bacterium. The zinc role is associated with other biological functions, including biofilm formation, motility, and potential cross-resistance to antimicrobial agents.

3.4 Summary

Several metals are of high significance in rubber industry where the most important are iron and zinc. Many other metals are used in the preparation of natural rubber in the form of additives, which play important role for quality of end rubber product. One of the widest applications of rubber is the tire industry which has large impact on environment not only in the phase of production but especially in the phase of consumption and waste product utilization. The iron is mainly used as a metal cord to strengthen and to make more resilient. Zinc plays twofold role, it is used as an activator of vulcanization process and improves the rubber thermal properties which are important during tire use.

The text presented some published techniques aimed at removal and utilization of metals from waste rubber materials like spent tires. The other important area of metal processing in the field of rubber industry is the waste metal treatment by the rubber-based adsorbent. Used rubber materials, especially waste tires, are proposed to be used as a valuable subtract to produce novelty metal adsorbents. Several methods of metal removal by waste tire-prepared adsorbent are described and discussed.

The impact of metal content is also subject of interest in the agricultural field. One of the proposed applications is to use rubber debris as a soil substitute in the plant growth systems. Several techniques and analyses of the metal impact and possibility of plant contamination, especially zinc, and the plant development are depicted in the text. The impact on microbial life is presented on the examples of *Salmonella* bacteria.

The problem of rubber metal content and its utilization from waste rubber material is still subject of research. The importance of this field becomes higher due to greater environmental awareness and law enforcements.

References

[1] Bowers JE. Natural rubber-producing plants for the United States. Beltsville, Md: National Agricultural Library, Totowa, New Jersey: Rowman and Littlefield, USA, 1990.

[2] Saendee P, Tangboriboonrat P. Latex interpenetrating polymer networks of epoxidised natural rubber/poly(methyl methacrylate): an insight into the mechanism of epoxidation. Colloid Polym Sci. 2006;284:634–43.

[3] Choi SS. Strain and anisotropic effects on migration behaviors of antiozonants in carbon black-filled natural rubber vulcanizates. Polym Test. 2002;21:741–44.

[4] Eskandari M, Arastoopour H. Studying the pulverization mechanisms of low-cross-link-density natural rubber with and without carbon black. Powder Technol. 2009;189:454–61.

[5] Rattanasom N, Prasertsri S, Ruangritnumchai T. Comparison of the mechanical properties at similar hardness level of natural rubber filled with various reinforcing-fillers. Polym Test. 2009;28:8–12.

[6] Utara S, Jantachum P, Sukkaneewat B. Effect of surface modification of silicon carbide nanoparticles on the properties of nanocomposites based on epoxidized natural rubber/natural rubber blends. J Appl Polym Sci. 2017;134(45289):1–14.

[7] Drozdov AD, Dorfmann A. The nonlinear viscoelastic response of carbon black-filled natural rubbers. Int J Solids Struct. 2002;39:5699–717.
[8] Mansilla MA, Marzocca AJ, Macchi C, Somoza A. Natural rubber/styrene-butadiene rubber blends prepared by solution mixing: influence of vulcanization temperature using a Semi-EV sulfur curing system on the microstructural properties. Polym Test. 2017;63:150–57.
[9] Bhowmick AK, Mukhopadhyay R, De SK. High temperature vulcanization of elastomers. Rubber Chem Technol. 1979;52:725–34.
[10] Nasir M, The GK. The effects of various types of crosslinks on the physical properties of natural rubber. Eur Polym J. 1988;24:733–36.
[11] Kumaran MG, Mukhopadhyay R, De SK. Network structure of lignin-filled natural rubber vulcanizate. J Polym Sci B Polym Lett Ed. 1979;17:399–404.
[12] Morrison NJ, Porter M. Temperature effects on the stability of intermediates and crosslinks in sulfur vulcanization. Rubber Chem Technol. 1984;57:63–85.
[13] McSwkeeney GP, Morrison NJ. The thermal stability of monosulfidic crosslinks in natural rubber. Rubber Chem Technol. 1983;56:337–43.
[14] Walter J Tire Technology International, Industry opinion, 2014, September, http://www.tire-technologyinternational.com/industry-blogs.php?BlogID=1127, last access: 31.08.2017.
[15] Department for Environment Food & Rural Affairs; 2010 to 2015 government policy: waste and recycling, Waste and recycling policy paper, 8 May 2015, https://www.gov.uk/government/publications/2010-to-2015-government-policy-waste-and-recycling/2010-to-2015-government-policy-waste-and-recycling#appendix-8-disposing-of-tyres; last access: 20.09.2017.
[16] European Environment Agency, EMEP/CORINAIR Emission Inventory Guidebook – 2007, https://www.eea.europa.eu//publications/EMEPCORINAIR5, last access: 20.09.2017.
[17] Adachi K, Tainosho Y. Soil environment affected by tire dust. In: Tazaki K, Ed. International Symposium of the Kanazawa University 21st-Century COE Program. 17–18 March 2003. Kanazawa, Japan: Water and Soil Environments, 2003:344–47.
[18] Map T. Trade statistics for international business development. Geneva, Switzerland: International Trade Centre, Market Analysis and Research Section, 2017. http://www.trade-map.org/Country_SelProduct_TS.aspx. last access: 29.09.2017.
[19] Mechant AA, Petrich MA. Pyrolysis of scrap tire and conversion of chars to activated carbon. AIChE J. 1993;39:1370–76.
[20] Savas BZ, Ahmad S, Fedroff D Transportation Research Record No. 1574, November 1996, National Research Council, Washington, DC, USA, 1996, 80–88.
[21] Mull MA, Stuart K, Yehia AA. Fracture resistance characterization of chemically modified crumb rubber asphalt pavement. J Mater Sci. 2002;37:557–66.
[22] Luo T, Isayev AI. Rubber/plastic blends based on devulcanized ground tire rubber. J Elastomers Plast. 1998;30:133–60.
[23] San Miguel G, Fowler GD, Sollars CJ. Adsorption of organic compounds from solution by activated carbons produced from waste tyre rubber. Sep Sci Technol. 2002;37:663–76.
[24] Hamadi NK, Swaminathan S, Chen XD. Adsorption of Paraquat dichloride from aqueous solution by activated carbon derived from used tires. J Hazard Mater. 2004;112:133–41.
[25] Tanthapanichakoon W, Ariyadejwanich P, Japthong P, Nakagawa K, Mukai SR, Tamon H. Adsorption–desorption characteristics of phenol and reactive dyes from aqueous solution on mesoporous activated carbon prepared from waste tires. Water Res. 2005;39:1347–53.
[26] Alexandre-Franco M, Fernández-González C, Macías-García A, Gómez-Serrano V. Uptake of lead by carbonaceous adsorbents developed from tire rubber. Adsorption. 2008;14:591–600.
[27] Alexandre-Franco M, Fernández-González C, Alfaro-Domínguez M, Gómez-Serrano V. Adsorption of cadmium on carbonaceous adsorbents developed from used tire rubber. J Environ Manage. 2011;92:2193–200.

[28] Manchón-Vizuete E, Macías-García A, Nadal Gisbert A, Fernández-González C, Gómez-Serrano V. Adsorption of mercury by carbonaceous adsorbents prepared from rubber of tyre wastes. J Hazard Mater. 2005;119:231–38.
[29] Hamadi NK, Chen XD, Farid MM, Lu MG. Adsorption kinetics for the removal of chromium(VI) from aqueous solution by adsorbents derived from used tyres and sawdust. Chem Eng J. 2001;84:95–105.
[30] Calisir F, Roman FR, Alamo L, Perales O, Arocha MA, Akman S. Removal of Cu(II) from aqueous solutions by recycled tire rubber. Desalination. 2009;249:515–18.
[31] Torretta V, Rada EC, Ragazzi M, Trulli E, Istrate EA, Cioca LI. Treatment and disposal of tyres: two EU approaches. A review. Waste Manage. 2015;45:152–60.
[32] Thomas BS, Gupta RC. A comprehensive review on the applications of waste tire rubber in cement concrete. Renew Sust Energ Rev. 2016;54:1323–3.
[33] Al-Lal AM, Bolonio D, Llamas A, Lapuerta M, Canoira L. Desulfurization of pyrolysis fuels obtained from waste: lube oils, tires and plastics. Fuel. 2015;150:208–16.
[34] Zhang L, Zhou B, Duan P, Wang F, Xu Y. Hydrothermal conversion of scrap tire to liquid fuel. Chem Eng J. 2016;285:157–63.
[35] Bičáková O, Straka P. Co-pyrolysis of waste tire/coal mixtures for smokeless fuel, maltenes and hydrogen-rich gas production. Energ Convers Manage. 2016;116:203–13.
[36] Mui EL, Cheung WH, Valixm M, McKay G. Dye adsorption onto activated carbons from tyre rubber waste using surface coverage analysis. J Colloid Interface Sci. 2010;347:290–300.
[37] Acevedo B, Rocha RP, Pereira MF, Figueiredo JL, Barriocanal C. Adsorption of dyes by ACs prepared from waste tyre reinforcing fibre. Effect of texture, surface chemistry and pH. J Colloid Interface Sci. 2015;459:189–98.
[38] Acevedo B, Barriocanal C, Lupul I, Gryglewicz G. Properties and performance of mesoporous activated carbons from scrap tyres, bituminous wastes and coal. Fuel. 2015;151:83–90.
[39] Gupta VK, Nayak A, Agarwal S, Tyagi I. Potential of activated carbon from waste rubber tire for the adsorption of phenolics: effect of pre-treatment conditions. J Colloid Interface Sci. 2014;417:420–30.
[40] Cataldo F, Ursini O, Angelini G. Surface oxidation of rubber crumb with ozone. Polym Degrad Stab. 2010;95:803–10.
[41] Cao XW, Luo J, Cao Y, Yin XC, He GJ, Peng XF, et al. Structure and properties of deeply oxidized waster rubber crumb through long time ozonization. Polym Degrad Stab. 2014;109:1–6.
[42] Ganjian E, Khorami M, Maghsoudi AA. Scrap-tyre-rubber replacement for aggregate and filler in concrete. Constr Build Mater. 2009;23:1828–36.
[43] Thomas BS, Gupta RC. Properties of high strength concrete containing scrap tire rubber. J Clean Prod. 2016;113:86–92.
[44] Acevedo B, Barriocanal C. Texture and surface chemistry of activated carbons obtained from tyre wastes. Fuel Process Technol. 2015;134:275–839.
[45] Gupta VK, Suhas NA, Agarwal S, Chaudhary M, Tyagi I. Removal of Ni(II) ions from water using scrap tire. J Mol Liq. 2014;190:215–22.
[46] Gupta VK, Gupta B, Rastogi A, Agarwal S, Nayak A. Pesticides removal from waste water by activated carbon prepared from waste rubber tire. Water Res. 2011;45:4047–55.
[47] Gupta VK, Nayak A, Agarwal S. Performance evaluation and application of oxygen enriched waste rubber tire adsorbent for the removal of hazardous aniline derivatives from waste water. Chem Eng J. 2012;203:447–57.
[48] Gupta VK, Ganjali MR, Nayak A, Bhushan B, Agarwal S. Enhanced heavy metals removal and recovery by mesoporous adsorbent prepared from waste rubber tire. Chem Eng J. 2012;197:330–42.

[49] Saleh TA, Al-Saadi AA, Gupta VK. Carbonaceous adsorbent prepared from waste tires: experimental and computational evaluations of organic dye methyl orange. J Mol Liq. 2014;191:85–91.
[50] Saleh TA, Gupta VK, Al-Saadi AA. Adsorption of lead ions from aqueous solution using porous carbon derived from rubber tires: experimental and computational study. J Colloid Interface Sci. 2013;396:264–69.
[51] Al-Saadi AA, Saleh TA, Gupta VK. Spectroscopic and computational evaluation of cadmium adsorption using activated carbon produced from rubber tires. J Mol Liq. 2013;188:136–42.
[52] Gupta VK, Gupta B, Rastogi A, Agarwal S, Nayak A. A comparative investigation on adsorption performances of mesoporous activated carbon prepared from waste rubber tire and activated carbon for a hazardous azo dye – Acid Blue 113. J Hazard Mater. 2011;186:891–901.
[53] Gupta VK, Ali I, Saleh TA, Siddiqui MN, Agarwal S. Chromium removal from water by activated carbon developed from waste rubber tires. Environ Sci Pollut Res. 2013;20:1261–68.
[54] Amari T, Themelis NJ, Wernick IK. Resource recovery from used rubber tires. In Resour Policy. 1999;25:179–88.
[55] Allred RE, Coons NC, Finley DJ, Shoemaker JM, Wilder RL, Wilder JD. Large-scale recycling process for scrap tires and rubber products. In: Proceedings of 32nd SAMPE Europe International Technical Conference & Forum 2011 (SEICO 11): New Material Characteristics to Cover New Application Needs, Paris, France, 2011.
[56] Fukuzaki N, Yanaka T, Urushiyama Y. Effects of studded tires on roadside airborne dust pollution in Niigata, Japan. Atmos Environ. 1967;20:377–86.
[57] Adachi K, Tainosho Y. Characterization of heavy metal particles embedded in tire dust. Environ Int. 2004;30:1009–17.
[58] http://www.metso.com/. Accessed: 31 Aug 2017.
[59] Raclavská H, Corsaro A, Juchelková D, Sassmanová V, Frantík J. Effect of temperature on the enrichment and volatility of 18 elements during pyrolysis of biomass, coal, and tires. Fuel Process Technol. 2015;131:330–37.
[60] Queneau PB, Hogsett RF, Beckstead LW, Barchers DE. Processing of petroleum coke for recovery of vanadium and nickel. Hydrometallurgy. 1989;22:3–24.
[61] Tsuboi I, Kunugita E, Komasawa I. Recovery and purification of boron from coal fly ash. J Chem Eng Jpn. 1990;23:480–48.
[62] Biswas RK, Wakihara M, Taniguchi M. Recovery of vanadium and molybdenum from heavy oil desulphurization waste catalyst. Hydrometallurgy. 1985;14:219–30.
[63] Sharma KD. An approach to reduce magnesium from zinc electrolyte with recovery of zinc from disposed residue of an effluent treatment plant. Hydrometallurgy. 1990;24:407–15.
[64] Xue Z, Hua Z, Yao N, Chen S. Separation and recovery of nickel and cadmium from spent Cd–Ni storage batteries and their process wastes. Sep Sci Technol. 1992;27:213–21.
[65] Walter J. Zinc about it. Tire Technol Int. 2009;March:18–19.
[66] Smolders E, Degryse F. Fate and effect of zinc from tire debris in soil. Environ Sci Technol. 2002;36:3706–10.
[67] Muschack W. Pollution of street run-off by traffic and local conditions. Sci Total Environ. 1990;93:419–31.
[68] Falahi-Ardakani A. Contamination of environment with heavy metals emitted from automotives. Ecotoxicol Environ Saf. 1984;8:152–61.
[69] Lagerwerff JV, Specht AW. Contamination of roadside soil and vegetation with cadmium, nickel, lead, and zinc. Environ Sci Technol. 1970;4:583–86.
[70] Christensen ER, Guinn VP. Zinc from automobile tires in urban runoff. J Environ Eng Division. 1979;105:165–68.
[71] Dannis ML. Rubber dust from the normal wear of tires. Rubber Chem Technol. 1974;47:1011–37.

[72] Pierson WR, Brachaczek WW. Airborne particulate debris from rubber tires. Rubber Chem Technol. 1974;47:1275–99.

[73] Kinoshita T, Yamaguchi K, Akita S, Nii S, Kawaizumi F, Takahashi K. Hydrometallurgical recovery of zinc from ashes of automobile tire wastes. Chemosphere. 2005;59:1105–11.

[74] Baba AA, Adekola FA. Dissolution kinetics and zinc(II) recovery from spent automobile tyres by solvent extraction with Cyanex®272. World Rev Sci Technol Sustainable Develop (WRSTSD). 2011;8:182–95.

[75] Smedley PL, Kinniburgh DG. A review of the source, behaviour and distribution of arsenic in natural waters. Appl Geochem. 2002;17:517–68.

[76] Friis RH. The Praeger Handbook of Environmental Health. Santa Barbra, CA: Praeger ABC-CLIO, LLC, 2012.

[77] Dias JM, Mcm AF, Almeida MF, Rivera-Utrilla J, Sánchez-Polo M. Waste materials for activated carbon preparation and its use in aqueous-phase treatment: a review. J Environ Manage. 2007;85:833–46.

[78] Ioannidou O, Zabaniotou A. Agricultural residues as precursors for activated carbon production – a review. Renew Sust Energ Rev. 2007;11:1966–2005.

[79] Daikopoulos C, Georgiou Y, Bourlinos AB, Baikousi M, Karakassides MA, Zboril R, et al. Arsenite remediation by an amine-rich graphitic carbon nitride synthesized by a novel low-temperature method. Chem Eng J. 2014;256:347–55.

[80] Mouzourakis E, Georgiou Y, Louloudi M, Konstantinou I, Deligiannakis Y. Recycled-tire pyrolytic carbon made functional: a high-arsenite [As(III)] uptake material PyrC350®. J Hazard Mater. 2017;326:177–86.

[81] Seristatidou E, Mavrogiorgou A, Konstantinou I, Louloudi M, Deligiannakis Y. Recycled carbon (RC) materials made functional: an efficient heterogeneous Mn-RC catalyst. J Mol Catal A. 2015;403:84–92.

[82] Ghimire KN, Inoue K, Makino K, Miyajima T. Adsorptive removal of arsenic using orange juice residue. Sep Sci Technol. 2002;37:2785–99.

[83] Baikousi M, Daikopoulos C, Georgiou Y, Bourlinos A, Zbořil R, Deligiannakis Y, et al. Novel ordered mesoporous carbon with innate functionalities and superior heavy metal uptake. J Phys Chem C. 2013;117:16961–71.

[84] Mishra AK, Ramaprabhu S. Functionalized graphene sheets for arsenic removal and desalination of sea water. Desalination. 2011;282:39–45.

[85] Mishra A, Ramaprabhu S. The role of functionalised multiwalled carbon nanotubes based supercapacitor for arsenic removal and desalination of sea water. J Exp Nanosci. 2012;7:85–97.

[86] Bostick BC, Fendorf S, Manning BA. Arsenite adsorption on galena (PbS) and sphalerite (ZnS). Geochim Cosmochim Acta. 2003;67:895–907.

[87] Rönngren L, Sjöberg S, Sun ZX, Forsling W. Surface reactions in aqueous metal sulfide systems: 5. The complexation of sulfide ions at the ZnS-H_2O and PbS-H_2O interfaces. J Colloid Interface Sci. 1994;162:227–35.

[88] Shen G, Xu Y, Liu B. Preparation and adsorption properties of magnetic mesoporous Fe_3C/carbon aerogel for arsenic removal from water. Desalin Water Treat. 2016;57:24467–75.

[89] Newman MC, Alberts JJ, Greenhut VA. Geochemical factors complicating the use of aufwuchs to monitor bioaccumulation of arsenic, cadmium, chromium, copper and zinc. Water Res. 1985;19:1157–65.

[90] Amy GL, Chen HW, Dinzo A, Brandhuber P. Adsorbent treatment technologies for arsenic removal. Denver, Colorado, United States: American Water Works Association, Awwa Research Foundation, 2005.

[91] Kaminsky W. Chemical recycling of mixed plastics of pyrolysis. Adv Polym Technol. 1995;14:337–44.

[92] Freundlich HM. Over the adsorption in solution. J Phys Chem. 1906;57(385471):1100–07.
[93] Namasivayam C, Kadirvelu K. Uptake of mercury(II) from wastewater by activated carbon from an unwanted agricultural solid by-product: coirpith. Carbon. 1999;37:79–84.
[94] Yardim MF, Budinova T, Ekinci E, Petrov N, Razvigorova M, Minkova V. Removal of mercury(II) from aqueous solution by activated carbon obtained from furfural. Chemosphere. 2003;52:835–4.
[95] Kadirvelu K, Kavipriya M, Karthika C, Vennilamani N, Pattabhi S. Mercury(II) adsorption by activated carbon made from sago waste. Carbon. 2004;42:745–52.
[96] Zhang FS, Nriagu JO, Itoh H. Mercury removal from water using activated carbons derived from organic sewage sludge. Water Res. 2005;39:389–95.
[97] Cestari AR, Vieira EF, Lopes EC, Da Silva RG. Kinetics and equilibrium parameters of Hg(II) adsorption on silica–dithizone. J Colloid Interface Sci. 2004;272:271–76.
[98] Meng X, Hua Z, Dermatas D, Wang W, Kuo HY. Immobilization of mercury(II) in contaminated soil with used tire rubber. J Hazard Mater. 1998;57:231–41.
[99] Danwanichakul P, Dechojarasrri D, Meesumrit S, Swangwareesakul S. Influence of sulfur-crosslinking in vulcanized rubber chips on mercury(II) removal from contaminated water. J Hazard Mater. 2008;154:1–8.
[100] Kurniawan TA, Chan GY, Lo WH, Babel S. Comparisons of low-cost adsorbents for treating wastewaters laden with heavy metals. Sci Total Environ. 2006;366:409–26.
[101] Arief VO, Trilestari K, Sunarso J, Indraswati N, Ismadj S. Recent progress on biosorption of heavy metals from liquids using low cost biosorbents: characterization, biosorption parameters and mechanism studies. Clean Soil Air Water. 2008;36:937–62.
[102] Ceribasi IH, Yetis U. Biosorption of Ni(II) and Pb(II) by *Phanerochaete chrysosporium* from a binary metal system – kinetics. Water SA. 2001;27:15–20.
[103] Dimpe KM, Ngila JC, Nomngongo PN. Application of waste tyre-based activated carbon for the removal of heavy metals in wastewater. Cogent Eng. 2017;4(1330912):1–11.
[104] Fung PP, Cheung WH, McKay G. Systematic analysis of carbon dioxide activation of waste tire by factorial design. Chin J Chem Eng. 2012;20:497–504.
[105] Al-Asheh S, Banat F. Adsorption of copper ions on to tyre rubber. Adsorpt Sci Technol. 2000;18:685–700.
[106] Zavvar MH, Abdorrahman H, Vahdat J. Removal of Cu(II) from wastewater by waste tire rubber ash. J Serb Chem Soc. 2010;75:845–53.
[107] Maroufi S, Nekouei RK, Sahajwalla V. Thermal isolation of rare earth oxides from Nd–fe–B magnets using carbon from waste tyres. ACS Sustain Chem Eng. 2017;5:6201–08.
[108] Xu T, Li M, Zhang C. Reclamation of Nd, Dy and Co oxides from NdFeB Scrap. Chin Rare Earth. 2004;25:31–34.
[109] Yang Y, Walton A, Sheridan R, Güth K, Gauß R, Gutfleisch O, et al. REE Recovery from end-of-life NdFeB permanent magnet scrap: a critical review. J Sustain Metall. 2017;3:122–49.
[110] Wang Y, Liu Y, Guo J, Wang S, Weng G. Recovery of rare earth metals from NdFeB waste materials using hydrochlioric acid hydrometal. China. 2006;25:195–97.
[111] Lyman JW, Palmer GR. Recycling of neodymium iron boron magnet scrap. High Temp Mater Process. 1993;11:175–88.
[112] Saito T, Sato H, Motegi T. Recovery of rare earths from sludges containing rare-earth elements. J Alloys Compd. 2006;425:145–47.
[113] Uda T. Recovery of rare earths from magnet sludge by $FeCl_2$. Mater Trans. 2002;43:55–62.
[114] Murase K, Machida K, Adachi GJ. Recovery of rare metals from scrap of rare earth intermetallic material by chemical vapour transport. J Alloys Compd. 1995;217:218–25.
[115] Morales-Muñoz S, Luque-García JL, Luque de Castro MD. Demetalization of oils resulting from recycled tires by liquid–liquid extraction using modified superheated water: an environmentally friendly approach. Talanta. 2004;63:391–96.

[116] Zuloaga O, Fitzpatrick LJ, Etxebarria N, Dean JR. Influence of solvent and soil type on the pressurised fluid extraction of PAHs. J Environ Monit. 2000;2:634–38.
[117] Fitzpatrick LJ, Zuloaga O, Etxebarria N, Dean JR. Environmental applications of pressurized fluid extraction. Rev Anal Chem. 2000;19:75–122.
[118] Hartonen K, Inkala K, Kangas M, Riekkola ML. Extraction of polychlorinated biphenyls with water under subcritical conditions. J Chromatogr A. 1997;785:219–26.
[119] Hawthorne SB, Grabanski CB, Martin E, Mille DJ. Comparisons of Soxhlet extraction, pressurized liquid extraction, supercritical fluid extraction and subcritical water extraction for environmental solids: recovery, selectivity and effects on sample matrix. J Chromatogr A. 2000;892:421–33.
[120] Morales-Muñoz S, Luque-García JL, Luque de Castro MD. Acidified pressurized hot water for the continuous extraction of cadmium and lead from plant materials prior to ETAAS. Spectrochim Acta Part B. 2003;58:159–65.
[121] Fernández-Pérez V. Luque de Castro MD, Micelle formation for improvement of continuous subcritical water extraction of polycyclic aromatic hydrocarbons in soil prior to high-performance liquid chromatography–fluorescence detection. J Chromatogr A. 2000;902:357–67.
[122] Morales-Muñoz S, Luque-García JL, Luque de Castro MD. Pressurised liquid–liquid extraction. An approach to the removal of inorganic non-metal species from used industrial oils. Chemosphere. 2004;56:943–47.
[123] Fernández-Pérez V, Jiménez-Carmona MN, Luque de Castro MD. Continuous liquid–liquid extraction using modified subcritical water for the demetalisation of used industrial oils. Anal Chim Acta. 2001;433:47–52.
[124] Khoshgoftarmanesh AH, Taheri S, Shariatmadari H, Ghaziaskar SH, Chaney RL. Using acid-washing leachates of ground tire rubber and its ash as a zinc source for hydroponics-grown tomato. J Residuals Sci Technol. 2010;7:69–72.
[125] Johnson CM, Stout PR, Broyer TC, Carlton AB. Comparative chlorine requirements of different plant species. Plant Soil. 1957;8:337–53.
[126] Bowman DC, Evans RY, Dodge LL. Growth of *Chrysanthemum* with ground automobile tires used as container soil amendment. Hortic Sci. 1994;29:774–76.
[127] Handreck KA. Zinc toxicity from tire rubber in soilless potting media. Commun Soil Sci Plant Anal. 1996;27:2615–23.
[128] Crampton M, Ryan A, Eckert C, Baker KH, Herson DS. Effects of leachate from crumb rubber and zinc in green roofs on the survival, growth, and resistance characteristics of *Salmonella enterica* subsp. *enterica* serovar Typhimurium. Appl Environ Microbiol. 2014;80:2804–10.

Katarzyna Staszak and Karolina Wieszczycka

4 Energy industry

Abstract: The potential sources of metals from energy industries are discussed. The discussion is organized based on two main metal-contains wastes from power plants: ashes, slags from combustion process and spent catalysts from selective catalytic NO_x reduction process with ammonia, known as SCR. The compositions, methods of metals recovery, based mainly on leaching process, and their further application are presented. Solid coal combustion wastes are sources of various compounds such as silica, alumina, iron oxide, and calcium. In the case of the spent SCR catalysts mainly two metals are considered: vanadium and tungsten – basic components of industrial ones.

Keywords: ash, metal recovery, selective catalytic NOx reduction (SCR), spent catalyst

4.1 Introduction

The energy industry is both energy production and its processing and transmission. Considering this type of industry in terms of waste the mainly focus is on the power industry. Power production is very important branch of industry. Its level decides on horizontal social and economic development of each country. Depending on the sources and energy carries the power industry can by divided into renewable (solar radiation, biomass, winds, waves and tides, hydropower and geothermal energy) and non-renewable (wood, peat, coal, brown coal, crude oil, natural gas) ones. The wastes that are produced in power plants depend of the energy source and used technologies. For example coal includes 20 to 60 % of carbon, 1–3 % of hydrogen, 0.3–3 % of sulfur, 1–2 % of nitrogen and 5–12 % of oxygen. The exact amounts of each element depend on the type and rank of coal, how the coal was formed millions of years ago and the location or mines from which the coal is sourced [1]. It is even more difficult to define the composition for example biomass, which is, after all, used as a source of energy.

An interesting report on the energy market is the annual report presented by the U.S. Energy Information Administration. The latest one "International Energy Outlook 2017" presents the viewpoint for international energy markets through

This article has previously been published in the journal *Physical Sciences Reviews*. Please cite as: Staszak, K., Wieszczycka, K. Energy industry. *Physical Sciences Reviews* [Online] **2017**, *3* (5). DOI: 10.1515/psr-2018-0023

https://doi.org/10.1515/9783110547061-004

2040 [2]. The data presented in this report shows that the total world energy consumption will rise from 575 quadrillion British thermal units (Btu) in 2015 to 736 quadrillion Btu in 2040. The highest growth will occur in countries outside of the Organization for Economic Cooperation and Development (OECD), especially China and India. Energy consumption in non-OECD countries increases 41 % between 2015 and 2040 in contrast to a 9 % increase in OECD countries. What is important, increase of world consumption of marketed energy is observed for all fuel sources, except coal, where demand is constant. Coal is increasingly replaced by natural gas, renewables, and nuclear power (in the case of China). As expected the world's fastest-growing energy source is renewable source, the second is nuclear power with consumption increasing by an average 2.3 % and 1.5 % per year between 2015 and 2040, respectively. Still, the main source of energy is and will be fossil fuels, with assumption 77 % of energy use in 2040, with the largest source of world energy consumption – liquid fuels, mostly petroleum-based. The real date from to 2015 show that World energy consumption by energy source, in quadrillion Btu is as follows: petroleum and other liquids (180), coal (155), natural gas (125), renewables (70) and nuclear (25). As predicted by the year 2040, this order will be minor changes as follows: petroleum and other liquids (225), natural gas (180), coal (155), renewables (125) and nuclear (40). The predictive growth of natural gas usage on power plants is associated with its advantages: low capital costs, favorable heat rates, and relatively low fuel cost. The share of individual sources varies when only net electricity generation is considered. In this case coal dominates with 40 % contribution in 2015, and 30 % in 2040. The next ones are renewables (20 and 30 %), natural gas (20 and 28 %), nuclear (15 and 10 %), petroleum and other liquids (5 and 2 %), with contributions in 2015 and 2040, respectively. Thus, in 2040 renewables provide the same share of word electricity generation as coal. Among the renewables the highest share has got hydropower, the lowest geothermal. This change in contribution of energy source in total world energy consumption and production, especially move way from coal and growth in the use of non-CO_2-emitting sources of energy, such as wind and solar, caused, very important for ecological point of view, decreasing of carbon dioxide emission to the atmosphere.

The data presented in the International Energy Outlook 2017 report, as well as in others references [3–5], indicate the continuous development of combustion technologies for saving energy and for suppressing environmental impact. New technologies to reduce pollution are also proposed and applied. For example, during the combustion technology the NO_x is formation. To reduce its emission following solutions are used: reduction of temperature during the combust process, reduction of nitrogen in fuels, end-of-pipe (post-combustion) treatment for removing NO_x by converting it to N_2 [6]. Similar procedures are applied to limit the other air pollution in the energy industry, such as CO_2, CO, SO_x and PPM.

4.2 Potential sources of wastes

4.2.1 Catalyst using in pollution reduction in energy industry

4.2.1.1 Basic principle of selective catalytic reduction process

According to the environmental regulations NO_x should be removed from gases emitted into the atmosphere in all industry processes. The need to reduce emissions is due to the fact that nitrogen oxides are responsible for the production of acid rains, formation of ozone in the troposphere and respiratory problems [7]. They are responsible for bronchitis, pneumonia, viral infections and hay fever. The main source of nitrogen oxides emission is transport (about 46 % in the 15 EU countries in 2000 [8]) and fuel combustion from stationary sources, primarily from power stations, industrial heaters and cogeneration plants. The most important method of reduction of NO_x emission is their decomposition to N_2 and H_2O through a selective catalytic reduction process with ammonia, known as SCR [9]. The mechanism of this decomposition is based on several reactions with NH_3 as reducing agent [10]:

$$4NH_3 + 4NO + O_2 \rightarrow 4N_2 + 6H_2O \quad (4.1)$$

$$2NH_3 + NO + NO_2 \rightarrow 2N_2 + 3H_2O \quad (4.2)$$

$$8NH_3 + 6NO_2 \rightarrow 7N_2 + 12H_2O \quad (4.3)$$

Wherein, approximately 90 % of NO_x in the flue gas from coal-fired boilers is nitric oxide (NO) [11]. Therefore the reaction (1) dominates in this process. Moreover the SCR technology is proposed for the combined removal of nitrogen and sulfide oxides (sulfur present in the coal is oxidized to SO_2 in the boiler), reduction of NO_x with oxidation of dioxins and furans or C and CO in thermal power plants, industrial boilers and cogeneration units. In the presence of sulfur, the SCR process is based on the following reactions:

$$2SO_2 + O_2 \rightarrow 2SO_3 \quad (4.4)$$

$$SO_3 + NH_3 + H_2O \rightarrow NH_4HSO_4 \quad (4.5)$$

$$SO_3 + 2NH_3 + H_2O \rightarrow (NH_4)_2SO_4 \quad (4.6)$$

To avoid the deposition and accumulation of the ammonium sulfates onto the catalyst, and consequently deactivation of catalyst, the reaction occurs in the presence of large excess of water, and temperature higher than 300°C [12]. Moreover minimize the amount of unreacted NH_3 in the flue gas is desirable. This quantity is referred to as NH_3 slip and, in general, must be held below 5 ppm and preferably 2 to 3 ppm. At the same time it should be noted that normally, without ammonium, almost all of the SO_3 converts to H_2SO_4 as flue gas is cooled in the air preheater [13]. The emissions of SO_3 from a boiler depend on coal sulfur content, combustion conditions, flue gas characteristics, and air pollution devices used in the process. Formation and emission of SO_3/H_2SO_4 cause environmental and plant operation problems, mainly due to corrosive properties. The commonly practical technology to reduce of SO_3, if the SCR process is not used, is the injection of alkali in the flue gas stream into or after furnace with efficiency between 40 and 90 %.

4.2.1.2 Catalysts in SCR systems

The SCR system is well-known process, patented in the U.S. by Englehard Corporation in 1957 and used in industry since 1970s. It was first installed in Japan, in 1985 was applied in Europe. This technology commonly removes 80 %–90 % of NO_x emissions in coal-fired power plants. The efficiency depends on coal characteristics, flue gas composition, temperature, reagent injection and distribution, and catalyst formulation and activity. Three types of commercial catalysts have been developed for SCR systems: noble metals, metal oxides and zeolites. The catalysts based on noble metals are very active in the selective reduction of NO_x. Unfortunately, they also oxidize NH_3. That is a reason that this type of catalyst are used only in specific solution of commercial catalyst system combine high NO_x reduction and CO oxidation activities at low temperature and for low temperature and natural gas applications [10]. The good examples are the automotive catalytic converters based on Pt-Rh or Pd [14]. Metal oxides catalysts are mainly based on vanadium supported on titania in the anatase form, and promoted with tungsten or molybdenum (V_2O_5-WO_3 $(MoO_3)/TiO_2$). This type of catalyst is proposed for NO_x reduction and SO_2 oxidation. Because V_2O_5 is responsible for the activity in the reduction of nitrogen oxides and undesired oxidation of sulfur(IV) oxide, its content is low, especially in the presence of high amount of SO_2 (below 1 % w/w). Addition of WO_3 or MoO_3 (10 and 6 % w/w, respectively) helps to increase the acidity, the activity and thermal stability of the catalyst and to limit the oxidation of SO_2 [15]. In the case of occurrence of As compounds in the exhaust gas it is better to use of MoO_3 because it prevents catalyst deactivation [16]. Metal oxide catalysts work generally in temperature between 300 and 400°C. For higher temperatures the zeolite catalysts with transition metal ions (e. g. Fe(III)) are proposed [17, 18], such as Fe/ZSM-5, Fe^{3+}-exchanged TiO_2-pillared clay, Fe_2O_3–WO_3/ZrO_2 and Fe_2O_3/TiO_2 [19–21]. Small amount of noble metal (Pt, Rh, or Pd, order with decreasing of SCR activity) to the Fe-ZSM-5 catalyst decrease the necessary temperature for selective catalytic reduction of NO with ammonia from 300–400 °C to 200–300 °C [22]. The catalyst with addition of

1% of Cu is also proposed [23–25]. An overview of the selective catalytic reduction of NO_x by metal-exchanged zeolites is described in detail, in work [26]. The analysis of chemistry of this process, including undesired side-reactions and aspects of the reaction mechanism over zeolites and the active sites involved, is presented.

The two leading shapes of SCR catalyst used are honeycomb and plate. The honeycomb form usually is an extruded ceramic (titanium oxide) with the catalyst either incorporated throughout the structure (homogeneous) or coated on the substrate. In plate geometry, the support material is generally coated with a catalyst.

Review of technological processes show that typical commercial SCR catalyst is V_2O_5-WO_3/TiO_2-based catalyst. It has several advantages as low cost, high NO_x removal efficiency and good sulfur resistance. Disadvantages, such as high starting temperature, narrow active temperature window, low N_2 selectivity at high temperature and high conversion of SO_2 to SO_3, cause that new solutions are still looking for new catalyst composition [27, 28]. Moreover, vanadium pentoxide shows toxicity to environment and humans (such as causing irritation to skin, eyes and respiratory tract) [29]. A promising metal oxides catalyst is CeO_2/TiO_2-based catalyst with high activity of nitrogen oxides reduction with ammonia in a broader temperature window [30–33], low-temperature activity, selectivity and durability Sb-promoted V_2O_5/TiO_2 catalyst [34, 35] and manganese-based catalysts (MnO_2-TiO_2, MnO_2-CeO_2-TiO_2 and MnO_2-Fe_2O_3-CeO_2-TiO_2) [36–38]. Especially MnO_2-Fe_2O_3-CeO_2-TiO_2 catalyst exhibit good low-temperature activity in reduction of NO with NH_3, low N_2O escape, low oxidation to NH_3, good resistance to SO_2 poisoning and good activity stability [39]. Modification of vanadium-based catalyst is also proposed in the literature. Germanium addition to CeO_2-WO_3 catalyst allows obtaining catalyst with very high NO_x conversion at a wide temperature range [40]. Moreover catalyst with $V_2O_3(SO_4)_2$ as active component are successfully tested [41].

The catalytic activity of SCR catalysts decrease gradually during the operation time due to the poison from the complex chemical compositions in flue gas, fouling, thermal degradation, vapor compound formation, vapor–solid reactions between vapor and catalyst surface as well as solid-solid reactions between catalyst carrier and promoter, and attrition/crushing [11, 42–45]. Generally, the lifetime of SCR catalyst is 2–3 years in typical process of coal or oil combustion [46, 47], and much smaller in plants burning biomass [48]. The reason of such difference of life-time is much higher content of alkali metals in the biomass in comparison to the typical fossil fuels.

If it possible the regeneration of catalyst are made, not only to reduce the cost of operating SCR units, but also to reduce the amount landfill disposal. The regeneration of used catalysts will be a preferable option because it could prolong the lifetime of catalysts and thus save metal resources. Unfortunately, after several times of regeneration [42, 49], the SCR catalyst comes to the end of its life. This results in a significant amount of waste each year. Only in Taiwan 1000 tons per year spent SCR catalyst is generated [50], in China in 2018 it is estimated on 38 000 tons [51]. To avoid the worst solution – landfill disposal, it is suggested to recover metals from spent catalysts. This is due not only to the ecological but also to the economic issue. The spent catalysts are not only hazardous waste but also precious natural resources.

The serious problem in coal-burning power plants is the emission of mercury. It should be noted that the heavy metals in exhaust gases mostly exist in solid forms but emitted mercury with high vapor pressure exists in a gas form, causing its removal at the source more difficult [52]. The global emission of mercury is estimated on the range from 1000 to 6000 tons/yr [53], with the largest source – coal-burning plants. Mercury in coal-burning flue gas exists in three forms: elemental (Hg^0), oxidized (Hg^{2+}), and particle-bound (Hg(p)). Mercury is released during coal combustion, part of it is oxidized by excess amount of chlorine in coal ($HgCl_2$). Both forms of mercury have ability to adsorb on solid surface, such as fly ash particles, and form the particle-bound mercury. In conventional wet scrubber process the oxidized mercury can be easily removed, due to its dissolution in water. The elemental Hg is not soluble in water and others methods should be used to reduce its emission. One of the solutions is adsorption of mercury on activated carbon, alone or modified by impregnation with sulfur, chlorine and iodine [54, 55]. Unfortunately, this method is effective only for high concentration of Hg. Since mercury on the second oxidation stage is easy to remove, application of catalyst to oxidize elemental mercury is proposed. Similar to described above SCR process, the V_2O_5/TiO_2 catalysts are mainly used [52, 56–59], here in the presence of hydrochloric acid, according to the following mechanism:

$$2Hg + 4HCl + O_2 \rightarrow HgCl_2 + H_2O \qquad (4.7)$$

To increase the Hg^0 catalytic efficiency the modification of SCR V_2O_5/TiO_2 catalyst by noble (Au, Pd [60]), rare earth (Ru [61]) and other metals (Mn, Cu, Cr [61–63]) are proposed. Elemental mercury catalytic oxidation efficiency is high (over 90 %). Thus this method can be used for Hg emission control.

Typical compositions of fresh and spent catalyst using in selective catalytic reduction processes are presented in Table 4.1. In the first three pairs the difference between composition of fresh and spent catalysts is shown. Compared with the fresh catalysts, spent catalysts show lower vanadium contents. Moreover, the sum of others elements concentrations such as Al, Si, Ca, Fe, K, and Na increase to varying degrees in spent catalysts. This indicates that different elements such as Fe and Al accumulate in catalyst during their work. Evidently, the spent catalyst could not to be directed reused by blending with fresh catalyst, because the deposited alkali metals exert an adverse effect on catalytic activity [64]. The toxicity of alkali oxides to catalysts followed with sequence: $Cs_2O > Rb_2O > K_2O > Na_2O > Li_2O$ [65, 66]. It is

Table 4.1: Composition of SCR catalysts, in wt%.

Type/producer	V	W	Ti	Others metals	Other	Ref
Coal power plant catalyst (V_2O_5-WO_3/TiO_2)/NDA	3.3*	3.0*	100*	**Ca** (3.8), **Na** (4.1), **Al** (78.7)*	Si (11.1), S (2.6)*	[46]
Coal power plant catalyst (V_2O_5-WO_3/TiO_2)/NDA/time in line 350 000 h	3.1*	3.2*	100*	**Ca** (5.0), **Na** (4.0), **Al** (74.2)*	Si (23.0), S (79.2)*	[46]
Commercial honeycomb monolith catalyst/Zhejiang Tuna Environmental Science & Technology Co., Ltd.	0.49	4.16	49.98	**Ca** (1.06), **Na** (0.02), **K** (0.04)	Si (2.33), S (0.26), As (0), O (41.66)	[69]
Supra/time in line – more than 30,000 h in a low grade coal fired power plant (Inner Mongolia, China)	0.30	3.61	51.98	**Ca** (0.95), **Na** (0.09), **K** (0.03)	Si (1.82), S (0.28), As (1.14), O (39.1)	[69]
V_2O_5–WO_3/TiO_2 catalyst/Jiangsu Longyuan Catalyst Co., Ltd, Wuxi, China	0.39	3.48	55.04	**Ca** (0.84), **Na** (0.07), **K** (0.02), **Al** (0.44), **Fe** (0.04)	Si (0.87)	[81]
V_2O_5–WO_3/TiO_2 catalyst/Jiangsu Longyuan Catalyst Co., Ltd, Wuxi, China/35,000 h in-line	0.31	4.57	51.88	**Ca** (0.90), **Na** (0.13), **K** (0.24), **Al** (0.92), **Fe** (0.24)	Si (0.92)	[81]
NDA/Haldor Topsøe A/S	1–7.3 as V_2O_5	5–13 as WO_3	NDA as TiO_2	NDA	NDA	[42, 70]
Thermal power plant in China catalyst/NDA	0.29 (0.52 as V_2O_5)	3.00 (3.78 as WO_3)	36.00 (60.07 as TiO_2)	**Ca** (1.50), **K** (0.11), **As** (0.01), **Al, Fe** (NDA)	P (0.30), Si (NDA)	[47]
Spent SCR catalyst/NDA	1.57	6.37	33.40	**Ca, Al** (NDA)	Si (2.81), P (NDA)	[50]

(continued)

Table 4.1: (*continued*)

Type/producer	V	W	Ti	Others metals	Other	Ref
SCR honeycomb monolith catalyst/ENEL Produzione (Livorno, Italy)	0.55	9.0	NDA	**Al** (NDA)	Si (NDA)	[68]
SCR honeycomb monolith catalyst/ENEL Produzione (Livorno, Italy)	1.8	8.5	NDA	**Al** (NDA)	Si (NDA)	[68]
Coal-fired power plant catalyst from southwest China/NDA	1 as V_2O_5	NDA as WO_3	NDA as TiO_2	NDA	NDA	[71]
Samchunpo thermoelectric power plant catalyst in Korea/NDA	1.23 as V_2O_5	7.73 as WO_3	70.9 as TiO_2	**CaO** (2.45), **MgO** (0.55), $\mathbf{Al_2O_3}$ (5.57), $\mathbf{MoO_3}$ (0.10), $\mathbf{Fe_2O_3}$ (0.77)	etc. (0.10)	[72, 74]
Commercial V_2O_5 –WO_3/TiO_2 catalyst/ Institute of Process Engineering, Chinese Academy of Sciences	0.74 as V_2O_5	4.61 as WO_3	87.25 as TiO_2	**CaO** (1.38), $\mathbf{Al_2O_3}$ (0.96), $\mathbf{Fe_2O_3}$ (0.08)	SiO_2 (3.74), SO_3 (0.84), P_2O_5 (0.13)	[77]
Commercial V_2O_5–WO_3/TiO_2 catalyst/ Institute of Process Engineering, Chinese Academy of Sciences	1.20 as V_2O_5	4.69 as WO_3	88.51 as TiO_2	**CaO** (1.43), **MgO** (0.08), $\mathbf{Al_2O_3}$ (0.59), $\mathbf{Fe_2O_3}$ (0.11), $\mathbf{K_2O}$ (0.20), $\mathbf{ZrO_2}$ (0.08)	SiO_2 (2.82), SO_3 (0.09), P_2O_5 (0.20)	[86]
Commercial honeycomb monolith catalyst	1.86 as V_2O_5	4.39 as WO_3	49.84 as TiO_2	**Ca, Mg, Na, K, Mo** (NDA), $\mathbf{Al_2O_3}$ (25.05)	SiO_2, P_2O_5 (NDA)	[73]

NDA – no data available; white row – fresh catalyst, gray row – spent catalyst; * – The element molar ratios (the value of Ti is given as 100).

assumed that SCR catalysts were primarily deactivated by chemical poisoning with alkali oxides, as was presented in works [67, 68]. The sent catalysts need to be regenerated, though by washing (mainly by water and nonorganic acid) to remove the disturbing compounds [42, 49, 64].

As it is seen these catalysts are mainly composed of TiO_2. The small composition of V and W (less than 10 % of each component) does not seem at first attractive but V and W account for more than 40 % of the total metal price [74]. Therefore the recovery of vanadium and tungsten from spent catalyst is studied widely. These methods are based mainly on soda roasting and acid leaching [75]. These methods are usually efficient because in typical SCR catalyst the vanadium in the form of V_2O_5 is dispersed on TiO_2 support, but their interaction is week [76]. In soda roasting, NaOH, Na_2CO_3, and NaCl are used as leaching agents. The main idea of these processes is to selectively transfer metal ions from solid waste into aqueous solutions. Afterward, further traditional purification techniques, such as sediment, precipitation, solvent extraction, ion exchange or crystallization, are proposed in the literature [50, 77]. The basic of these processes in convert of V_2O_5 and WO_3 (water insoluble states) from spent catalyst to the forms of water-soluble states $NaVO_3$ and Na_2WO_4 according to the reactions:

$$V_2O_5 + 2NaOH \rightarrow 2NaVO_3 + H_2O \tag{4.8}$$

$$WO_3 + 2NaOH \rightarrow Na_2WO_4 + H_2O \tag{4.9}$$

$$V_2O_5 + Na_2CO_3 \rightarrow 2NaVO_3 + CO_2 \tag{4.10}$$

$$WO_3 + Na_2CO_3 \rightarrow Na_2WO_4 + CO_2 \tag{4.11}$$

Because main component of SCR catalyst is TiO_2 the following reaction also occurs during the leaching process:

$$TiO_2 + 2NaOH \rightarrow Na_2TiO_3 + H_2O \tag{4.12}$$

Because titanium (IV) oxide is barely soluble, in leaching solution mainly vanadium (soluble) and tungsten (easy soluble) ions exist. To separate V and W ions in the leaching solution the calcium precipitation process are proposed with the following mechanism:

$$Ca(OH)_2 + Na_2WO_4 \rightarrow CaWO_4 + 2NaOH \tag{4.13}$$

$$Ca(OH)_2 + 2NaVO_3 \rightarrow Ca(VO_3)_2 + 2NaOH \tag{4.14}$$

$$CaCl_2 + Na_2WO_4 \rightarrow CaWO_4 + 2NaCl \quad (4.15)$$

$$CaCl_2 + 2NaVO_3 \rightarrow Ca(VO_3)_2 + 2NaCl \quad (4.16)$$

Besides precipitation the separation of tungsten from vanadium with the anion exchange resin is proposed [50]. In this process the difference in the number of negative charges between V and W in alkaline solution is utilized (see eqs (4.8) and (4.9)). Typical procedure of recovery metals from catalyst include besides the leaching process, also the initial step with catalyst preparation (grinding, sieving) and the separation, mainly by filtration, the leaching solution after process. Generally in the literature the conventional stirring leaching method of spent catalyst is proposed. Zhang et al. [78] showed that leaching can be intensified by ultrasonic wave. The obtained results indicated that the leaching efficiency of V and W enhanced with ultrasound energy was 76.92 % and 69.87 %, respectively in very short time – 3 min. These results are higher by 23.08 % and 2.59 %, respectively, compared with the conventional stirring leaching result obtained at the same time. In the optimal conditions (2 h ultrasonic leaching time, liquid-solid ratio of 12:1, and ultrasonic power of 500 W) the recovery of V and W reached 89.01 % and 96.05 %, respectively.

The conditions of various leaching processes combined with others methods for metals separation and main conclusions are presented in Table 4.2.

Some investigations are dedicated not only to the recovery of metals from spent SCR catalysts, but also to their re-use, mainly as new catalysts. Interesting alternative to utilize spent SCR catalysts is their reuse as new ones after oxalic acid leaching and impregnating with V_2O_5 [81]. This procedure allows obtaining newly synthesized catalyst with deNO_x activity comparable to fresh catalyst (91 % NO conversion at 300°C), and with good resistance to SO_2 and H_2O. Moreover, because vanadium and tungsten oxides are known as a component of photocatalyst, it is proposed to use spent SCR catalyst as a substrate to prepare photocatalytic materials [82–85]. After classical leaching of V and W from waste V_2O_5–WO_3/TiO_2 catalyst the appropriate reagents are added, for example $Bi(NO_3)_3$ solution to obtain Bi_2WO_6/$BiVO_4$ photocatalyst [77] or $Zn(NO_3)_2$ to obtain $Zn_3(VO_4)_2$/$ZnWO_4$ photocatalyst [86]. The main assumptions, experimental conditions and conclusions are summarized in Table 4.3.

It should be mentioned that alternative processes for reduction of NO_x emission are also applied as selective non-catalytic reduction (SNCR) or SCONO$_X$™. In SNCR process the reduction of NO_x to N_2 is obtained in the presence of oxygen by reaction with amine-based reagents, either ammonia (NH_3) or urea, $CO(NH_2)_2$ at 800–1000°C. The higher temperature is needed for urea. SNCR does not require a catalyst, but the efficiency is much smaller than in typical SCR technology and ranges from 20 to 70 %. Moreover SNCR applications are limited to smaller boilers because of difficulties in

Table 4.2: Examples of vanadium and tungsten recovery from spent SCR catalysts by leaching processes.

Metal contents (initial conc.)	Process condition	Results	Ref
0.29 % V, 3.00 % W	– Leaching agent 0.1–5 M H_2SO_4 – Experimental conditions: temp. in a range of 30 – 80 °C, time from 5 to 120 min, Solid/liquid ratio from 1/5 to 1/20.	– Vanadium recovery yield increases with increases in temperature and sulfuric acid concentration but decreases with an increase in solid to liquid ratio. – Vanadium removal: from 20 to 60 % depending on the operation conditions.	[47]
1.57 % V, 6.37 % W	– Leaching agent: NaOH – Optimal conditions: NaOH concentration of 0.3 kg/kg of catalyst, pulp density 3 %, leaching temp. 70°C, particle size 74 µm, time 30 min. – Separation of V and W through base anion exchange resin (Amberlite IRA900).	– Vanadium and tungsten removal: 87 and 91 %, respectively after leaching process. – Final concentration of W after leaching and ion exchange process: 8.4 g/L with 98 % purity.	[50]
V, W (NDA)	– Leaching agent: NaOH. – Optimal conditions: 7.5 M NaOH, temp. 100 °C.	– Vanadium and tungsten removal: 92.94 and 97.30 %, respectively.	[79]
1.23 % V_2O_5 7.73 % WO_3	– Leaching agent: 1–4 M NaOH. – Leaching process is carried out with autoclave (1 liter, ≤80 bar) reactor to obtain the high yield rate.	– Temperature in the range of 200~300 °C is proportional to the leaching efficiency of vanadium and tungsten.	[80]
1.23 % V_2O_5 7.73 % WO_3	– Leaching agent: NaOH with or without addition of Na_2CO_3. – Leaching reaction time: 1, 2, 6, 12, and 24 h. – Temp: 200, 250, 300, and 350 °C with a heating rate of 200 °C/h. – Precipitation process: $Ca(OH)_2$. – The optimal leaching reaction conditions of 300 °C, L/S = 20, 2 M NaOH, and 0.2 M Na_2CO_3 as additive.	– Depending on the reaction temperature, the recovery of W increased from 64.8 to 96.0 %; V whit ranged from 86.3 to 90.5 %. – With 2 M NaOH the lowest recovery (88.2 %) of V was obtained whereas the highest recovery (96.0 %) of W was obtained. – Addition of 0.2 M Na_2CO_3: the leaching percentage of V was low at 86.6 % but that of W was high at 99.9 %. – Vanadium and tungsten removal: 86.6 and 99.9 %, respectively.	[74]
1.23 % V_2O_5, 7.73 % WO_3	– Leaching agent: Na_2CO_3. – Soda roasting process: temp. 850 °C, 120 min. – Water leaching: temp. 40 °C, 30 min.	– Vanadium and tungsten removal: 46 % and 92 %, respectively in optimal condition. – The W removal increases from 30 to 80 %, while V decreased 55 to 30 % with roasting temperature increasing from 600 to 900°C. – Procedure: After soda roasting process the water leaching.	[72]

Table 4.3: Examples of-use of spent SCR catalysts.

Metal contents (initial conc.)	Process condition	Results	Ref
0.74 % V_2O_5, 4.61 % WO_3, 87.25 % TiO_2	– Leaching agent: NaOH. – The optimal conditions: 5 M NaOH, solid to liquid ratio S/L 1:5, time 3 h, temp. 120 °C (continuous air stirring, 10 mL/min, 0.05 MPa).	– The percent leaching of V increases from 70 to 91.50 % and W from 38.50 to 69.70 %, while Ti changes only from 0.99 to 0.67 % with increase NaOH concentration from 1 to 5 M. – With temperature increased from 353 to 393 K, the percentage of vanadium and tungsten were both increased. – Vanadium and tungsten removal: 96.50 and 72.30 %, respectively. – By directly addition of the $Bi(NO_3)_3$ solution to the leaching solution the $Bi_2WO_6/BiVO_4$ photocatalyst can be prepared. – The percentage of photocatalytic degradation of $BiVO_4/Bi_2WO_6$ on methyl blue is 73 %.	[77]
1.20 % V_2O_5, 4.69 % WO_3	– Leaching agent: NaOH with addition of Na_3PO_4. – Precipitant agents: NaOH for Fe, Zr, Mg ions; Na_3PO_4 for Ca and $Mg(NO_3)_2$ for Si. – The $Zn_3(VO_4)_2/ZnWO_4$ photocatalyst is prepared by a hydrothermal-sintering process. – Operation conditions for photocatalyst preparation: addition of $Zn(NO_3)_2 \cdot 6H_2O$, pH 9.0 (correction by $NH_3{\cdot}H_2O$), temp. 180 °C, time 24 h. Eventually the hydrothermal product is calcinated at 680 °C for 6 h.	– First step: recovery of metals from spent catalyst with leaching efficiency: V_2O_5 (96.3 %), WO_3 (60.3 %), Al_2O_3 (80.62 %), SiO_2 (78.59 %), TiO_2 (0.2 %), CaO (3.27 – due to precipitation $Ca_3(PO_4)_2$). – Second step: removal of Ca, Al (as $AlOH_3$) Mg (as $MgOH_2$) and Si (as $MgSiO_3$) from leaching solution by selective precipitation. – Residual ratio after purification: V (77.5 %), W (90.3 %), Al (1 %), Si (95 %). – Concentrations in leaching (L) and purified solution (P), in mmol/L V W Ti Ca Al Si L 12.71 24.40 $4.43 \cdot 10^{-3}$ 1.67 9.28 73.81 P 9.84 22.02 $2.33 \cdot 10^{-3}$ 0.97 0.06 0.80 – The total recycle efficiencies for vanadium and tungsten are 74.6 % and 54.4 %, respectively. – The new photocatalysts $Zn_3(VO_4)_2/ZnWO_4$ are obtained. – The percentage of photocatalytic degradation of catalysts on methyl blue is over 90 %.	[86]

(continued)

Table 4.3: (*continued*)

Metal contents (initial conc.)	Process condition	Results	Ref
145 mg/L V_2O_5, 546 mg/L WO_3	– Leaching agent: Na_2CO_3. – The tungsten(VI) can be separated from mixed solution with small impurities of vanadium(V) by Fe_3O_4 under certain pH condition.	– The maximum separation factor tungsten(VI) over vanadium(V) is found at pH 1~2, with 85.83 % purity of W(VI). – Vanadium ions adsorb onto Fe_3O_4 nanoparticles with maximum adsorptive ion capacity at pH 3. – The newly obtained material is used as a heterogeneous Fenton-like catalyst.	[87]
1.20 % V_2O_5, 4.69 % WO_3	– Leaching agent: 1.5 M oxalic acid ($H_2C_2O_4 \cdot H_2O$). – Leaching solution was co-impregnated with NH_4VO_3 dissolved in oxalic acid at 80°C, time 12 h. Then the solution was dried at 105°C for 6 h and calcined at 600°C in air for 3 h. – Composition of mainly elements after leaching (L) is different in comparison to spent catalyst (S), results in % wt:	– The leaching process with oxalic acid allows removing the alkali metals form spent catalyst. – Impregnation with V_2O_5 allows obtaining a new catalyst with deNO_x activity comparable to fresh catalyst.	[81]

	V	W	Ti	Ca	Al	Si	Fe	K	Na
S	0.31	0.13	4.57	51.9	0.90	0.92	0.92	0.24	0.24
L	0.12	0.03	3.94	52.4	0.75	0.52	1.02	0.07	0.04

achieving uniform distribution of reagent in the flue gas stream [6, 88–90]. In SCONO$_x$™ technology a potassium carbonate (K_2CO_3) coated catalyst are used. In this process catalytic oxidation carbon-monoxide (CO) to carbon-dioxide (CO_2), and nitric oxide (NO) to nitrogen-dioxide (NO_2) occurs. The CO_2 is exhausted while the NO_2 absorbs onto the catalyst to form potassium nitrites (KNO_2) and potassium nitrates (KNO_3). Several types of catalyst are proposed in the literature Pt–K_2CO_3/ZrO_2 [91], Pt–x$K_2CO_3/K_2Ti_8O_{17}$ [92] with Pt content about 1–2%. Although this technology is characterized by high efficiency of NO_x removal, it is not used in the large scale power plant. For this reason, the recovery of metals from these catalysts has not been described in the literature, and the compositions of these catalysts are only in experimental phase.

4.2.2 Coal combustion wastes

Coal-burning power plants produce substantial quantities of coal ash, which is divided on fly ash and bottom ash. Fly ash is defined as the portion of ash produced in coal combustion that has a sufficiently small particle size to be carried away from the boiler in the flue gas. In a typical power plant fly ash particles are captured in electrostatic precipitators or baghouses (baghouses can also be used to collect ash with bags that filter the fly ash out of the flue gas stream).

Bottom ash is formed when ash particles soften, melted and accumulates on the boiler walls and against steam tubes in the boiler. The created clinker eventually is blown off by jets of air or falls to hoppers located at the base of the furnace where they are collected and ground to a sand size gradation. Some bottom ash is transported to storage dry, but most is dewatered prior to unloading and transport to construction sites or storage stockpiles.

Boiler slag is also formed during burning process, but only when wet-bottom boilers are used. There are two types of wet-bottom boilers: the slag-tap boiler and the cyclone boiler. The slag-tap boiler burns pulverized coal and the cyclone boiler burns crushed coal. In each type, the bottom ash is kept in a molten state and tapped off as a liquid. Both boiler types have a solid base with an orifice that can be opened to permit the molten ash that has collected at the base to flow into the ash hopper below. The ash hopper in wet-bottom furnaces contains quenching water. When the molten slag comes in contact with the quenching water, it fractures instantly, crystallizes, and forms pellets-boiler slag.

The properties of ash are dependent on the composition of the parent coal, conditions during combustion, efficiency of emission control devices, storage and handling of the by-products, and climate. In the case of the fly ash the chemical properties are influenced to a great extent by the properties of the coal being burned and the techniques used for capturing and storage of ash. There are basically four types, or ranks, of coal, each vary in heating value and chemical composition. These types are: anthracite, bituminous, sub-bituminous and lignite. The principal components of bituminous, sub-bituminous, and lignite coal fly ash are silica, alumina, iron oxide, and calcium. However, lignite and sub-bituminous coal fly ash has a higher CaO, MgO, and SO_3 content and lower LOI than fly ash from bituminous coals (Table 4.4). Most of the major elements tend to exist in the relatively stable particle cores rather than on the particle surfaces where chemical and physical interactions occur more readily Consequently, concentrations of core elements with respect to particle size tend to be constant [93, 94].

Table 4.4: Chemical composition of coal flays ash produced from different coal.

Component (wt.%)	Coal Type			
	Bituminous [95]	Sub-bituminous [95]	Lignite [95]	Anthracitic [96]
SiO_2	20–60	40–60	15–45	50–57
Al_2O_3	5–35	20–30	10–25	30–37
Fe_2O_3	10–40	4–10	4–15	3–10
CaO	1–12	5–30	15–40	1–2
MgO	0–5	1–6	3–10	0–1
SO_3	0–4	0–2	0–10	0–1
Na_2O	0–4	0–2	0–6	1–3
K_2O	0–3	0–4	0–4	1–3
LOI	0–15	0–3	0–5	*

* Not determined.

Bottom ash particles are much coarser than fly ash. The grain size typically ranges from fine sand to gravel in size. The chemical composition of bottom ash is similar to that of fly ash but contains higher concentration of carbon. Bottom ash tends to be relatively more inert because the particles are larger and more fused than fly ash. Since these particles are highly fused, they tend to show less pozzolanic activity and are less suited as a binder constituent in cement or concrete products. However, bottom ash can be used as a concrete aggregate or for several other civil engineering applications where sand, gravel and crushed stone are used. Table 4.5 shows the chemical composition of the example bottom ash obtained by burning bituminous, sub-bituminous, lignite and anthracitic coal.

The chemical composition of **boiler slag** is generally similar to that of bottom ash generated from the same of coal type. This type of waste is composed principally of silica, alumina and iron with smaller amounts of calcium, magnesium, sulfate and other compounds. Boiler slag occurs predominantly at a diameter in the range of 5 to 0.5 mm and has a smooth texture. However, it can also achieve vesicular or porous structure and that depends on the quenching procedure, as well as the organic and mineral properties of the coal used, fluid dynamics, ash

Table 4.5: Chemical composition of coal bottom ash produced from different coal.

Component (wt.%)				Coal type
	Bituminous	Sub-bituminous	Lignite	Anthracitic
	[97]	[98]	[99]	[100]
SiO_2	52.5	60.9	41.7	45.1
Al_2O_3	22.3	14.5	17.1	34.1
Fe_2O_3	11.2	18.6	6.6	4.2
CaO	3.5	1.4	22.5	8.7
MgO	2.5	0.9	4.9	1.4
SO_3	**	0.3	0.4	3.1
Na_2O	0.7	0.2	1.4	0.5
K_2O	2.7	0.6	0.4	1
Ti_2O	0.8	1.1	*	1.5
P_2O_3	0.2	0.9	*	0.1
LOI	2.9	**	1.13	**

*Presented as a sum of Ti_2O, P_2O_3 and other ingredients = 3.83 wt.%; ** Not determined.

transport phenomena, vaporization and condensation of ash species, and so one [101]. For example, boiler slags generated from the burning of lignite or sub-bituminous coals are more porous than that of the bituminous coals, slag from pulverized fuel boiler is composed mainly of irregular, often porous, glassy forms, which state aluminosilicate glass of variable composition with local accumulations of Ca, Mg and Fe oxides.

Sulfur dioxide is pollutant produced by coal-fired power plants. To control SO_2 emission various methods can be applied [102]: (i) improvement of fuel quality by using the low-sulfur coal, (ii) appropriate methods of coal preparation by producer, they allow reducing the sulfur content of 10–30 %, depending on the coal quality and the technology [103], (iii) boiler modernization, (iv) changes in technology, i. e. natural gas instead of coal, such changes are observed now in energy sector worldwide, (v) flue gas desulfurization. The last opportunity, known as FGD, is the widest applied in the energy industry, and can be classified into four groups: (i) wet-scrubber, (ii) spray-dry scrubber, (iii) dry-scrubber and (iv) combined SO_2/NO_X removal process technologies. Each main category is divided into several subcategories depending on specific chemical reactions and flow conditions.

The most often, the wet scrubber technologies with gypsum producing, with about 62 % contribution, are applied in energy sector using slurry of alkaline sorbent, usually limestone or lime, or seawater to scrub gases [104]. The main task of this solution is oxidation of sulfur dioxide and its removing from flue gas by production of gypsum as a saleable product. Thus, there are four processes in absorbers and slurry tanks as follows: limestone dissolution, SO_2 absorption, SO_3^{2-} oxidation and gypsum crystallization with the overall reaction [104, 105]:

$$CaCO_{3(s)} + SO_{2(g)} + \frac{1}{2}O_{2(g)} + 2H_2O_{(l)} \rightarrow CaSO_4 \times 2H_2O_{(s)} + CO_2 \quad (4.17)$$

The wet FGD system is well-known technology applied in industrial from early 1970s in United States and Japan, and expanded rapidly in the 1980s in Europe [106]. It is characterized by high desulfurization performance and low operating cost.

It is worth to mention that in wet flue gas desulfurization technologies the effective capturing of oxidized mercury can occur. In this case Hg, as a one of the elements of coal (about 0.1 ppm), may be incorporated into the FGD slurry and its solid byproducts including synthetic gypsum [107]. Mercury removal efficiencies in this process range from 50 % to 75 %, or higher 85 %–90 %, when the units are preceded by selective catalytic reduction (SCR) [108]. The mercury content must be controlled, because the gypsum from FGD process is often used as building materials and there is s possibility to emission from them [109, 110]. The mercury content depends of the used coal and technologies applied in power plants. For example the Hg concentration in FGD gypsum samples is 0.15 and 0.31 mg/kg for two 1200 MW Spanish power stations equipped with a wet limestone-based with forced oxidation FGD facility, for mixture of local sub-bituminous coals (60 %) close to lignite and bituminous coals (40 %) or mixture of anthracites (80 %) and petroleum–coke (20 %), respectively [111].

Moreover, in FDG systems, the elements such As, Se, F, Cl, and B can be presented. The most of them is removed in the aqueous effluent (filtered water) or gypsum, while B and Se remain in the flue gases [112, 113]. Exemplary in work [114] authors presents the detailed composition of FDG gypsum, in mg/kg: As (2.6), Cl (307), F (778), Hg (0.212), S (175,379), Se (4.0), Ba (9.21), Cd (0.04), Cr (2.8), Mo (0.11), Ni (5.7), Pb (0.64), Sb (0.06), Zn (7.9). In dry and/or semi-dry technologies SO_2 reacts with an alkaline sorbent typically $Ca(OH)_2$ or CaO. This method is the second most popular FGD system worldwide (11 %) and it proposed for relatively small power plants with low S contents in coal (<1.5 wt%) [115].

4.2.2.1 Coal combustion wastes reuses

The most important use of coal fly ash is as partial substitute for cement in concrete. In 2015 the total amount of fly ash produced in the United States was 44.37 million tons out of which 54 % were used again in other sectors [116], comparable production has been observed in Europe with the reusing of fly ash in 50 % [117], while in Canada of the approximately 4.7 million tonnes of fly-ash produced annually by coal-fired power generating stations from which only 31 % is recycled [118]. Coal fly ash mostly comprises of silicas, aluminas and calcium compounds, therefore, it is extensively used in concrete either as a separately batched material being an addition or as an ingredient in blended cement. Utilization of fly ash in concrete is also an effective energy conservation measure that can improve the economy of concrete construction. According to "Standard Specification for Coal Fly Ash and Raw or Calcined Natural Pozzolan for Use in Concrete" (American Society for Testing and Materials (ASTM), C618 [119]) fly ashes are classified into two categories – Class F, produced from anthracite or bituminous coal, and Class C fly ash produced from sub-bituminous or lignite coal. The difference in the source coal causes difference in the amount of calcium and the silica, alumina, and iron content in the ash. In Class F fly ash, total calcium typically ranges from 1 to 12%, mostly in the form of calcium hydroxide and calcium. In contrast, Class C fly ash may have reported calcium oxide contents as high as 20–40%. Another difference is that the amount of alkalis, which are generally higher in the Class C fly ash than in the Class F. Table 4.6 presents the composition of example fly ash type Class C and Class F collected from coal combustion plants.

Table 4.6: Composition of example fly ash type Class C and Class F.

Component (wt.%)	Muskogee Plant Muskogee Oklahoma [120]	Manjung Plant, Perak Malaysia [121]	Power plants in West Bengal India [122]	Appalachian Power Company Clinch River Plant [120]
	Class C	Class C	Class F	Class F
SiO_2	30.1	38.8	58.9	59.4
Al_2O_3	16.6	14.7	19.1	21.4
Fe_2O_3	7.0	19.5	10.2	8.8
CaO	36.9	18.1	2.6	3.8
MgO	1.4	3.3	0.7	0.2
SO_3	4.4	1.5	-	0.0
Na_2O	0.6	1.7	0.6	0.0
K_2O	0.3	1.8	3.1	4.6
Ti_2O	2.7	1.0	1.3	1.8

LOI is a very important factor to determine the quality of fly ash for use as admixture in concrete. This parameter values are primarily residual carbon material, which may adversely affect the fly ash applying. Especially in the production of consistent durable concrete, a low and constant LOI is desirable. Problem is when activated carbon powder is used in power plant with air quality control systems to remove mercury from combustion gases. This additive mainly negatively influences on determination of air content in concrete [123].

Environmental protection and technical and economic conditions force co-combustion of fossil fuels with biomass or waste fuels in power plants. The most often co-fired biomasses include wood, bark, wood chips, corn straw, rapeseed straw, olive pulp, olive kernel, and herbaceous biomass [1, 2]. Co-fired wastes also generated fly ashes and slags [124–126], which composition is comparable with that obtained from coal combustion process. The similar composition of fly ash is a result of a low ash content of wood pellets (0.02–0.5 wt%) compared with lignite (10 wt%). All fly ashes contained approximately 44 wt% SiO_2, 21 % Al_2O_3, 14 % CaO and 4 % Fe_2O_3, putting them in the class C. Moreover, the LOI of co-fired biomasses ash is much lower than the 6 wt% (upper limit according to C618) and ranged from 0.4 to 0.9 wt%. The European Union has approved the use of fly ash derived from the co-combustion of coal with wood, straw, cultivated biomass, municipal solid waste and paper sludge as a cement ingredient; however, the regulation is very restricted. Standard EN 450–1:2012 allows using the fly ash form co-combustion in concrete manufacturing, but it introduces the requirement for minimum content of coal in fuel mixture (80 % by mass) and maximum content of ash from co-combusted materials (10 %) [127]. Moreover, the fly ash should contain less than 5 wt% carbon, 5 wt% total alkali, and 0.1 wt% chloride [128]. Optical microscopy images of coal and biomass fly ashes are presented in Figure 4.1.

Figure 4.1: Optical microscopy images of coal (A) and biomass (B) fly ashes (Keyence Digital Microscope VHX-6000 Series).

Other major uses for fly ash include constructing structural fills and embankments, waste stabilization and solidification, mine reclamation, and use as raw feed in cement manufacturing. Bottom ash is often used as aggregate, replacing sand and gravel, as well as an ingredient in manufacturing concrete blocks or as a raw feed in cement manufacture (Figure 4.2). This waste is also considered as structural fills and embankments and mine reclamation. In 2006 in Europe, about 3.1 million tonnes of bottom ash were used in the construction industry. Out of this 48 % was used as a fine aggregate in concrete blocks, 34 % in road construction and about 16 % in cement and concrete. Dense and hard boiler slag is used as a component in surface coatings of asphalt in road paving. Slag can also be used as blasting grit and roofing granules and of which about 55 % was used in road construction, e. g. as a drainage layer. Another uses are as aggregates in concrete and grouts [129, 130]. The utilization rate of flay ash, button ash and boiler slag increased greatly over last twenty years reaching a high level. In the United State the utilization rate in 2005 was 41 % of fly ash, 42.9 % of button ash and in 77.4 % for boiler slag [131], while in 2015 was 54.24 % for fly ash, 40.1 % for button ash and in 83.8 % for boiler slag [116]. Similar effect has been observed in Europe: the utilization rate in 2003 was 47 % of fly ash, 44 % of button ash and in 100 % for boiler slag, while in 2006, the rate was 49.1 % and 51.0 % for fly and button ash, respectively, and the utilization rate of boiler slag was also 100 %.

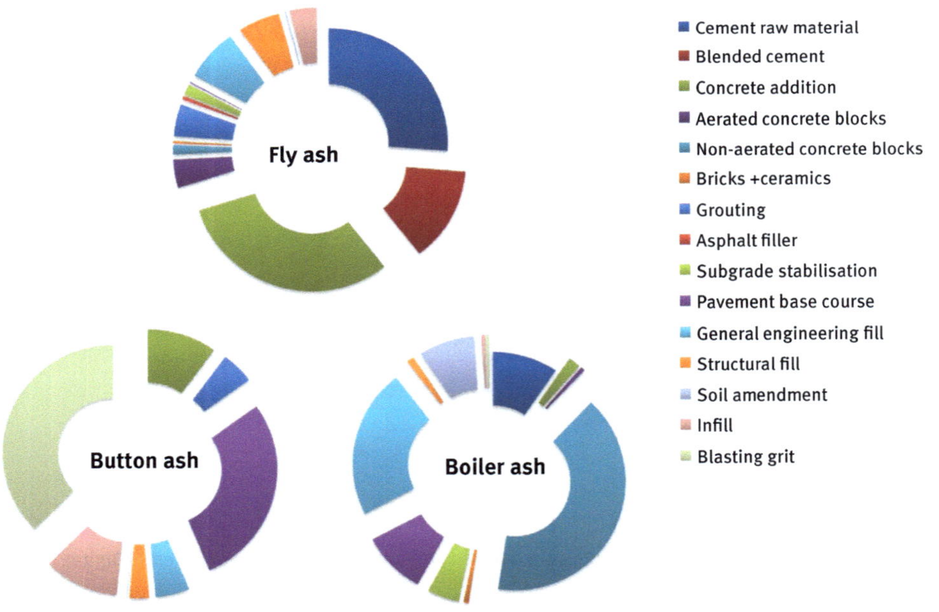

Figure 4.2: Utilization of fly ash, bottom ash and boiler slag in the construction industry in Europe in 2006 [117].

The residue of the generated waste is being dumped in landfills or open lands, causing economic and environmental problems [132, 133]. Utilization of dumped fly ash not only in construction industries is the most popular research goal over last years. Especially, the using of fly ash as sorbents of organic and inorganic compounds has been studied by numerous research centers. Vandenbusch and Sell have studied fly ashes as sorbent organic materials from a municipal waste treatment facility effluent [134] and the results indicated the ability of ash to remove 90 % of boron from the effluent. Potassium-fly ash sorbents have been investigated by Sanna and Maroto-Valer for high-temperature CO_2 sorption [135]. The coal fly ash before additional treatment seems to be a suitable sorbent material, however with low capacity (e.g. 8 mg Cu(II)/g) [136]. In another studies the fly ashes after hydrothermal treatment displayed a much better capacity to remove Cu(II), Ni(II), Cd(II) and Pb(II) from aqueous solutions (e. g. 1111 mg Pb(II)/g) [137, 138]. Another cost-effective and environmentally friendly method for recycling this waste is the using of fly ash in heterogeneous catalysis [139]. Of course, this type of application required the surface area enhance of the fly ash, which can be achieved by thermal or chemical treatment. An example of this application is steam and nitric acid activation of fly ash, which has been used then as a support for Cu, Ni, Ru, V and Fe catalysts to catalytic reduction of NO_x using NH_3 [140–142]. A Ni loaded fly ash was also tested for CO_2 reforming of methane [143]. Khatri et al. [144] synthesized a solid acid catalyst for benzylation reactions by loading sulfated zirconia on chemically activated fly ash. Saputra et al. [145] have employed fly ash as a support for the synthesis of Co oxide based catalyst which has ability to degrade phenol in the presence of Oxone. Zhang et al. [146] used activated fly ash as a heterogeneous Fenton-like catalyst for nitrophenol removal from water. The potassium impregnated fly ash based catalyst was used in biodiesel production displaying maximum oil conversion (86.13 %) [147]. Also using the synthesized sodalite as a catalyst for transesterification of soy oil, the very high conversion has been obtained (96 wt% at 65 °C) [148]. Efficient transesterification of soybean oil Chakraborty et al. have obtained using a fly-ash-supported CaO catalyst. The studies exhibited high catalytic performance in converting soybean oil into fuel-grade biodiesel with fatty acid methyl ester content of 97 % [149]. Fly-ash-supported CaO has also been employed as effective catalysts for the base catalyzed Knoevenagel reaction [150]. Sutarno and Arryanto [151] have synthesized zeolites from fly ash and used them as a catalysts for hydrocracking of heavy petroleum distillates.

4.2.2.2 Trace elements leachability from ash derived from combustion process

Increased coal consumption in large coal-fired power plants implies increased mobilization of potentially toxic trace elements. Especially trace elements, including As, B, Cd, Pb, Mo, Ni, Se, and Zn, are important constituents due to their high concentrations in fly ash and environmental significance. In another studies Ruane et al. [152] have also found that coal fly ash leachate samples were highly variable, but were generally high in B, Fe, Ca, Al, and sulfate, and often exceeded the drinking water quality levels (i.e. cadmium, chromium, iron, manganese, lead and pH). It was also indicated that the rate of leaching was affected by the fly ash particle size, agitation of the mixture, pH, and temperature of the leachate solution [153, 154]. The ultimate impact of each trace element will depend on its form in the ash and the toxicity, mobility, and availability of this particular form in agricultural ecosystems. As Markowski and Filby have indicated, As, Se, Mo, Cd, and Zn are associated with particle surface layers, Cr, Cu and Pb are distributed between the core and the surface, while Ni exists in the particle core. Furthermore, As, Se, Cd, Zn, Ni, Cr, and Pb increase markedly in concentration with decreasing particle size [155]. Similar analysis Izquierdo and Querol have presented in their review [156], wherein they have demonstrated that the elements enriched in the core of fly ash particles are not directly exposed to leaching, whilst surface-associated elements are more leachable in an aqueous environment. Also the acidity of the leaching system has been found to play a dominant role in the leachability of most elements contained in fly ash, while the alkalinity enhances the mobility of elements with tendencies to form oxyanionic species (e. g. Cr, Mo, V and W). Rest of elements such as Be, Cd, Co, Cu, Fe, Mg, Mn, Ni, Pb, REE, Si, Sn, Th, Tl, U and Zn show the minimum solubility in the pH 7–10 region and can be considered as being of low concern in alkaline fly ash under environmental leaching conditions. However if the pH decreases their leachability rises even though to an environmentally hazardous level [156].

In another research, it has indicated that the trace element affinities in coal exert a considerable influence on the behavior of these elements during combustion. Elements with sulfide and organic affinities are oxidized during coal combustion and consequently they may show volatile behavior because of the temperature rise accompanying oxidation or because of reaction with Cl, F, Na or S compounds which can induce volatility or capturing as oxide or sulfate [157]. Chadwick et al. [158] have also indicated an important influence of the organic affinity of uranium on its volatile behavior during combustion. It has been shown that uranium has partial organic affinity, which is responsible for its partial volatilization and, similar to Mn, V, Cr, Ni, Ln, Ga, Nd, As, Sb, Sn, Zn, Se, Pb and Hg, its volatility is inversely proportional to particle size what causes concentration in the smallest ash particle size fractions. Elements, such as, Mg, Na, K, Mo, Ce, Rb, Cs and Nb appear to have a smaller fraction volatilized during coal combustion [159]. It was also indicated that the volatility of trace elements increased from larger to smaller particle size and establishes an inverse relationship of volatility and particle size [160]. Coal contains significant amount of various trace metals, and after combustion these metal concentrations in fly ash are much higher than that of in parent coal. Concentration ranges for selected elements, along with trace element concentrations in ash samples from different power plants currently operating are listed in Table 4.7.

Table 4.7: Trace element analysis of coal and various combustion waste.

Element	Coal	Bottom ash [95]	Coal	Fly ash [162]	Bottom ash	Fly ash [162]	Fly ash** [124]
V	20	130	75.1	243	209	278	328
Cr	12	160	27.8	86	75.3	172	191
Mn	24	360	18.6	50	61	*	174
Ni	4	70	15.1	44	41.4	*	118
Cu	6	20	38.8	105	75.6	118	119
Zn	27	–	14.1	47	15.2	175	232
Ga	10	30	*	*	*	*	59
As	12	35	1.1	2.44	1.1	97	161
Se	3	2	*	*	*	13	13
Br	14	–	*	*	*	*	2.0
Rb	23	75	*	*	*	*	131
Sr	120	1200	*	*	*	1647	802
Zr	*	*	*	*	*	*	279
B	16	–	36.2	77	41.4	*	–
Ba	150	1200	*	*	*	1999	1296
La	5	70	*	*	*	*	69
Ce	*	*	*	*	*	*	157.5
Pb	6	70	13.2	35.3	15.3	109	77
Th	1.4	20	*	*	*	37	26.6
U	0.7	–	*	*	*	18	8.1
Co	3	20	13	38	33	48	*

* not determined; ** Co-combustion fly ash.

4.2.2.3 Removal of metals from fly ash

Most coal fly ashes currently produced in power plants contain leachable heavy metals and enhancement in utilization rate of fly ash requires removal of heavy metals prior to utilization. Also with the exhaustion of natural mineral resources, there is a need to recover desired metals from ashes and slags. Microbiological and chemical leaching has been proposed to remove effectively metals ions from the combustion waste. Biohydrometallurgical process uses a natural ability of microorganisms to dissolve metals present in the waste seems to be the most environmentally preferred. Example of a successful bioleaching is Al and Fe removal from coal fly ash by *Thiobacillus thiooxidans* (*T. thiooxidans*) bacteria (Table 4.8) [163]. These very good results of leaching (comparable to chemical leaching) require adaptation of the bacteria in coal fly ash. Leaching of metals from fly ash by *Yarrowia lipolytica* has been also considered and the results were compared with that obtained from chemical leaching. It has been was shown that Yeast cells are most effective in the copper leaching (59.4 %), slightly lower results have been obtained for zinc, nickel and chromium ions, while most of Al and Si remained in

Table 4.8: Examples of heavy metal recovery from fly ash by bioleaching methods (CFA – coal fly ash, MSWFA-fly ash from municipal solid waste).

Fly ash type	Bioleaching condition	Results	Ref
CFA	Bioleaching using *Acidithiobacillus ferrooxidans* (ZYR-1) fly ash pulp density 10 % w/v, 100 mL sucrose medium; 7-day pre-culturing; 28°C at 200 rpm	Bioleaching of Al in 25 % and Fe in 15–22 %.	[163]
MSWFA	Bioleaching using *Aspergillus niger* (AS 3.879) 20 g/L fly ash, 100 mL sucrose medium; two-day pre-culturing; 216 h, 30°C at 140 rpm	Maximum metals extraction of one-step bioleaching was: 98.7 % of Mn, 87.6 % of Cd, 69.7 % of Cu, 68.5 % of Zn, 42.1 % of Cr, 31.7 % of Fe, and 36.5 % of Pb.	[167]
MSWFA	Bioleaching using *Aspergillus niger* (AS 3.879) fly ash pulp density 1 % w/v, 100 mL sucrose medium; two-day pre-culturing; 30°C at 120 rpm pH 2.5–5.8	Al, Fe and Zn removal was accompanied by increase in citric acid concentration excreted by *A. niger*. Maximum metals extraction were 97 % of Al 98 % of Zn and 56 % of Fe.	[168]
MSWFA	Bioleaching using *Thiobacillusthiooxidans* (TM-32) fly ash pulp density 1 % w/v, 100 mL sucrose medium; fife-day pre-culturing; 28°C at 120 rpm pH 1.7	Maximum metals extraction: Cd in 100 %, Cr in 96 %, Cu in 73 %, Zn in 59 % and As in 42 %.	[166]
MSWFA	Bioleaching using *Acidithiobacillus ferrooxidans* (ATCC23270) fly ash pulp density 1 % w/v, 100 mL sucrose medium; fife-day pre-culturing; 28°C at 120 rpm pH 1.7	Maximum metals extraction: 100 % of Cr, Zn and Cd, 77 % of Cu and 49 % of As	
MSWFA	Bioleaching using *Thiobacillus thiooxidans* (TM-32) and *Acidithiobacillus ferrooxidans* (ATCC23270) fly ash pulp density 1 % w/v, 100 mL sucrose medium; fife-day pre-culturing; 28°C at 120 rpm pH 1.7	Maximum metals extraction: 100 % of Cr and Cd, 78 % of Zn, 67 % of Cu and 36 % of As.	

solid state [164]. Other bioleaching examples have focused on fly ash from municipal solid waste treatment. Bosshard et al. have investigated the bioleaching of fly ash from municipal waste incineration by *Aspergillus niger*. The using mixed culture of iron and sulfur oxidizing bacteria enables a significant recovery of Cd, Zn, Cu and Pb (81 %, 66 %, 57 % and 52 %, respectively), Mn and Al are extracted in approximately 30 %, and in less than 10 % Cr, Fe, and Ni [165]. Ishigaki et al. [166] have also investigated the metal leaching from fly ash municipal solid waste using cultures of a sulfur- and iron-oxidizing bacterium individual and as a mixture. The obtained results have shown that high metal leachability is exhibited in the 1 % ash culture, e g. 67 % and 78 % for Cu and Zn, respectively, and 100 % for Cr and Cd. Conducting process in the 3 % ash cultures the results are comparably.

Chemical solubilization of fly ash metals includes the use of acidic, neutral and alkaline reagents but effectiveness of the process depends on the leachability of the metals in leaching solution. Even though the chemical leaching seems to be more aggressive than bioleaching the chemical treatment of metals from combustion residues is a very slow process and the solid–liquid phase equilibrium may not be attained even with long leaching times. Considering acidic reagents, the leaching force is as follows: aqua regia > HCl > HNO_3 > H_2SO_4. In the case of H_2SO_4 very low leachability is observed which resulted from the formation of insoluble metal-sulfate species. The dissolution in this medium can be enhanced by carrying out the process in thermal and high-pressure conditions [169, 170].

Poor extractable lead in H_2SO_4 as well as in dilute HCl or HNO_3 and dissolution of Pb(II) requires concentrated HCl or HNO_3 [171]. Concentrated acidic solution is also required by Fe, Mn and Si, while in the diluted solution (e. g. 0.1 M H_2SO_4, HCl or HNO_3) all of elements display low leachabilities (1–7 %) [172–174]. Efficient chemical treatment has been observed for As using both diluted and concentrated acidic solutions of H_2SO_4, HCl and HNO_3. Organic acids as leaching reagents have been also considered, from which the most efficient are oxalic, citric and gluconic acids. In the case of Cu the leaching is with almost 100 % yield, regardless on type of organic acid and its concentrations [173, 175]. Al, Zn, Pb, Mn and Fe requires concentrated organic acid solutions to extract up to 100 % of Al, 70 % of Zn, 50 % of Fe, 80 % of Mn and 50 % of Pb [175]. In another study, it was found that 20 wt.% acetic acid dissolves 62 wt.% Zn and 94 wt.% Pb while leaching using 10 wt.% HCl dissolves from fly ash 62.5 wt.% Zn and 39.5 wt.% Pb [176]. Similar to the mineral acid leaching, a long-time treating of fly ash with acetic acid causes a decrease in Pb extraction [177]. Special predisposition to dissolution with organic acids Hg exhibits [178]. This element is highly extractable in acetic acid as opposed to H_2SO_4 and Na_2CO_3.

Alkaline leaching considered to be cost-effective process has been proposed by various researchers. The effectiveness of alkali leaching is a result of the solubility of the V, Al Zn and Pb at alkali condition. For example, considering NaOH, hydroxides react with metals having the amphoteric nature like Pd, Zn, Mo, Cr and V result in the formation of dissolved in the aqueous solutions complex ion species (e. g. $HV_2O_5^-$, $V_2O_7^{4-}$) [179]. Ammonia as lixiviant is also known to favor extraction of Cu, Ni owing to the formation of metal-amina complexes [180]. Ammonia forms relatively strong complexes with mercury compared to most other cationic elements and, therefore, may change the leaching characteristics of mercury. Wang et al. [181] have confirmed that mercury leaching by neutral and acidic solution is not significant [182], however, ammonia addition increased mercury leaching in the alkaline pH range, due to the formation of leachability mercury-ammonia complexes.

Unfortunately, there are conditions or elements for which the alkaline leaching is not an ideal method of dissolution. For example, arsenic has shown poor extractability in alkaline solutions [183], Fe forms insoluble $FeSO_4 \cdot 7H_2O$ [184] and low solubility of Zn in the presence of Pb is mainly a result of gelatinous precipitates formation [176]. The main examples of chemical treatment of fly ashes with experimental conditions and results are summarized in Table 4.9.

Table 4.9: Examples of heavy metal recovery from fly ash by chemical leaching (CFA – coal fly ash, MSWFA-fly ash from municipal solid waste).

Fly ash type	Leaching agent	Maximum metals recovery	Ref
Mineral acid leaching			
CFA	H_2SO_4 (0.1 M)	As leaching in 88–94 %	[183]
MSWFA	H_2SO_4 (0.1 M)	100 % of Ca, 58 % of Zn, 52.4 % of Cu, 28 % of Pb and Fe in 6 %	[175]
MSWFA	H_2SO_4 (1.5 M)	100 % of Cu and Zn, 98 % of Mn, 82 % of Mg, 81 % of Fe, 58 % of Si and Pb in 5 %	[174]
CFA	HNO_3 (0.1 M)	As leaching in 83–90 %	[183]
MSWFA	HNO_3 (0.1 M)	100 % of Ca, 57 % of Zn, 53 % of Cu, 42 % of Al and 25 % of Pb. Fe was dissolved only in 2 %	[173]
MSWFA	HCl (0.1 M)	100 % of Ca, 57 % of Zn, 50 % of Cu, 43 % of Al and 15 % of Pb. Fe was dissolved in 3 %	[173]
MSWFA	HCl (3 M)	100 % of Cu and Pb, 77 % of Zn, 76 % of Al, 71 % of Mn, 63 % of Ca and Fe and Si in 44 %	[174]
Organic acid leaching			
MSWFA	Citric acid (0.1 M)	100 % of Al and Cu, 97 % of Pb, 93 % of Ca and Fe in 67 %	[173]
CFA	Citric acid (0.3 M)	Cr leaching in 28 %	[185]
MSWFA	Citric acid (0.5 M)	100 % of Cu, 80 % of Al, 70 % of Zn, 50 % of Fe and Pb in 30 %	[175]
MSWFA	Oxalic acid (0.1 M)	47–45 % of Fe, Cu and Zn, while Pb was dissolved in 3 %	[173]
MSWFA	Oxalic acid (0.5 M)	100 % of Cu, 85 % of Al, 68 % of Zn, 60 % of Fe and Pb in 10 %	[175]
MSWFA	Gluconic acid (0.5 M)	100 % of Cu, 95 % of Al, 80 % of Fe, 72 % of Zn and Pb in 50 %	[175]
Alkaline leaching			
MSWFA Primary	NaOH (3 M)	Pb in 84 % and Zn in 25 %	[176]
MSWFA Secondary	NaOH (3 M)	Pb in 46 % and Zn in 29 %	[176]
CFA	NaOH (0.1 M)	As in 35–65 %	[183]
CFA	NH_3 (4N)	Ni in 59 % V in 10 % and Fe precipitate as hematite	[184]
CFA	NH_3 (1 g/L)	Cd and Cu in up to 0.5 %	[180]

4.3 Conclusions

The potential sources of metals in energy industry were presented, with the focus on two main metal-contains wastes from power plants – the solid wastes from coal combustion and spent SCR catalysts. Utilization of fly ash, bottom ash and boiler slag, especially in the construction industry, was discussed. The problem of metal removal from such wastes generated in power plants was also presented. Analysis of the latest literature in this field indicates that recovery of valuable metals such as vanadium and tungsten from waste SCR catalyst provides significant important social, environmental and economic benefits. The methods of leaching of these metals, as well as the re-use applications of spent catalysts were presented and discussed.

References

[1] Zamaraev KI. Catalytic science and technology for environmental issues. Catal Today. 1997;35:3–13.
[2] International Energy Outlook 2017, #IEO2017. U.S. Energy Information Administration, September 14, 2017. Available at: www.eia.gov/ieo.
[3] Madejski P, Janda T, Modliński N, Daniel N. A combustion process optimization and numerical analysis for the low emission operation of pulverized coal-fired boiler. In: Kyprianidis KG, Skvaril J, editors. Developments in combustion technology. Rijeka, Croatia: InTech, 2016:33–75.
[4] Beér JM. Combustion technology developments in power generation in response to environmental challenges. Prog Energy Combust Sci. 2000;26:301–27.
[5] Wall T, Liu Y, Spero C, Elliott L, Khare S, Rathnam R, et al. An overview on oxyfuel coal combustion – state of the art research and technology development. Chem Eng Res Design. 2009;87:1003–16.
[6] Mahmoudi S, Baeyens J, Seville JP. NO_x formation and selective non-catalytic reduction (SNCR) in a fluidized bed combustor of biomass. Biomass Bioenergy. 2010;34:1393–409.
[7] Kim SW, Heckel A, McKeen SA, Frost GJ, Hsie EY, Trainer MK, et al. Power plant NOx emission reductions and their impact on air quality. Geophys Res Lett. 2006;33(L2281):1–5.
[8] Klingstedt F, Arve K, Eränen K, Murzin DY. Toward improved catalytic low-temperature NO_x removal in diesel-powered vehicles. Acc Chem Res. 2006;39:273–82.
[9] Phil HH, Reddy MP, Kumar PA, Ju LK, Hyo JS. SO_2 resistant antimony promoted V_2O_5/TiO_2 catalyst for NH_3-SCR of NO_x at low temperatures. Appl Catal B. 2008;78:301–08.
[10] Forzatti P. Present status and perspectives in de-NO_x SCR catalysis. Appl Catal A. 2001;222:221–36.
[11] Schreifels JJ, Wang S, Hao J. Design and operational considerations for selective catalytic reduction technologies at coal-fired boilers. Front Energy. 2012;6:98–105.
[12] Matsuda S, Kamo T, Kato A, Nakajima F, Kumura T, Kuroda H. Deposition of ammonium bisulfate in the selective catalytic reduction of nitrogen oxides with ammonia. Ind Eng Chem Prod Res Develop. 1982;21:48–52.
[13] Srivastava RK, Miller CA, Erickson C, Jambhekar R. Emissions of sulfur trioxide from coal-fired power plants. J Air Waste Manag Assoc. 2004;54:750–62.
[14] Kim CH, Woo SI, Jeon SH. Recovery of platinum-group metals from recycled automotive catalytic converters by carbochlorination. Ind Eng Chem Res. 2000;39:1185–92.
[15] Lietti L, Nova I, Forzatti P. Selective catalytic reduction (SCR) of NO by NH_3 over TiO_2-supported $V_2O_5–WO_3$ and $V_2O_5–MoO_3$ catalysts. Top Catal. 2000;11:111–22.

[16] Hums E. A catalytically highly-active, arsenic oxide resistant V-Mo-O phase – results of studying intermediates of the deactivation process of V_2O_5-MoO_3-TiO_2 (ANATASE) $DeNO_x$ catalysts. Res Chem Intermed. 1993;19:419–41.

[17] Chen JP, Hausladen MC, Yang RT. Delaminated Fe_2O_3-pillared clay: its preparation, characterization, and activities for selective catalytic reduction of NO by NH_3. J Catal. 1995;151:135–46.

[18] Li J, Chang H, Ma L, Hao J, Yang RT. Low-temperature selective catalytic reduction of NO_x with NH_3 over metal oxide and zeolite catalysts – A review. Catal Today. 2011;175:147–56.

[19] Yang S, Li J, Wang C, Chen J, Ma L, Chang H, et al. Fe–Ti spinel for the selective catalytic reduction of NO with NH_3: mechanism and structure–activity relationship. Appl Catal B. 2012;117–118:73–80.

[20] Liu F, He H. Selective catalytic reduction of NO with NH_3 over manganese substituted iron titanate catalyst: reaction mechanism and H_2O/SO_2 inhibition mechanism study. Catal Today. 2010;153:70–76.

[21] Apostolescu N, Geiger B, Hizbullah K, Jan MT, Kureti S, Reichert D, et al. Selective catalytic reduction of nitrogen oxides by ammonia on iron oxide catalysts. Appl Catal B. 2006;62:104–14.

[22] Qi G, Yang R. Low-temperature SCR of NO with NH_3 over noble metal promoted Fe-ZSM-5 catalysts. Catal Lett. 2005;100:243–46.

[23] Kumar A, Smith MA, Kamasamudram K, Currier NW, Yezerets A. Chemical $deSO_x$: an effective way to recover Cu-zeolite SCR catalysts from sulfur poisoning. Catal Today. 2016;267:10–16.

[24] Putluru SS, Riisager A, Fehrmann R. Alkali resistant Cu/zeolite $deNO_x$ catalysts for flue gas cleaning in biomass fired applications. Appl Catal B. 2011;101:183–88.

[25] Oord R, Ten Have IC, Arends JM, Hendriks FC, Schmidt J, Lezcano-Gonzalez I, et al. Enhanced activity of desilicated Cu-SSZ-13 for the selective catalytic reduction of NO_x and its comparison with steamed Cu-SSZ-13. Catal Sci Technol. 2017;7:3851–62.

[26] Brandenberger S, Kröcher O, Tissler A, Althoff R. The state of the art in selective catalytic reduction of NO_x by ammonia using metal -exchanged zeolite catalysts. Catal Rev. 2008;50:492–531.

[27] Busca G, Lietti L, Ramis G, Berti F. Chemical and mechanistic aspects of the selective catalytic reduction of NOx by ammonia over oxide catalysts: A review. Appl Catal B. 1998;18:1–36.

[28] Roy S, Hegde MS, Madras G. Catalysis for NO_x abatement. Appl Energ. 2009;86:2283–97.

[29] Domingo JL. Vanadium and tungsten derivatives as antidiabetic agents. Biol Trace Elem Res. 2002;88:97–112.

[30] Chen L, Li J, Ge M, Zhu R. Enhanced activity of tungsten modified CeO_2/TiO_2 for selective catalytic reduction of NO_x with ammonia. Catal Today. 2010;153:77–83.

[31] Gao X, Jiang Y, Fu U, Zhong Y, Luo Z, Cen K. Preparation and characterization of CeO_2/TiO_2 catalysts for selective catalytic reduction of NO with NH_3. Catal Commun. 2010;11:465–69.

[32] Gao X, Jiang Y, Zhong Y, Luo Z, Cen K. The activity and characterization of CeO_2-TiO_2 catalysts prepared by the sol-gel method for selective catalytic reduction of NO with NH_3. J Hazard Mater. 2010;174:734–39.

[33] Chang H, Ma L, Yang S, Li Y, Chen L, Wang W, et al. Comparison of preparation methods for ceria catalyst and the effect of surface and bulk sulfates on its activity toward NH_3-SCR. J Hazard Mater. 2013;262:782–88.

[34] Kumar PA, Jeong YE, Gautam S, Ha HP, Lee KJ, Chae KH. XANES and DRIFTS study of sulfated Sb/V/Ce/TiO_2 catalysts for NH_3-SCR. Chem Eng J. 2015;275:142–51.

[35] Lee KJ, Kumar PA, Maqbool MS, Rao KN, Song KH, Ha HP. Ceria added Sb-V_2O_5/TiO_2 catalysts for low temperature NH_3 SCR: physico-chemical properties and catalytic activity. Appl Catal B. 2013;142–143:705–17.

[36] Sheng Z, Hu Y, Xue J, Wang X, Liao W. SO_2 poisoning and regeneration of Mn-Ce/TiO_2 catalyst for low temperature NO_x reduction with NH_3. J Rare Earth. 2012;30:676–82.

[37] Yan D, Yu Y, Huang X, Liu S, Liu Y. Poisoning effect of SO_2 on Mn-Ce/TiO_2 catalysts for NO reduction by NH_3 at low temperature. J Fuel Chem Technol. 2016;44:232–38.
[38] Shen B, Zhang X, Ma H, Yao Y, Liu T. A comparative study of Mn/CeO_2, Mn/ZrO_2 and Mn/Ce-ZrO_2 for low temperature selective catalytic reduction of NO with NH_3 in the presence of SO_2 and H_2O. J Environ Sci. 2013;25:791–800.
[39] Yu J, Guo F, Wang Y, Zhu J, Liu Y, Su F, et al. Sulfur poisoning resistant mesoporous Mn-base catalyst for low-temperature SCR of NO with NH_3. Appl Catal B. 2010;95:160–68.
[40] Chang H, Li J, Yuan J, Chen L, Dai Y, Arandiyan H, et al. Ge, Mn-doped CeO_2–WO_3 catalysts for NH_3–SCR of NO_x: Effects of SO_2 and H_2 regeneration. Catal Today. 2013;201:139–44.
[41] Jing W, Guo Q, Hou Y, Ma G, Han X, Huang Z. Catalytic role of vanadium(V) sulfate on activated carbon for SO_2 oxidation and NH_3-SCR of NO at low temperatures. Catal Commun. 2014;56:23–26.
[42] Shang X, Hu G, He C, Zhao J, Zhang F, Xu Y, et al. Regeneration of full-scale commercial honeycomb monolith catalyst (V_2O_5–WO_3/TiO_2) used in coal-fired power plant. J Ind Eng Chem. 2012;18:513–19.
[43] Tang F, Xu B, Shi H, Qiu J, Fan Y. The poisoning effect of Na^+ and Ca^{2+} ions doped on the V_2O_5/TiO_2 catalysts for selective catalytic reduction of NO by NH_3. Appl Catal B. 2010;94:71–76.
[44] Klimczak M, Kern P, Heinzelmann T, Lucas M, Claus P. High-throughput study of the effects of inorganic additives and poisons on NH_3-SCR catalysts – part I: V_2O_5–WO_3/TiO_2 catalysts. Appl Catal B. 2010;95:39–47.
[45] Niksa S, Krishnakumar B, Ghoreish F. Analytical management of SCR catalyst lifetimes and multipollutant performance. J Air Waste Manag Assoc. 2016;66:215–23.
[46] Yu Y, He C, Chen J, Meng X. Deactivation mechanism of de-NO_x catalyst (V_2O_5-WO_3/TiO_2) used in coal fired power plant. J Fuel Chem Technol. 2012;40:1359–65.
[47] Li Q, Liu Z, Liu Q. Kinetics of vanadium leaching from a spent industrial V_2O_5/TiO_2 catalyst by sulfuric acid. Ind Eng Chem Res. 2014;53:2956–62.
[48] Zheng Y, Jensen AD, Johnsson JE. Laboratory investigation of selective catalytic reduction catalysts: deactivation by potassium compounds and catalyst regeneration. Ind Eng Chem Res. 2004;43:941–47.
[49] Khodayari R, Odenbrand CU. Regeneration of commercial SCR catalysts by washing and sulphation: effect of sulphate groups on the activity. Appl Catal B. 2001;33:277–91.
[50] Wu WC, Tsai TY, Shen YH. Tungsten recovery from spent SCR catalyst using alkaline leaching and ion exchange. Minerals. 2016;6(107):1–12.
[51] Zeng R. Reclamation and recycling of SCR waste catalyzer. China Environ Protect Ind. 2013;2:39–42.
[52] Lee W, Bae GN. Removal of elemental mercury (Hg(0)) by nanosized V_2O_5/TiO_2 catalysts. Environ Sci Technol. 2009;43:1522–27.
[53] Mercury Study Report to Congress. U. S. Government Printing Office: Washington, DC, 1997.
[54] Vidic RD, Siler DP. Vapor-phase elemental mercury adsorption by activated carbon impregnated with chloride and chelating agents. Carbon. 2001;39:3–14.
[55] Granite E, Pennline H, Hargis R. Novel sorbents for mercury removal from flue gas. Ind Eng Chem Res. 2000;39:1020–29.
[56] Li Y, Murphy P, Wu C, Powers K, Bonjongo J. Development of silica/vanadia/titania catalysts for removal of elemental mercury from coal-combustion flue gas. Environ Sci Technol. 2008;42:5304–09.
[57] Kamata H, Ueno SI, Naito T, Yukimura A. Mercury oxidation over the $V_2O_5(WO_3)$/TiO_2 commercial SCR catalyst. Ind Eng Chem Res. 2008;47:8136–41.
[58] Senior CL. Oxidation of mercury across selective catalytic reduction catalysts in coal–fired power plants. J Air Waste Manag Assoc. 2005;56:23–31.

[59] Niksa S, Fujiwara N. A predictive mechanism for mercury oxidation on selective catalytic reduction catalysts under coal-derived flue gas. J Air Waste Manag Assoc. 2005;55:1866–75.
[60] Zhao YX, Mann MD, Pavlish JH, Mibeck BA, Dunham GE, Olson ES. Application of gold catalyst for mercury oxidation by chlorine. Environ Sci Technol. 2006;40:1603–08.
[61] Yan N, Chen W, Chen J, Qu Z, Guo Y, Yang S, et al. Significance of RuO_2 modified SCR catalyst for elemental mercury oxidation in coal-fired flue gas. Environ Sci Technol. 2011;45:5725–30.
[62] Li JF, Yan NQ, Qu Z, Qiao SH, Yang SJ, Guo YF, et al. Catalytic oxidation of elemental mercury over the modified catalyst Mn/alpha-Al_2O_3 at lower temperatures. Environ Sci Technol. 2010;44:426–31.
[63] Qiao SH, Chen J, Li JF, Qu Z, Liu P, Yan NQ, et al. Adsorption and catalytic oxidation of gaseous elemental mercury in flue gas over MnO_x/aumina. Ind Eng Chem Res. 2009;48:3317–22.
[64] Gao F, Tang X, Yi H, Zhao S, Zhang T, Li D, et al. The poisoning and regeneration effect of alkali metals deposed over commercial V_2O_5-WO_3/TiO_2 catalysts on SCR of NO by NH_3. Chin Sci Bull. 2014;59:3966–72.
[65] Chen L, Li JH, Ge MF. The poisoning effect of alkali metals doping over nano V_2O_5-WO_3/TiO_2 catalysts on selective catalytic reduction of NO_x by NH_3. Chem Eng J. 2011;170:531–37.
[66] Larsson AC, Einvall J, Andersson A, Sanati M. Targeting by comparison with laboratory experiments the SCR catalyst deactivation process by potassium and zinc salts in a large-scale biomass combustion boiler. Energy Fuels. 2006;20:1398–405.
[67] Kamata H, Takahashi K, Odenbrand CU. The Role of K_2O in the selective reduction of NO with NH_3 over a V_2O_5-WO_3/TiO_2 commercial selective catalytic reduction catalyst. J Mol Catal A. 1999;139:189–98.
[68] Lisi L, Lasorella G, Malloggi S, Russo G. Single and combined deactivating effect of alkali metals and HCl on commercial SCR catalysts. Appl Catal B. 2004;50:251–58.
[69] Xue Y, Zhang Y, Zhang Y, Zheng S, Zhang Y, Jin W. Electrochemical detoxification and recovery of spent SCR catalyst by in-situ generated reactive oxygen species in alkaline media. Chem Eng J. 2017;325:544–53.
[70] Zheng Y, Jensen AD, Johnsson JE. Deactivation of V_2O_5-WO_3-TiO_2 SCR catalyst at a biomass-fired combined heat and power plant. Appl Catal B. 2005;60:253–64.
[71] Peng Y, Li J, Si W, Luo J, Wang Y, Fu J, et al. Deactivation and regeneration of a commercial SCR catalyst: comparison with alkali metals and arsenic. Appl Catal B. 2015;168–169:195–202.
[72] Kim H, Lee J, Kim J. Leaching of vanadium and tungsten from spent SCR catalysts for De-NO_x by soda roasting and water leaching method. J Korean Inst Resour Recycl. 2012;21:65–73.
[73] Yu Y, He C, Chen J, Yin L, Qiu T, Meng X. Regeneration of deactivated commercial SCR catalyst by alkali washing. Catal Commun. 2013;39:78–81. ISSN 1566-7367.
[74] Kim JW, Lee WG, Hwang IS, Lee JY, Han C. Recovery of tungsten from spent selective catalytic reduction catalysts by pressure leaching. J Ind Eng Chem. 2015;28:73–77.
[75] Jia Y, Ding W, Ding X, Chen Y, Zhong Q. The leaching of tungsten and vanadium from spent selective catalytic reduction catalyst. Huanjing Kexue Xuebao/Acta Scientiae Circumstantiae. 2016;36:3759–64.
[76] García-Bordejé E, Lázaro MJ, Moliner R, Galindo JF, Sotres J, Baró AM. Structure of vanadium oxide supported on mesoporous carbon-coated monoliths and relationship with its catalytic performance in the SCR of NO at low temperatures. J Catal. 2004;223:395–403.
[77] Huo Y, Chang Z, Li W, Liu S, Dong B. Reuse and valorization of vanadium and tungsten from waste V_2O_5–WO_3/TiO_2 SCR catalyst. Waste Biomass Valorization. 2015;6:159–65.
[78] Zhang C, Liu J, Yang X, Wu F, Xia Q, Li Z. Ultrasound assisted enhancement in vanadium and tungsten leaching from waste SCR catalyst. Gongneng Cailiao/J Functional Mater. 2015;46:20063–67.

[79] Tang D, Song H, Liu D, Wu W, Zheng C, Gao X. Study on leaching kinetics of extracting vanadium and tungsten by sodium hydroxide from spent SCR catalyst. Chin J Environ Eng. 2017;11:1093–100.
[80] Choi IH, Kang JS, Kumar JR, Mun GH, Lee JY. Extraction of tungsten and vanadium from spent SCR catalyst with caustic soda by autoclave process. EARTH 2015 – Proceedings of the 13th International Symposium on East Asian Resources Recycling Technology, 2015:354–56.
[81] Qi C, Bao W, Wang L, Li H, Wu W. Study of the V_2O_5-WO_3/TiO_2 catalyst synthesized from waste catalyst on selective catalytic reduction of NO_x by NH_3. Catalysts. 2017;7(110):1–13.
[82] Fu H, Lin J, Zhang L, Zhu Y. Photocatalytic activities of a novel $ZnWO_4$ catalyst prepared by a hydrothermal process. Appl Catal A. 2006;306:58–67.
[83] Huang G, Zhang C, Zhu Y. $ZnWO_4$ photocatalyst with high activity for degradation of organic contaminants. J Alloys Compd. 2007;432:269–76.
[84] Guo H, Guo D, Zheng Z, Wen W, Chen J. Hydrothermal synthesis and visible light photocatalytic activities of $Zn_3(VO_4)_2$ nanorods. J Mater Res. 2014;29:2934–41.
[85] Hoyos DA, Echavarría A, Saldarriaga C. Synthesis and structure of a porous zinc vanadate, $Zn_3(VO_4)_2{\cdot}3H_2O$. J Mater Sci. 2011;36:5515–18.
[86] Wang Y, Chang Z, Blamo BJ, Wu X, Liu S, Li W. Preparation of photocatalytic $Zn_3(VO_4)_2/ZnWO_4$ from waste V_2O_5–WO_3/TiO_2 SCR catalyst. Waste Biomass Valorization. 2017;8:2423–30.
[87] Sha X, Ma W, Meng F, Wang R, Fuping T, Wei L. Adsorption of vanadium (V) from SCR catalyst leaching solution and application in Methyl Orange. Water Environ Res. 2016;88:2219–27.
[88] Farcy B, Abou-Taouk A, Vervisch L, Domingo P, Perret N. Two approaches of chemistry downsizing for simulating selective non catalytic reduction $DeNO_x$ process. Fuel. 2014;118:291–99.
[89] Modliński N. Numerical simulation of SNCR (selective non-catalytic reduction) process in coal fired grate boiler. Energy. 2015;92:67–76.
[90] Liang L, Hui S, Pan S, Shang T, Liu C, Wang D. Influence of mixing, oxygen and residence time on the SNCR process. Fuel. 2014;120:38–45.
[91] Hou N, Zhang Y, Meng M. Carbonate-based lean-burn NO_x trap catalysts Pt–K_2CO_3/ZrO_2 with large NO_x storage capacity and high reduction efficiency. J Phys Chem C. 2013;117:4089–97.
[92] Zhang Y, Meng M, Dai F, Ding T, You R. States and function of potassium carbonate species in the polytitanate nanobelt supported catalysts used for efficient NO_x storage and reduction. J Phys Chem C. 2013;117:23691–700.
[93] Hansen LD, Fisher GL. Elemental distribution in coal fly ash particles. Environ Sci Technol. 1980;14:1111–7.7.
[94] Kauffer N, Litchman D. Comparison of micron and submicron fly ash particles using scanning electron microscopy and x-ray elemental analysis. Environ Sci Technol. 1984;18:54–57.
[95] Roussak O, Gesser HD. Applied chemistry, a textbook for engineers and technologists. New York, USA: Springer, 2013.
[96] Geological Survey Professional Paper. U.S. Government Printing Office, 1949.
[97] Uzarowicz Ł, Zagórski Z. Mineralogy and chemical composition of technogenic soils (Technosols) developed from fly ash and bottom ash from selected thermal power stations in Poland. Soil Science Annual. 2016;66:82–91.
[98] Wee HL, Wua H, Zhanga D, French D. The effect of combustion conditions on mineral matter transformation and ash deposition in a utility boiler fired with a sub-bituminous coal. Proc Combustion Inst. 2005;30:2981–89.
[99] Ghafoorit N, Bucholc J. Investigation of lignite-based bottom ash for structural concrete. J Mater Civ Eng. 1996;8:128–37.
[100] Duan L Partitioning behaviour of trace elements with particle size distribution in fly ash under air and oxy-fuel combustion in 5th international meeting on oxy-fuel combustion research network, Wuhan, China, 2015. Available at: http://www.ieaghg.org/docs/General_Docs/5oxy

%20presentations/Capacity%20Building%20Course/07%20-%20L.%20Duan%20(Southeast %20University).pdf. Accessed: 23 Aug 2017.
[101] Borio RW, Levasseur AA. Overview of coal ash deposition in boilers. In 188th National Meeting of the American Chemical Society, Philadelphia, 1984:26–31.
[102] Kamiński J. Technologies and costs of SO_2-emissions reduction for the energy sector. Appl Energy. 2003;75:165–72.
[103] Luttrell GH, Kohmuench JN, Yoon RH. An evaluation of coal preparation technologies for controlling trace element emissions. Fuel Process Technol. 2000;65–66:407–22.
[104] Nygaard HG, Kiil S, Johnsson JE, Jensen JN, Hansen J, Fogh F, et al. Full-scale measurements of SO_2 gas phase concentrations and slurry compositions in a wet flue gas desulphurisation spray absorber. Fuel. 2004;83:1151–64.
[105] Zhong Y, Gao X, Huo W, Luo ZY, Ni MJ, Cen KF. A model for performance optimization of wet flue gas desulfurization systems of power plants. Fuel Process Technol. 2008;89:1025–32.
[106] Córdoba P. Status of Flue Gas Desulphurisation (FGD) systems from coal-fired power plants: overview of the physic-chemical control processes of wet limestone FGDs. Fuel. 2015;144:274–86.
[107] Kairies CL, Schroeder KT, Cardone CR. Mercury in gypsum produced from flue gas desulfurization. Fuel. 2006;85:2530–36.
[108] Srivastava RK, Hutson N, Martin B, Princiotta F, Staudt J. Control of mercury emissions from coal-fired electric utility boilers: an overview of the status of mercury control technologies. Environ Sci Technol. 2006;40:1385–93.
[109] Zou R, Zeng X, Luo G, Qiu Y, Zhang B, Xu Y, et al. Mercury stability of byproducts from wet flue gas desulfurization devices. Fuel. 2016;186:215–21.
[110] Ochoa-González R, Díaz-Somoano M, Martínez-Tarazona MR. Control of Hg^0 re-emission from gypsum slurries by means of additives in typical wet scrubber conditions. Fuel. 2013;105:112–18.
[111] Rallo M, Lopez-Anton MA, Perry R, Maroto-Valer MM. Mercury speciation in gypsums produced from flue gas desulfurization by temperature programmed decomposition. Fuel. 2010;89:2157–59.
[112] Clarke L. The fate of trace elements during combustion and gasification: an overview. Fuel. 1993;72:731–36.
[113] Cheng C, Hack P, Chu P, Chang YN, Lin TY, Ko CS, et al. Partitioning of mercury, arsenic, selenium, boron, and chloride in a full-scale coal combustion process equipped with selective catalytic reduction, electrostatic precipitation, and flue gas desulfurization systems. Energy Fuels. 2009;23:4805–16.
[114] Álvarez-Ayuso E, Querol X, Tomás A. Environmental impact of coal combustion desulphurisation plant: abatement capacity of desulphurisation process and environmental characterization of combustion by-products. Chemosphere. 2006;665:2009–17.
[115] Zheng Y, Kiil S, Johnsson JE, Zhong Q. Use of spray dry absorption product in wet flue gas desulphurization plants: pilot-scale experiments. Fuel. 2002;81:1899–905.
[116] 2015 Coal Combustion Product (CCP). Production and use survey report, American coal ash association. Available at: https://www.acaa-usa.org/Portals/9/Files/PDFs/2015-Survey_Results_Table.pdf. Accessed: 23 Aug 2017.
[117] ECOBA 2006 CCP Production & Use Survey (Europe (EU 15). ECOBA – European Coal Combustion Products Association. Available at: https://www.acaa-usa.org/Portals/9/Files/PDFs/ECOBA_Stat_2006_EU15.pdf. Accessed: 23 Aug 2017.
[118] Natural Resources Canada. Production and use of Coal Combustion Products (CCPs), 2008–2010 and 2009–2011, March, 2013
[119] ASTM C618-17. Standard specification for coal fly ash and raw or calcined natural pozzolan for use in concrete. West Conshohocken, PA: ASTM International, 2017.
[120] Lopez-Flores F. Flyash and effect of partial cement replacement by flyash. Joint Highway Research Project JHRP-82-11, School of Civil Engineering, Purdue University, 13 July 1982.

[121] Cheng-Yong H, Yun-Ming L, Abdullah MM, Hussin K. Thermal resistance variations of fly ash geopolymers: foaming responses. Scientific Reports 7, Article number: 45355, 2017.
[122] Dutta BK, Khanra S, Mallick D. Leaching of elements from coal fly ash: assessment of its potential for use in filling abandoned coal mines. Fuel. 2009;88:1314–23.
[123] Gao YM, Shim HS, Hurt RH, Suuberg EM, Yang MY. Effects of carbon on air entrainment in fly ash concrete: the role of soot and carbon black. Energy Fuels. 1997;11:457–62.
[124] Hower J, Robertson J. Chemistry and petrology of fly ash derived from the co-combustion of western United States coal and tire-derived fuel. Fuel Process Technol. 2004;85:359–77.
[125] Barbosa R, Lapa N, Boavida D, Lopes H, Gulyurtlu I, Mendes B. Co-combustion of coal and sewage sludge: chemical and ecotoxicological properties of ash. J Hazard Mater. 2009;170:902–09.
[126] Jukić M, Ćurković L, Šabarić J, Kerolli-Mustafa M. Fractionation of heavy metals in fly ash from wood biomass using the BCR sequential extraction procedure. Bull Environ Contam Toxicol. 2017;99:524–29.
[127] EN 450-1:2012, Fly ash for concrete. Definition, specifications and conformity criteria.
[128] Vom Berg W, Feuerborn HJ. Present situation and perspective of CCP management in Europe. Lexington, Kentucky, USA: World of Coal Ash (WOCA), 2005.
[129] Ramme BW, Tharaniyil MP. We energies Coal combustion products utilization handbook, 3rd ed. Wisconsin, USA : We Energies, 2013.
[130] Vardaka G, Thomaidis K, Leptokaridis C, Tsimas S. Use of steel slag as coarse aggregate for the production of pervious concrete. J Sustain Dev Energy Water Environ Syst. 2014;2:30–40.
[131] 2005 Coal Combustion Product (CCP). Production and use survey report, American coal ash association. Available at: https://www.acaa-usa.org/Portals/9/Files/PDFs/2005-Survey_Results_Table.pdf. Accessed: 25 Aug 2017.
[132] Page AL, Elseewi A, Straughan IR. Physical and chemical properties of flyash from coal-fired power plants with reference to environmental impacts. In: Gunther F, Gunther J, editors. Residue reviews. New York: Springer, 1979. p. 83–120.
[133] Ghio AJ, Silbajoris R, Carson JL, Samet JM. Biological effects of oil fly ash. Environ Health Perspect. 2002;110(Suppl 1):89.
[134] Vandenbusch MB, Sell NJ. Fly ash as a sorbent for the removal of biologically resistant organic matter. Res Conser Recycl. 1992;6:95–116.
[135] Sanna A, Maroto-Valer MM. Potassium-based sorbents from fly ash for high-temperature CO_2 capture. Environ Sci Pollut Res Int. 2016;23:22242–52.
[136] González A, Moreno N, Navia R, Querol X. Development of a non-conventional sorbent from fly ash and its potential use in acid wastewater neutralization and heavy metal removal. Chem Eng J. 2011;166:896–905.
[137] Harja M, Buema G, Sutiman DM, Cretescu I. Removal of heavy metal ions from aqueous solutions using low-cost sorbents obtained from ash. Chem Pap. 2013;67:497–508.
[138] Remenárová L, Pipíška M, Florková E, Hornik M, Rozloznik M, Augustin J. Zeolites from coal fly ash as efficient sorbents for cadmium ions. Clean Techn Environ Policy. 2014;16:1551–64.
[139] Wang SB. Application of solid ash based catalysts in heterogeneous catalysis. Environ Sci Technol. 2008;42:7055–63.
[140] Xuan X, Yue C, Li S, Yao Q. Selective catalytic reduction of NO by ammonia with fly ash catalyst. Fuel. 2003;82:575–79.
[141] Saracco G, Specchia S, Specchia V. Catalytically modified fly-ash filters for NO_x reduction with NH_3. Chem Engin Sci. 1996;51:5289–97.
[142] Li L, Wang S, Zhu Z, Yao X, Yan Z. Catalytic decomposition of ammonia over fly ash supported Ru catalysts. Fuel Process Technol. 2008;89:1106–12.

[143] Chen WJ, Sheu FR, Savage RL. Catalytic activity of coal ash on steammethane reforming and water-gas shift reactions. Fuel Process Technol. 1987;16:279–88.
[144] Khatri C, Mishra MK, Rani A. Synthesis and characterization of fly ash supported sulfated zirconia catalyst for benzylation reactions. Fuel Process Technol. 2010;91:1288–95.
[145] Saputra E, Muhammad S, Sun HQ, Anga HM, Tadéa MO, Wang SB. Red mud and fly ash supported Co catalysts for phenol oxidation. Catal Today. 2012;190:68–72.
[146] Zhang AL, Wang NN, Zhou JT, Jiang P, Liu GF. Heterogeneous Fenton-like catalytic removal of p-nitrophenol in water using acid-activated fly ash. J Hazard Mater. 2012;201–202:68–73.
[147] Babajide O, Petrik L, Musyoka N, Amigun B, Ameer F. Use of coal fly ash as a catalyst in the production of biodiesel. Petroleum Coal. 2010;52:261–72.
[148] Manique MC, Lacerda LV, Alves AK, Bergmann CP. Biodiesel production using coal fly ash-derived sodalite as a heterogeneous catalyst. Fuel. 2017;190:268–73.
[149] Chakraborty R, Bepari S, Banerjee A. Transesterification of soybean oil catalyzed by fly ash and egg shell derived solid catalysts. Chem Eng J. 2010;165:798–805.
[150] Jain D, Khatri C, Rani A. Fly ash supported calcium oxide as recyclable solid base catalyst for Knoevenagel condensation reaction. Fuel Process Technol. 2010;91:1015–21.
[151] Sutarno S, Arryanto Y. Synthesis of faujasite from fly ash and its applications for hydrocracking of petroleum distillates. Bull Chem React Eng Catal. 2007;2:45–51.
[152] Ruane R, Milligan D, Young R, Chu R, Olem H. Aquatic effects of wet ash disposal and wet limestone scrubber systems. Water Sci Technol. 1983;15:149–53.
[153] Keyser TR, Natusch DF, Evans CA, Linton RW. Characterizing the surface of environmental particles. Environ Sci Technol. 1978;12:768–73.
[154] Eisenberg SH, Tittlebaum ME, Eaton HC, Soroezak MM. Chemical characteristics of selected fly ash leachates. J Environ Sci Health. 1986;21:383–402.
[155] Markowski GR, Filby R. Trace element concentration as a function of particle size in fly ash from a pulverised coal utility boiler. Environ Sci Technol. 1985;19:796–04.
[156] Izquierdo M, Querol X. Leaching behaviour of elements from coal combustion fly ash: an overview. Int J Coal Geol. 2012;94:54–66.
[157] Querol X, Fernhndez-Turiel JL, Lbpez-Soler A. Trace elements in coal and their behaviour during combustion in a large power station. Fuel. 1995;74:331–43.
[158] Chadwick MJ, Highton NH, Lindman N. Environmental impacts of coal mining and utilisation. Oxford: Pergamon Press, 1987:171–217.
[159] Iyer R. The surface chemistry of leaching coal fly ash. J Hazard Mater. 2002;93:321–29.
[160] Fischer GL, Chang DP, Brummer M. Fly ash collected from electrostatic precipitators, micro-crystalline structures and the mystery of spheres. Science. 1976;192:553–55.
[161] Tang Q, Liu G, Zhou C, Sun R. Distribution of trace elements in feed coal and combustion residues from two coal-fired power plants at Huainan, Anhui, China. Fuel. 2013;107:315–22.
[162] Moreno N, Querol X, Andrés JM, Stanton K, Towler M, Nugteren H, et al. Physico-chemical characteristics of European pulverized coal combustion fly ashes. Fuel. 2005;84:1351–63.
[163] Seidel A, Zimmels Y, Armon R. Mechanism of bioleaching of coal fly ash by Thiobacillus thiooxidans. Chem Eng J. 2001;83:123–30.
[164] Bankar A, Winey M, Prakash D, Kumar AR, Gosavi S, Kapadnis B, et al. Bioleaching of fly ash by the tropical marine yeast, *Yarrowia lipolytica* NCIM 3589. Appl Biochem Biotechnol. 2012;168:2205–17.
[165] Bosshard PP, Bachofen R, Brandl H. Metal leaching of fly ash from municipal waste incineration by *Aspergillus niger*. Environ Sci Technol. 1996;30:3066–70.
[166] Ishigaki T, Nakanishi A, Tateda M, Ike M, Fujita M. Bioleaching of metal from municipal waste incineration fly ash using a mixed culture of sulfur-oxidizing and iron-oxidizing bacteria. Chemosphere. 2005;60:1087–94.

[167] Yang J, Wang Q, Wang Q, Wu T. Comparisons of onestep and two-step bioleaching for heavy metals removed from municipal solid waste incineration fly ash. Environ Eng Sci. 2008;25:783–89.

[168] Xu TJ, Ting YP. Fungal bioleaching of incineration fly ash: metal extraction and modeling growth kinetics. Enzyme Microb Technol. 2009;44:323–28.

[169] Li LS, Wu YS, Liu YY, Zhai YC. Extraction of alumina from coal fly ash with sulfuric acid leaching method. Chin J Process Eng. 2011;11:254–58.

[170] Wu C-Y, Yu HF, Zhang HF. Extraction of aluminum by pressure acid-leaching method from coal fly ash. Trans Nonferrous Met Soc China. 2012;22:2282–88.

[171] Katsuura H, Inoue T, Hiraoka M, Sakai S. Full-scale plant study on fly ash treatment by the acid extraction process. Waste Manag. 1996;16:491–99.

[172] Huang K, Inoue K, Harada H, Kawakita H, Keisuke O. Leaching behavior of heavy metals with hydrochloric acid from fly ash generated in municipal waste incineration plants. Trans Nonferrous Met Soc China. 2011;21:1422–27.

[173] Huang K, Inoue K, Harada H, Kawakita H, Ohto K. Leaching of heavy metals by citric acid from fly ash generated in municipal waste incineration plants. J Mater Cycles Waste Manag. 2011;13:118–26.

[174] Fedje KK, Ekberg C, Skarnemark G, Steenari BM. Removal of hazardous metals from MSW fly ash – an evaluation of ash leaching methods. J Hazard Mater. 2010;173:310–17.

[175] Wu HY, Ting YP. Metal extraction from municipal solid waste (MSW) incinerator fly ash – chemical leaching and fungal bioleaching. Enzyme Microb Technol. 2006;38:839–47.

[176] Nagib S, Inoue K. Recovery of lead and zinc from fly ash generated from municipal incineration plants by means of acid and/or alkaline leaching. Hydrometallurgy. 2000;56:269–92.

[177] Yang GC, Tsai CM. A study on heavy metal extractability and subsequent recovery by electrolysis for a municipal incinerator fly ash. J Hazard Mater. 1998;58:103–20.

[178] Kazonich G, Kim AG, Dahlberg MD, editors. Comparison of leaching results for three high mercury fly ash samples. In:Proceedings of the Air Quality IV: Mercury, Trace Elements, and Particulate Matter Conference, 2003.

[179] Mizutani S, Yoshida T, Sakai SI, Takatsuki H. Release of metals from MSW I fly ash and availability in alkali condition. Waste Manag. 1996;16:537–44.

[180] Wang J, Ban H, Teng X, Wang H, Ladwig K. Impacts of pH and ammonia on the leaching of Cu(II) and Cd(II) from coal fly ash. Chemosphere. 2006;64:1892–98.

[181] Wang J, Wang T, Mallhi H, Liu Y, Ban H, Ladwig K. The role of ammonia on mercury leaching from coal fly ash. Chemosphere. 2007;69:1586–92.

[182] Gustin MS, Ladwig K. An assessment of the significance of mercury release from coal fly ash. J Air Waste Manage Assoc. 2004;54:320–30.

[183] Xu YH, Nakajima T, Ohki A. Leaching of arsenic from coal fly ashes 1. Leaching behavior of arsenic and mechanism study. Toxicol Environ Chem. 2001;81:55–68.

[184] Tsai SL, Tsai MS. A study of the extraction of vanadium and nickel in oil-fired fly ash. Resour Conserv Recycl. 1998;22:163–76.

[185] Pangayao D, Gallardo S, editors. Leaching of chromium from coal ash using citric acid, oxalic acid and gluconic acid by batch leaching procedure. In: IEEE 2014 International Conference on Humanoid, Nanotechnology, Information Technology, Communication and Control, Environment and Management (HNICEM) 93. Torma AE, 2–14.

M. Teresa A. Reis and M. Rosinda C. Ismael

5 Electroplating wastes

Abstract: Electroplating and other metal finishing industries, like every metal and metallurgical industry, are associated with the generation of waste. Spent electroplating baths, waste pickle liquors, etching solutions and rinse waters from electroplating units and steel finishing operations are complex solutions containing acids and several hazardous metals. It is compulsory, for environmental and economic reasons, to treat these solutions for recovering acid, metals and/or reusing these solutions/waters. This article is a review on the characterization of the wastes that are generated in electroplating industry, steel processing and copper etching in printed circuit boards manufacture, as well as on the treatment and regeneration methods of such streams. Various techniques, such as neutralization, crystallization, evaporation, pyrohydrolysis, electrodialysis, ion exchange, classical solvent extraction and membrane-based solvent extraction are presented, their advantages and disadvantages being scrutinized.

Keywords: Acid regeneration, Electroplating bath, Etching solution, Metals recovery, Spent pickling solution, Waste water treatment

5.1 Introduction

Wastes from electroplating and other metal finishing operations are considered hazardous and may represent a loss of raw materials. As well known, the consumption of water and metals is enormous. Actually, electroplating units use water in the electrolytic bath, washing and rinsing of the electroplated articles. Steel finishing operations such as washing, pickling, cleaning, rinsing, etc. also use a huge amount of water. Waste pickle liquor from the pickling units and bleed streams from the electroplating units are of major concern due to the high concentration of metals and acid that are in solution [1]. For instance, hot-dip galvanizing is the most extensively used surface treatment to protect steel pieces against corrosion; steelwork plants in the EU produce 300,000 m^3/year of spent pickling solutions [2, 3]. Hence, the recovery of metals and acid from these wastes must be performed due to environmental reasons, besides the economic benefit.

The development of modern electronic equipment gave rise to a rapid evolution of the printed circuit boards industry. The etching of copper from printed circuit

This article has previously been published in the journal *Physical Sciences Reviews*. Please cite as: Reis, T. A., Ismael, M. R. C. Electroplating wastes. *Physical Sciences Reviews* [Online] **2018**, *3*. DOI: 10.1515/psr-2018-0024

https://doi.org/10.1515/9783110547061-005

boards (PCBs) is an important process in their fabrication. Etching is carried out to completely remove the base copper and leave only the circuit pattern. Spent etching solutions with a high content of copper are then generated and need to be regenerated for environmental and economic reasons too.

It is noteworthy that the waste waters from industries like electroplating, galvanizing and steel industries, as well as the PCBs, are also a potential source of pollution. These hazardous wastes may cause severe environmental problems and health disorders and ought to be treated before being discharged to the aquatic medium [4].

It is worth mentioning that the surface treatment industry is not a major source of emissions to air. However, some emissions such as NO_x and acids from pickling operations, chromium mist released from Cr(VI) plating and ammonia from copper etching may locally important. Essentially, emissions are primarily to water, and about 300,000 tonnes of hazardous wastes are produced per year, mainly as sludge from waste water treatment or spent process solutions [5].

This article focuses on the characterization of the wastes that are generated in surface treatment industry, such as electroplating, steel processing and copper etching in PCBs manufacture, as well as on the treatment methods of such streams.

5.2 Characterization of the wastes

The surfaces treatment is an essential and integral part of many industries whose implementation increased considerably in the last decades. In this type of treatment, a metallic coating is applied to the surface to improve surface protection against the environment (e. g. corrosion), to give special properties to the surface (e. g. reflectivity) or just changing its appearance (decorative art). As example, zinc is widely used, as metal of sacrifice, to protect steel against corrosion; copper is an excellent conductor used in PCBs and communications equipment; chromium and nickel is used to increase the hardness and wear resistance of the surface.

The surface treatment process generally consists of three main steps: surface preparation/cleaning through degreasing (alkaline or acid agents) and/or removal of rust and oxides (pickling agent: H_2SO_4, HCl, HCl+HNO_3, etc.); plating or electroplating of metal and finishing with rinse water. This process is water intensive, which leads to the production of a huge volume of contaminated effluents/waste waters. It is obligatory, for economic and environmental reasons, to treat these solutions with the objective of recovering metals and/or reusing metal solutions/water. The selection of the treatment depends on the composition of the solution, taking special attention to the concentration, toxicity and market value of the metal(s). Therefore, it is important to characterize the different effluents/waste waters produced.

Cyanide solutions can be used for direct deposition of copper, zinc, silver and gold on surfaces. These solutions allow producing a finer grained metal deposit when compared with acid solution. In addition, cyanide-based plating solutions tend to be more tolerant of impurities than other solutions. This bath must be kept at high pH to avoid the formation of hydrogen cyanide gas. A typical zinc cyanide bath contains 25–60 g/L of zinc, 10–150 g/L NaCN and 25–140 g/L NaOH [6]. The zinc coating is of high quality besides the operating requirements for a cyanide zinc plating process are really minimal, when compared to the other zinc plating processes [7]. Despite this and due to cyanide toxicity and the stringent regulation against water pollution, these alkaline baths have been replaced by alkaline non-cyanide baths (6–14 g/L Zn, at 120 g/L NaOH). The metal deposition can also be performed in sulphuric acid solution.

Chromium plating process can be carried out by using Cr(VI) or Cr(III) solution. As it is well known, the Cr(VI) is highly irritant and toxic. Therefore, the Cr(III) solution is preferred as this ion is considerably less toxic than the hexavalent chromium. A typical plating chromium bath contains 250–450 g/L of chromate and 2.5–4 g/L H_2SO_4, depending on which bath is used, with less than 10% of the chromium acid used is deposited on the metal products [8, 9]. During the plating process, some impurities like Fe(III), Al(III), Cr(III), Cu(II) and Ni(II) are being accumulated in the bath attaining concentrations in the range 10–15 g/L. These solutions must be replaced for the reason that it leads to undesirable effects on plating process.

The most widely used nickel plating electrolyte is the Watts' bath, which contains nickel sulphate, nickel chloride, boric acid and organic additives to improve the surface properties and appearance of electro-deposited nickel [10, 11]. A typical composition of these baths includes 240–300 g/L $NiSO_4 \bullet 6H_2O$, 30–90 g/L $NiCl_2 \bullet 6H_2O$ and 30–45 g/L H_3BO_3 (pH 3.5–4.5).

The manufacture of PCBs, which are the basis of electronic sub-assemblies, involves a complex set of individual operations, some of them having environmental impact. In general, the base material is clad with copper, with a thin layer, the electronic circuit being screen-printed or photo-printed. After this, etching is carried out to completely remove the unmasked copper thus leaving only the circuit pattern. The etching solutions commonly used in the industry are ammoniacal and acidic etchants with cupric chloride; hydrogen peroxide/sulphuric acid are used for micro-etching and ferric chloride is now rarely used [5]. The PCBs that have been made resistant by deposited metals, i.e. tin, are almost exclusively processed in ammoniacal etchants [5]. The ammoniacal etching of metallic copper is described by the following global reaction (dissolution and oxidation):

$$Cu^\circ + 2\,NH_3 + 2\,NH_4Cl + 0.5\,O_2 \rightarrow Cu(NH_3)_4Cl_2 + H_2O \tag{5.1}$$

The concentration of copper in a typical spent etching solution (7–10 M NH_3/NH_4^+, 4–4.5 M Cl^-, pH 8.0–9.5) can attain a value as high as 150–160 g/L [12, 13]. Besides the spent etching solution, the PCBs industry also produces more dilute effluents as a result of rinse waters, the typical concentration of Cu(II) being ~5 g/L [5].

Copper present in spent etching solutions and rinse waters must be recovered due to economic and environmental reasons. Actually, economic benefits to the process are attained if the concentration of copper in the etching solution is reduced, leading to the reuse of the regenerated solution. The solvent extraction (SX) process is the most attractive option for the recovery of copper(II) from spent etching solutions. In fact, the Reference Document on Best Available Techniques (BAT), BREF, for the Surface Treatment of Metals and Plastics sector reports that liquid–liquid ion-exchange (IE) (i.e. SX) is the BAT for regenerating ammonia etching solution with copper recovery [5]. Further details and discussion on the use of SX will be presented in Section 5.3.7.

The typical cupric etchant contains ~180 g/L of copper as $CuCl_2$ and 1–3.5 M HCl. However, the presence of chloride ions as NaCl (200–300 g/L) allows decreasing the concentration of HCl to 0.1 M or lower, the etching being possible under a higher concentration of Cu(II) ions (up 290 g/L) [14]. Various processes are commercially available for the regeneration of $CuCl_2$ etchant. Chemical regeneration processes such as chlorine, sodium chlorate, hydrogen peroxide and oxygen mainly use chemical additives into waste etchant [15]. The cheapest regeneration process is that of chlorine, but the addition of water is a very negative aspect, besides the safety feature. All the above-mentioned chemical methods cause further problems such as the disposal of excess regenerated etchant. Electrolysis and electrodialysis processes solve this problem and allow recovering copper from the etchant. However, both techniques have the drawback of expensive equipment and materials and the release of H_2 in electrolysis is also a disadvantageous aspect [15]. It is worth mentioning that the regeneration method influences the composition of the etchant. For instance, after regeneration with $NaClO_3$, the etchant contains 200–300 g/L NaCl, 150–180 g/L Cu(II) and 0.1–1 M HCl, whereas it comprises $CuCl_2$ and HCl (1–3 M) after oxidation with H_2O_2.

In 2015, the world produced just over 1,600 million tonnes of steel [16, 17]. The last phase of steel production involves heat treatments like hot rolling, annealing or forging. These treatments are carried out at temperatures higher than 800°C in the presence of air. The oxygen of the air reacts with metals present on the steel surface to form oxides films, such as iron oxides (FeO), chromium oxide and manganese oxide [18, 19]. These oxides must be removed from steel surface before the galvanizing process. The most common method is hot-dip galvanizing, in which steel pieces are submerged in a bath of molten zinc. Iron on the surface of steel reacts with zinc to form a multi-layered coating of zinc-iron alloy. This alloy increases the resistance of steel against corrosion prolonging steel life by 20–40 times [20]. High-quality coating requires a clean surface, i.e. a surface free of scales and rust. This can be accomplished by immersion of the piece in sulphuric, hydrochloric or in $HF+HNO_3$ acid solution, the process being called pickling. Up to the middle of last century, sulphuric acid was the most widely used. In the meantime, H_2SO_4 has been substituted by HCl since, although it is more expensive, it leads to higher rate of dissolution at lower temperatures, better surface quality and more economical regeneration [21]. During this time, the pickling of stainless steel with hydrofluoric acid or mixtures for hydrofluoric acid and nitric acid has also been applied on industrial scale [22, 23]. The typical compositions of pickling solution and operating temperatures are listed in Table 5.1.

Table 5.1: Pickling acid solution.

Sulphuric acid solution	Hydrochloric acid solution	Nitric and hydrofluoric acid solution
200–250 g/L H_2SO_4 95–100°C [1]	150–200 g/L HCl 60–70°C [1, 20]	60–180 g/L HNO_3 and 10–80 g/L HF. 50°C [1, 22, 24, 25]

In the pickling process, the acid reacts with oxides solubilizing the metal. Therefore, the concentration of the acid in solution decreases and the metal concentration increases. When the concentrations of acid and metals in solution attain limiting values, the pickling process losses its efficacy and the solution must be replaced. For example, the concentration of iron in spent hydrochloric acid solution can attain 130 g/L and in HNO_3+HF solution can attain 30–40 g/L [22, 25].

It is worth noting that all the effluents above mentioned, spent baths, spent pickling solution and rinse water, are complex solutions containing acids and several hazardous metals such zinc, nickel, chromium, copper, iron, lead, etc. Table 5.2 alerts about the toxicity of these metals and lists the range of concentrations allowed in solution to be discharged in public sewer or surface waters in some EU countries [5]; Table 5.3 lists the composition of some real effluents. As can be noted, these solutions are environmentally inacceptable and require treatment. Some interesting reviews on the treatment of these type of effluents generated are published elsewhere [1, 3, 4, 26–30]. Treatment selection

Table 5.2: Toxicity and limit discharges to public sewer or surface waters.

Metal	Toxicity	Limit[1] [5], mg/L
Chromium	Cr(VI): highly toxic, ulcers in nasal membranes, lung cancer [31]	0.1–0.5 Cr(VI)
Copper	Necrosis and Wilson's disease [32, 33]	0.4–4
Nickel	Gastrointestinal distress, pulmonary fibrosis and skin dermatitis [34]	0.5–5
Lead	Anaemia, insomnia, headache, dizziness, irritability, weakness of muscles, hallucination and renal damages [35]	0.2–1
Zinc	Stomach cramps, skin irritations, vomiting, nausea and anaemia [4]	0.5–7

Note: [1]Range of concentrations between minimum and maximum limits in Belgium, France, Germany, England and Wales and Portugal.

Table 5.3: Characterization of spent bath and rinse water in surface treatment processes.

	Acid	Acid Conc., g/L	Zn, g/L	Fe, g/L	Other	Ref.
Spent pickling solution (SPS)	HNO_3 HF	120–160 60–80	–	35–45	5–20 Cr, 3–6 Ni, in g/L	[29]
Rinse water after pickling	HNO_3 HF	2 1	–	1	0.14 Cr, 0.07 Ni, in g/L	[36]
SPS	HCl	52	6.3	160	47 Cu, 96, 102 Ni, 56 Cr, in mg/L	[37]
SPS	HCl	58–230	82	96	173 Mn, 38 Al, 104 Pb, 19 Cr, 19 Ni, 27 Cd, in mg/L	[38]
SPS	HCl	36.5	122.3	95.0	–	[39]
SPS	HCl	9.1	70.2	92.2	–	[40]
Rinse water after pickling	HCl	not detected	0.71	7.28	0.01 g/L Cr	[40]
Zn electroplating bath	H_2SO_4	150–160	50–100	–	40–50 g/L $MgSO_4$, 10–15 g/L $MnSO_4$, 200 mg/L chloride, 300–400 mg/L Ca	[1]
Cu spent electro-plating bath	H_2SO_4	194	–	0.26	40 Cu, 9.6 Ni, 0.11 Bi, 0.0007 As, in g/L	[41]
Cu electroplating rinse water	–	pH 0.2–0.3	0.024	0.15	0.61 Ni, 2.9 Pb, 0.5 As, 0.45 Bi, 0.012 Sb, in mg/L	[1]
Cu spent electro-plating bath					26 Cu^+, 13 KOH, 12 KCN, 350 K_2CO_3, in g/L	[42]
Cr electroplating (Rinse water)	–	–	0.167	0.003	5.4 g/L Cr, 36 mg/L Ni, 99 mg/L SO_4	[43]
Cr electroplating (Rinse water)	–	pH 1.1–2.0	1.0	0.003–0.005	50 mg/L Cr(VI), 100 g/L Cl, 10 g/L SO_4	[44]
Cr electroplating (Rinse water)	–	pH 1.53	–	–	359 Cr(VI), 72.9 Cu, 51.7 Ni, in mg/L	[45]
Ni electroplating (Rinse water)	–	pH 1.4–6.6	–	–	1.6–2.5 g/L Ni, 3–6 mg/L Cr	[46]
Ni electroplating (Rinse water)	–	pH 6.5	–	–	215 Ni^{2+}, 40 Cl^-, 275 SO_4^{2-}, in mg/L	[47]
Ni electroplating (Rinse water)	–	pH 2.0	0.048–0.158	0.016–0.054	2.8–8.7 Ni, 0.3–1.2 Na, 0.3–1.2 B, 0.4–1.6 Cl, 4.6–18.5 SO_4, in mg/L	[11]
Electroplating rinse water	–	2.2–2.8	0.097–0.20	–	72–215 Ni, 22–48 Cr, 12–28 Cu, 5–10 Pb, 6–12 Ni, in mg/L	[48]

should have environmental issues as a priority. In the last decades, an increasingly stringent regulation has been established regarding the discharge of acids and metals to protect people and the environment. As stated before, the removal of acids and metals from solution should also result in an economic benefit, i.e. these products must be recovered or reused. It is worth mentioning that the decontaminated water obtained in the effluent treatment can be reused instead of being discharged to environment. The following sections summarize different techniques and processes to treat wastes from plating/electroplating industries.

5.3 Treatment and regeneration methods

5.3.1 Neutralization

The surface treatments like galvanization, chrome plating and nickel plating, etc. require a surface free of scales and metal oxides. This can be carried out using concentrated solutions of hydrochloric, sulphuric, nitric or phosphoric acid. Through use, the concentration of dissolved metals in the pickling solutions increases and the concentration of free acid also as the pickling efficiency decreases. To ensure an adequate pickling rate, it is necessary to maintain significant concentrations of free acids in solution. Therefore, the spent pickling solutions, to be substituted, are characterized by high concentrations of acids and metals like Fe, Cr, Ni, Zn, Cu, etc. [4, 29, 49, 50]. These solutions should be treated in some way before being discharged as waste, as neither metals nor acids are permitted in the effluents. Industrially, the neutralization of the acid and consequently the precipitation of the metal has been widely implemented either by small- or large-scale industries, as results of its relative simplicity and low cost [51–53]. The neutralization of the acid from spent pickling liquors can be carried out using caustic soda (NaOH), soda ash, high calcium limestone, dolomitic limestone, high calcium quicklime, high calcium hydrated lime ($Ca(OH)_2$), dolomitic quicklime, dolomitic hydrated lime, highly hydrated dolomitic lime and ammonia (NH_3OH).

The use of limes as neutralizing/precipitant agent normally requires a first step involving its dissolution in water. The resulting slurry, ($Ca(OH)_2$), is then fed to the process. The hydroxyl ion (OH^-) neutralizes the acid and precipitates metal hydroxides like $Fe(OH)_3$, $Zn(OH)_2$, $Cu(OH)_2$, $Cr(OH)_3$, $Ni(OH)_2$, etc. It is known that the ferric iron precipitates at pH below 5 while Zn, Cr, Cu and Ni precipitate at pH higher than 7 [50, 52, 54–56]. The process of neutralization of acid with lime and limestone from spent liquor containing iron is well documented by Hoak and Lewis [57–60]. The studies revealed that limestones vary widely in their rate of reaction with pickle liquor, and the rate depends upon particle size, chemical analysis and a specific reactivity peculiar to a particular limestone [57]. The authors also noticed that the rate of precipitation of iron was associated to the rate of oxidation of the ferrous hydrate. This process was improved by maintaining the liquor temperature at 75°C and providing an efficient aeration, whereby a substantial reduction in sludge volume was attained, settling was complete in less than an hour, and the vacuum filtration rate was increased markedly [58]. These findings are particularly interesting because they can be achieved with less reactive limestones like dolomite, which contains magnesia in its composition. The possibility of using different precipitants is an added value for the neutralization process. It is worth noting that this process is implemented in both large and small plants and the use of low-cost reagents, sometimes local reagents, is desirable.

The use of lime over sodium hydroxide presents as advantageous a better sedimentation and dewatering properties of the formed metal hydroxide also as a reduction of the concentration of fluoride in solution, as result of low solubility of calcium fluoride. The precipitation with NaOH is faster (3–4 times) [53, 61] and, also, allows reducing in ~35% of the volume of sludge [62]. Mughal et al. [56] evaluated the precipitation of Cr with NaOH from an electroplating waste water. The results revealed that the Cr precipitation was not pH dependent in the range of pH 9–10.5. The process attained an efficiency of 95% leaving less than 0.1 mg/L of Cr in the filtrate.

Chen et al. [50] used lime and fly ash to precipitate Cu, Cr, Pb and Zn from nitrate solution containing 100 ppm of each metal. The presence of fly ash increased the particle size of the precipitate and significantly improved the process of sedimentation of sludge. The removal of metal from solutions was efficient, the concentrations of chromium, copper, lead and zinc in effluents were reduced to 0.08, 0.14, 0.03 and 0.45 mg/L, respectively.

Forsell and Niklasson [63] developed a method for removing iron, chromium, nickel and fluoride ions from pickling steel liquor. In a first step, the pH of the solution is adjusted to 5 with slaked lime, which led to the precipitation of iron, chromium and fluoride ions. After filtration, the pH of the liquor was increased to the range 7–10 and nickel precipitates. This solid is not contaminated and can be processed to give nickel oxide.

Dufour and co-workers [49] carried out studies on the sequential separation of Cr, Fe, Mo and Ni from pickle liquor using KOH and KOH+NH_4OH as precipitants. They tested four stages of precipitation, namely at pH 3, 5, 7 and 9. Almost all the Mo precipitated in the first stage at pH 3. But, it was not possible to achieve a good separation of metals. In fact, the four precipitates were a mixture of all the metals. The best results were obtained at pH 7 with the production of a solid containing, mainly, Cr, Fe and Mo. The Ni remained in solution and was recovered in a second step by precipitation.

As noted, one the main drawbacks in the use of the neutralization process is the co-precipitation of metals [64] that leads to the production of a contaminated sludge. This one must be sent to disposal, which cost contributes appreciably to the overall cost of pickling the metal. Otherwise, it is well known the problems associated to disposal namely scarce spaces for dumping sludge and the possibility of sludge being leachates. It is a growing pressure to eliminate this type of residues and to recycle metals and acids [22]. It is worth noting that the sulphate and fluorine anions, from supernatant solution, precipitate as calcium sulphate and calcium difluoride, but the nitrate anions remain in solution, which is also a problem for effluent discharge. In Table 5.4, some advantages and disadvantages of the neutralization process are summarized.

Table 5.4: Characteristics of the neutralization process.

Advantages	Disadvantages
– Low operating costs – Simple technique and inexpensive equipment – Applied at both small and larger plants	– Slow kinetics of metal precipitation – Inefficient separation of metals – Poor settling properties of sludge – Expensive disposal of sludges: large volume of a relatively low density sludge; limited space for sludge disposal; landfill must be well-isolated to avoid the leaching of soluble component by rain water – Losses of reagents: pickle acids and metals

5.3.2 Crystallization

The removal of iron and chromium from spent liquors can be performed using the crystallization technique [65–67]. The solids obtained from the crystallization process are a by-product, since the main objective of the overall process is to recover the acids. Whatever, the production of any marketable product is always an economic benefit to the process. For example, the by-product ferrous chloride can be used in dye, ink and pigments or can be converted in ferric chloride [68] to be used as flocculants in waste water treatment. The by-product ferrous sulphate is used as fertilizer.

The $FeSO_4{\cdot}H_2O$, $FeSO_4{\cdot}7H_2O$ can be obtained from both H_2SO_4 and HCl liquors [21, 69]. In the latter, H_2SO_4 must be added to the liquor leading to the following reaction:

$$FeCl_2 + H_2SO_4 \rightarrow 2HCl + FeSO_4 \tag{5.2}$$

The formation and growing of crystal require a supersaturated solution, which can be achieved by cooling the solution [70, 71] or evaporating the water [21].

Kehrmann [69] patents a method of producing ferrous sulphate heptahydrate from an HCl pickling solution through the addition of sulphuric acid. Hydrochloric acid and sulphate monohydrate are formed. The ferrous sulphate monohydrate is then dehumidified and converted in a crystallization reactor (160–200°C) in the presence of excess water into ferrous sulphate heptahydrate. Ozdemir et al. [70] proposed as a first step of a crystallization process the addition of HCl to the spent liquors to lower the solubility of the ferrous chloride. This solution is then cooled to –40°C in a first crystallizer. The $FeCl_2$ crystals are separated and the liquor is cooled to –57°C leading to the formation of more crystals. These operating conditions are not easy to apply and are costly.

Solids like $FeF_3.3H_2O$, $K_2FeF_5.H_2O$ and $CrF_3.3H_2O$ are obtained from HCl+HF and HNO_3+HF liquors [24, 25, 66, 71, 72]. Hermoso et al. [71] used the crystallization method to remove iron and chromium from spent HNO_3/HF as fluorides (K_2FeF_5 and CrF_3). The best operating conditions are 60–65°C, pH 4.1–4.2 and free fluoride from 14% to 15% of alkali stream as potassium fluoride. The nickel remaining in the solution precipitates as pure (~100%) $Ni(OH)_2$ in another stage with an yield of 72%.

In the Outokumpu process the sulphuric acid is added to the pickling liquor and the mixture is fed to an evaporator (80°C; 5 kPa). Vapours of HF and HNO_3 are condensed and recycled to the pickling bath. The resulting slurry is fed to the crystallizer. The metal sulphates precipitate and the supernatant, H_2SO_4 solution, is recycled to the evaporator [53].

The PHAR (Pickling Hydrochloric Acid Regeneration) process has been developed for the regeneration of HCl from spent pickle liquors. In this process, the sulphuric acid and HCl spent liquor are continuously fed to reactor, leading to the formation of HCl and $FeSO_4$ (see reaction (2)). The aqueous solution proceeds to the crystallizer, by decreasing the crystallization temperature from −3°C to −8°C increases the iron removal efficiency from 72% to 85%. The $FeSO_4{\cdot}7H_2O$ compound is separated from the mother solution by vacuum filtration or centrifugation [21] and the mother solution returns to the pickling tanks. The by-product of this process, ferrous sulphate heptahydrate, is also a marketable commodity. The process does not generate waste water or other residual wastes that would require additional treatment or disposal [73].

Leonzio [19] has recently described a new process combining essentially the evaporation and crystallization to produce hydrochloric acid and iron-zinc sulphate heptahydrates from spent liquors. The process uses two heat pumps in replacement of industrial boilers. The author claimed, based on a simulation, the reduction on tonne of equivalent petrol, equal to 39%, and air pollutants, in particular carbon dioxide equal to 59%, according to the European Union climate package. The promising results need to be confirmed at an industrial scale.

5.3.3 Evaporation

The volatile acids HF and HNO_3 can be regenerated by evaporation from spent pickling solution by the so-called "sulphuric acid distillation". This methodology is applied in processes like Fluorex (Prosep Technologies [22, 74, 75]) and OPAR (Outokumpu Pickling Acid Recovery). In "sulphuric acid distillation", the picking solution is introduced in a reactor containing sulphuric acid (50–60%) at 80–90°C under vacuum conditions [22]. The sulphuric acid and nitrate/fluoride metal compounds react according to the following reactions:

$$2FeF_3 + 3H_2SO_4 \rightarrow 6HF\rightarrow + Fe_2(SO_4)_3 \tag{5.3}$$

$$2CrF_3 + 3H_2SO_4 \rightarrow 6HF\rightarrow + Cr_2(SO_4)_3 \tag{5.4}$$

$$Ni(NO_3)_2 + 3H_2SO_4 \rightarrow 2HNO_3\rightarrow + NiSO_4 \tag{5.5}$$

HF, HNO_3 and H_2O volatilize and proceed to the absorber. The acids are recovered at the bottom of the absorber and are recycled to the pickling bath.

The OPAR process crystalizes the metal sulphates in the reactor. Subsequently, they are dissolved in water and reprecipitated by neutralization with calcium hydroxide and spent slag. The resulting sludge is landfilled [76]. This fact associated with problems related to the difficulty of crystallization of Ni and Cr and to the very expensive equipment that is necessary for handling the highly corrosive solution, makes that this process has not achieved widespread acceptance [77].

The Fluorex process removes the sulphate solution from the reactor and replaces it with purified acid before the precipitation of the salts occurs. The sulphuric acid is separated from metals by using the acid purification unit (APU^{TM}) system (IE/sorption). The acid is recycled to the reactor and the metals in solution precipitate as hydroxides by using ammonia. The supernatant solution is treated with lime that leads to the regeneration of ammonia and the production of gypsum of high purity. According to Brown and Sheedy [75], this process combines the best features of both the lime and caustic neutralization processes, i.e. low operating costs, while producing recyclable metal hydroxide filter cake and gypsum, which can be readily and economically disposed off. This process revealed to be very efficient in what concerns to the total recovery of fluoride (94%) and nitrate (>99%).

5.3.4 Pyrohydrolysis

Pyrohydrolysis of chloride refers to the thermal decomposition of a metal chloride into metal oxides and hydrogen chloride gas at elevated temperatures. Pyrohydrolysis of iron chloride, occurring at temperatures above 400°C, can be expressed by the following equations:

$$4FeCl_2 + 4\ H_2O + O_2 \leftrightarrow 8\ HCl + 2\ Fe_2O_3 \tag{5.6}$$

$$2FeCl_3 + 3\ H_2O \leftrightarrow 6\ HCl + Fe_2O_3 \tag{5.7}$$

This process is implemented in approximately 400 industrial plants for the regeneration of HCl from spent liquor [78–81]. Although the method has proven to be an economic way of recovering HCl acid, it requires high investment that makes difficult its implementation in the medium- and small-scale plants [21]. It is worth noting that the equipment must be built in expensive corrosion resistant alloys, since the process deals with highly aggressive chloride compounds. Also, a dust collecting and gas cleaning system must be provided. In addition to high investment, the process has also presented high operational cost due to the high energy consumption in the pyrohydrolysers [1, 21, 82].

A simplified flowsheet for the pyrohydrolysis process of iron chloride using spray roaster or fluidized bed technology is shown in Figure 5.1. It is worth mentioning that

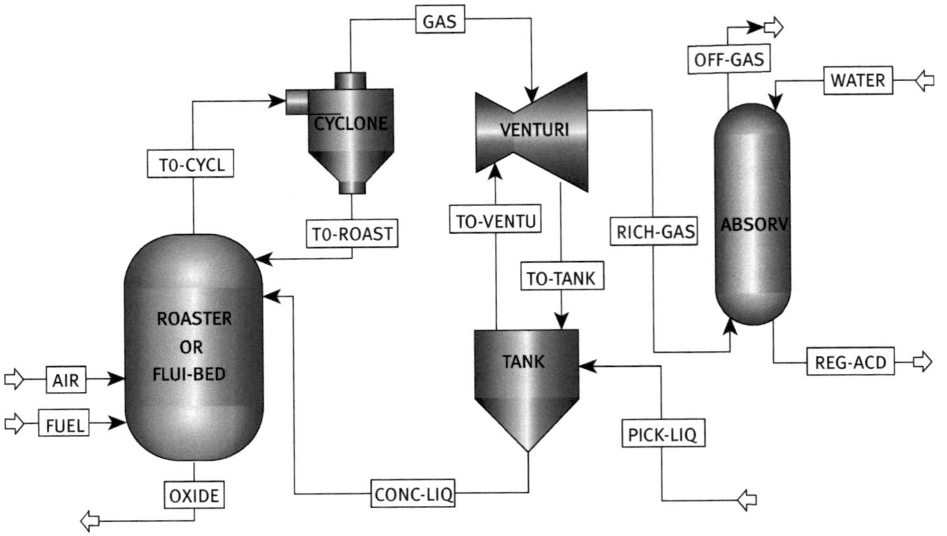

Figure 5.1: Generic flow diagram for pyrohydrolysis plants.

so far approximately 300 spray roasters and 100 fluidized beds have been built worldwide with a capacity from 0.3 up to 30 m^3/h of spent acid solution [78].

Figure 5.1 shows that the pyrohydrolysis process of iron chloride includes three main steps, a pre-concentration of waste pickle liquor, the pyrohydrolysis reaction and the absorption of HCl gas, which leads to the production of an aqueous hydrochloric acid solution [80, 81]. The pre-concentration of the pickle liquor is accomplished by promoting its contact with the hot exhaust gas from the pyrohydrolysis step in a venturi scrubber where heat and mass transfer occur. The concentrated liquor is sent back to the recycle tank while the venturi off-gas, rich in HCl and saturated in water vapor, proceeds to an adiabatic absorber. Water or acid rinse water from the pickling line (commonly 0.5% (w/w) HCl) is fed to absorber column; the regenerated HCl in aqueous solution is collected at the bottom of the column [81]. The pre-concentrated liquor from the recycle tank is fed to a spay roaster or a fluidized bed, where the iron chloride pyrohydrolysis occurs [78, 81–84]. When the spray roaster is used, the pre-concentrated liquor is sprayed into the top of the reactor in a countercurrent process with the combustion gases. Temperatures inside the reactor vary between 700–900°C (burner level) and 400–500°C (gas-off). FeO is collected at the bottom of the reactor while the gases leave the reactor through the top. With regards to pyrohydrolysis in a fluidized bed reactor, fuel and air are introduced directly into the bottom of the bed. The gases of combustion gas flow upwards and fluidize the bed. It is worth noting that the temperature is uniform throughout the reactor (700–900°C).The pre-concentrated liquor is fed into the bed, being rapidly evaporated and leading to the formation of an oxide layer that covers an existing oxide layer.

Table 5.5 presents some characteristics of the pyrohydrolysis process of iron chloride carried out in a spay roaster or in a fluidized bed [80].

Table 5.5: Pyrohydrolysis of iron chloride in spay roasters *versus* fluidized bed reactors.

Spay roaster reactor	**Fluidized bed reactor**
– Less intensive in both energy and electricity consumption.	– Lower volume of the reactor.
– The gas stream leaves the reactor at 400–500°C.	– The gas stream leaves reactor at reaction temperature (700–900°C).
– Oxide produced: lightweight and porous; must be produced when the oxide has market.	– Oxide produced: granular and heavy product; must be produced when the oxide is sent to disposal in landfills.

The pyrohydrolysis process was adapted to the HF/HNO_3 system, and is known as PYROMARS [22, 79]. This process involves a pre-concentration of the spent acid using an evaporator. The produced liquor is fed to the Ruthner-type spray roaster where the following reactions occur:

$$2FeF_3 + 3\,H_2O \leftrightarrow Fe_2O_3 + 6\,HF \tag{5.8}$$

$$2HNO_3 \leftrightarrow NO_2 + NO + O_2 + 6\,H_2O \tag{5.9}$$

$$NO_2 \leftrightarrow NO + 1/2O_2 \tag{5.10}$$

The FeO is collected at bottom of the reactor while the produced gases (water vapour, HF, HNO_3, NO, NO_2 and combustion gases) are sent to the absorber. The HF+HNO_3 regenerated acids in aqueous solution leave the absorber and proceed to the storage tank while the gases are fed to an oxidation column where NO is convert to NO_2. The NO_2 returns to the absorber where it reacts with water to produce HNO_3. The Andritz Metals claimed for this process recoveries of ~99% for HF acid and metals (Ni, Cr, Fe, etc.), and up to 80% for HNO_3.

5.3.5 Dialysis, electrodialysis and electrodialysis/electrolysis

The plating is a water-intensive process. In addition to pickling and plating baths, water is also used in final rinse steps to eliminate traces of baths from the metal surfaces. Nowadays, the reuse of water as well as the recovery of heavy metals and mineral acids is economical and environmentally obligatory. This can be accomplished using membranes separation techniques, such as diffusion dialysis and electrodialysis [30, 85, 86].

The diffusion dialysis membrane technology has been used in the recovery of acid such as HF, HNO_3, HCl and H_2SO_4 from spent plating, anodizing and acid pickling baths and also from metal finishing solutions [87–93]. In the diffusion dialysis, an IE membrane is placed between two solutions, i.e. between the solution to be treated and water/solution with a lower concentration of acid. The driving force for dialysis is the difference in chemical potential of the solute on either side of a membrane. As result, the acid molecules diffuse through the membrane in the direction of lowest chemical potential. Therefore, the components of the solution to be treated are distributed between one solution containing most of the acid and the other containing most of the dissolved metals (Fe, Ni, Zn, Cd, Cr). According to literature data, acid recovery is in the range of 80–90%, while the metal rejection rates keep of 70–90% [87, 94]. Although the separation is not completely efficient, the resulting solutions have enough quality to be recycled to the process, reducing the consumption of fresh chemicals. The diffusion dialysis presents as limitations its relatively low processing capability and efficiency. On the other hand, it is known as being environmental friendly and also as one of the cheapest membrane techniques due to its low operating and maintenance cost resulting from its operational simplicity [95].

Electrodialysis is a membrane separation process that uses an electric field as the driving force. The electrodialysis cell consists in an alternating set of anionic and cationic membranes placed between the cathode and the anode. The application of an electric field leads to the migration of cations to the cathode and anions to the anode, which is allowed or hindered by selective membranes. Nowadays, the most important large-scale application of electrodialysis is still water desalination. However, its applications in waste water/spent effluent treatment are gaining increasing importance, particularly in the galvanic industry and metal surface treatment processes because of the two solutions resulting from an electrodialysis process can be reused. The diluted solution may be used as rinse water and the concentrated solution may be sent to pickling/plating baths or to metal recovery process like electrolysis. The combination electrodialysis/electrolysis that allows recovering the metal (electrolysis) and producing a decontaminated effluent (electrodialysis) is being a subject of several studies. The recovery and concentration of metals such as zinc, nickel, copper, chromium and cadmium and the recovery of acids such as HCl, HNO_3, HF, etc. from waste water and spent pickling solutions are examples of application of electrodialysis and electrodialysis/electrolysis techniques [30, 86, 96–103].

Scarazzato [30] published an extensive review of the use of electrodialysis techniques in the electroplating industries. It was noted that the waste water treatment was the most cited application of electrodialysis in the plating industry, especially for copper, nickel and zinc recovery and for chromium(VI) removal. The authors refer that one main important issue to be improved for this application is the energy waste when dilute solutions are used. For such case, the use of hybrid techniques, such as electrodeionization, was the most evaluated alternative. Benvenuti et al. used the electrodialysis process to treat plating bath and rinse water in a nickel plating industry [99, 100]. The studies carried out at bench scale allowed establishing the optimal operating condition for the process. These parameters were applied at industrial scale to treat effluents containing 10–5,000 mg/L Ni. The concentrated solution attains 22 g/L Ni and can be used as a reinforcement of the plating bath. The dilute solution had enough quality to be used in the final rising steps. The authors claimed 20% of chemical savings for the nickel bath and 90% of reduction of the effluent volume sent to waste water treatment [100].

The application of electrodeionization process was assessed for Ni and pure water recovery from the electroplating industry [102, 103]. The process was performed in two stages, leading to a recovery of 99.8% of nickel present in the feed (initial concentration 50 mg/L Ni). In addition, a closed-circuit circulation of the concentrate stream was used to obtain a solution containing 11 g/L Ni. Water with enough quality to be used as make-up was also produced [103]. The use of electrolysis-electrodialysis-electrodeionization combination process revealed to be efficient to recover nickel from an aqueous solution containing up to 2,000 mg/L Ni. This process allowed recovering 100% of water and 99.8% of nickel, this one with purity as high as 93.9% [34].

Tran et al. [101] treated a nickel plating waste water using crystallization in fluidized pellet reactor and electrodialysis with bipolar membranes. The crystallization step allowed recovering 74% and 94.4% of calcium and nickel, respectively, from waste water containing 90–120 mg/L Ca, 40–60 mg/L Ni, 41.5–47.3 g/L Na, 75.9–88.6 g/L sulphates and vestiges of strontium, iron and chromium. After this pretreatment, it was possible to regenerate the acid (1.76 N) and the base (2.41 N) from the respective salt by electrodialysis. The authors point as advantageous the possibility of reducing water pollution and producing acid/base solution that can be reused.

Dydo et al. [11] studied the possibility of recovering and concentrating nickel from actual electroplating waste water. The results showed that is possible to attain 20.81 g/L of Ni in solutions after electrodialysis corresponding to a metal recovery of 85% and a concentration factor of two. This solution can be treated by the electrolysis method to recover nickel.

The application of electrodialysis on the treatment of aqueous solutions of copper has been the subject of several studies [97, 104–107]. Scarazzato et al. [97] concluded that electrodialysis (ED) is suitable for the treatment of copper baths. In one stage, it was possible to extract 99.7% of copper from the solution. Peng et al. [106] studied copper and water recovery from copper-electroplating waste water using electrolysis and electrodialysis processes. The results showed that it is possible to recover 99.5% of copper (99.7% of purity) and 100% of water, when the feed solution contained 100–1,000 mg/L of Cu. Cifuentes et al. [108–110] published several papers involving the recovery of copper and water from copper electrowinning solutions by electrodialysis. The authors concluded that ED is a very effective technique to remove of Cu and Fe from the aqueous solution, which allows the reuse of water of these electrolytes.

The widespread use of chromium in the last decades associated with the fact that chromium is a highly toxic pollutant which has forced its removal/recovery from spent effluents and waste waters. The use of electrodialysis to concentrate chromium in aqueous solution is well documented [8, 43, 111–113]. Chang et al. [45] studied the recovery of chromium from waste water using micellar-enhanced ultrafiltration and electrodialysis. The result pointed to a recovery of 85.3% of Cr, when the feed phase contained 359 mg/L Cr(VI), 72.9 mg/L Cu(II) and 52.7 mg/L Ni(II). Tor et al. [114] studied the simultaneous recovery of trivalent and hexavalent chromium ions through charged IE membranes by using three detachable cells. The recovery increased with increasing current density and decreased in the presence of co-existent ions (K^+ and Cu^{2+} for Cr(III); Cl^- and SO_4^{2-} for Cr(VI)) in the feed phase. Chaudhary et al. [115] treated a feed phase containing 100 ppm of Cr and 100 ppm of Cu by an electrolysis–dialysis process. The results showed that it is possible to achieve 92% regeneration of chromium and 90% recovery of copper. Chen et al. [116] tested two-stage electrodialysis system to concentrate and purify chromate from a low pH electroplating waste water. The first stage operates at pH 2.2, which allows concentrating the monovalent chromium ($HCrO_4^-$) four times. The second stage operates at pH 8.5 where the divalent ion $Cr_2O_4^{2-}$ predominates. This species is retained while monovalent cations present in solution pass through the membrane. Khan [117] tested successfully a three-compartment electrochemical membrane reactor based on the electrolysis and electrodialysis principles to treat aqueous solutions containing Cr/Cu/Ni. They justify the use of this membrane configuration as it allows managing three different tasks simultaneously: (1) removal and recovery of metal ions, (2) separation and recovery of chromic acid and (3) purification of static rinse water.

The electrodialysis presents low selectivity relatively to the ions with similar valences. Several authors tried to overcome this problem using complexing/ligand agents. Cherif et al. [118] used ethylenediamine tetraacetic acid (EDTA) to improve the separation between Zn and Ag. They noticed that the transport rate of uncomplexed Ag^+ ions was about three times higher than that of complexed zinc ions. EDTA was also applied in the separation of Ca^{2+}/Cd^{2+} [119], Co^{2+}/Ni^{2+} [120]. Frioui et al. [121] tested the use of various chemical ligands such as citrate, glycine, phosphate and EDTA on the separation of Ag/Zn and Cu/Cd. The best results (99%) were achieved, for both systems, with EDTA. Babilas and Dydo [122] tested the use of the lactic, malic and citric acid as chelating agents in the separation of Zn/Fe from an electroplating waste water. The use of citric acid leads to the best results, i.e. 92.36% for the retention of iron and 86.6% for the recovery of zinc.

The main drawbacks of electrodialysis are the low selectivity in solutions containing different metal ions with similar valences and membrane fouling, which leads to high energy consumption [34, 123]. The fouling can be minimized with pretreatment of feeds or by reversing periodically the direction of ion flow. The main advantages of the electrodialysis system are the separation without phase changes, no addition of chemicals, high permeability selectivity, low electrical resistance, good mechanical stability and high chemical stability [123, 124]. These advantages justify the high interest in the use of this technique.

5.3.6 Ion exchange

5.3.6.1 Spent pickling solutions (SPS)

As stated before, spray roasting is implemented worldwide in many industrial plants mainly for the treatment of stainless steel SPS. Nevertheless, the investment and operational costs associated with spray roasting in small hot-dip galvanizing plants are too high, besides other issues related to the composition of the solution mentioned before (see Section 5.3.4). Thus, the use of IE is gaining considerable noteworthiness under certain concentration ranges or together with other processes [125]. The advantages of IE for regenerating SPS are the high effectiveness, low operation cost, little equipment/space and application in industry [3, 126]. Actually, the use of IE is common in industry, particularly to recover free HCl. However, IE also produces a high volume of diluted solutions with large consumption of fresh water.

Marañón et al. [127] studied the IE treatment for removing iron from HCl-exhausted baths in an attempt to make the re-utilization of acid viable for industry while reducing the amount of waste generated. These authors noticed that cationic and chelating resins allowed removing Fe(II) present in the solution as a cation, while anionic resins were able to remove Fe(III) chlorocomplexes. Higher capacities were achieved when using anionic resins, especially with Lewatit MP 500 (up to 6 mg Fe/mL). Further work was carried out by the same authors [128] with the aim of removing Fe(III) and Zn(II) from SPS, using the anionic resins Lewatit MP 500 (macroporous) and Lewatit M 504 (gel). These two strongly basic anion exchange resins showed a similar behaviour in terms of the removal and recovery of zinc, which were very efficient when the concentration of Zn(II) in the pickling acid did not exceed 1 g/L. The best results were attained with a two-step operation, since it was possible to perform the sequential removal of Zn(II) and Fe(III) in two columns in series, while Fe(II) passed through the resin practically without retention. The maximum concentration recommended for iron exchange was 5 g/L. The elution of Fe(III) from the resin with water was very fast and complete, but the same did not occur with zinc chlorocomplex, thus needing a large volume of water. Even though this problem could be solved with the use of NH_4OH instead water as the regenerant, this process produces a large volume of effluents.

The application of anionic resins to separate Zn(II) and Fe(II) from HCl-exhausted baths, by zinc retention and Fe(II) rejection, has also been examined elsewhere [129]. Csicsovszki et al. [129] proposed a combined treatment with anion exchange and membrane electrowinning techniques to recover HCl from SPS. Acid and iron are recovered by membrane electrolysis with nickel electrode as a cathode, after zinc be separated with a quaternary ammine and eluted with 0.1 M HCl.

Alternatively, IE with anionic resins can be applied for retaining acid in a column (retardation process), the metals being eluted from the resin first as a waste or by-product [3]. The method was commercialized and applied as RECOFLO Acid Purification System and KOMParet Retardation System [3, 130]. Miesiąc [130] also proposed the removal of Zn(II) and Fe(II) from spent HCl solution using the acid retardation process. The best results were obtained for the strongly basic gel-form Lewatit VP OC 1071, the separation of HCl and Fe(II) being effective. Even though such techniques have their own drawbacks, (poor selectivity, dilution of recovered solutions, etc.), the uses of resin beds are attractive and applicable in metal finishing industry.

In addition, it should be mentioned that the recovery of HNO_3/HF from pickling liquors by IE has been widely used in steel processing plants by using APU technology (APU™) [3].

5.3.6.2 Electroplating solutions

IE has also been reported as a possible technology for the removal of heavy metals from electroplating baths and rinse waters. Thus, the removal and recovery of nickel, chromium, zinc, copper, etc. from such solutions has been studied [131–133]. Priya et al. [131] investigated the recovery and reuse of Ni(II) from rinse water of electroplating industries by using IE. These authors examined the adsorption of Ni(II) ion from a simulated rinse water on Amberlite IR-120 cation-exchange. The treated water could be recycled for rinsing solution, since the concentration of Ni(II) was depleted to values less than 1 mg/L. Furthermore, the adsorbed Ni(II) ion was efficiently desorbed with the use of HCl solution, which allows its reuse for making up the plating bath.

Papadopoulos et al. [134] studied the application of IE for the removal of nickel in a waste water stream (with pH 5.5–6.3), which was originated from a rinse bath of a metal surface treatment industry. The IE process using clinoptilolite showed to be quite effective for the removal of this metal. In addition, the combination of IE with the precipitation process was found more efficient than the individual IE, leading to a higher removal of nickel. Juang et al. [135] investigated the removal of Ni from synthetic electroplating waste water using a strongly acidic cation-exchange resin (i.e. Purolite NRW-100) in fixed beds. The breakthrough curves were satisfactorily predicted by applying the Thomas model, when the pH of the inlet solution pH was below 3.0. Dizge et al. [126] concluded that Ni(II) can be effectively removed from synthetic solutions and industrial effluents, using a Lewatit cation-exchange resin (Lewatit MonoPlus SP 112), which is strongly acidic. The maximum uptake of nickel was observed at pH 6.0, the efficiency of removal being low level at pH <3.0.

Revathi et al. [133] investigated the removal of copper, nickel and zinc ions from synthetic electroplating rinse water using the cationic exchange resin Ceralite IR 120. It was concluded that the effluent could be reused in the rinsing tanks after treatment by IE. Most of the copper sorbed in the resin was successfully eluted from the column by 1.5 M H_2SO_4. Copper present in the regenerated liquor was recovered as metal foil by electrowinning process [133]. Recently, Revathi et al. [136] examined the removal of Cu(II) ions from synthetic electroplating rinse water by using the same resin, Ceralite IR 120, but impregnated with polyethyleneimine. The modified IE resin was found to be a better sorbent for the removal of Cu(II) ions from the target solutions.

IE has also been applied for treating solutions in chromium plating. The study on the removal of Cr(VI) from chrome plating industry waste water by IE is published elsewhere [132]. The authors of this work reported that Cr(VI) can be efficiently removed by using Tulsion A-27 (MP) resins by IE column process. The regeneration of a Cr(III) electroplating bath by IE chelating resins based on iminodiacetic acid (IDA) functional groups was patented by Reynolds [137]. Gode and Pehlivan [138] analysed the behaviour of two chelating exchange resins, containing IDA groups (Lewatit TP 207 and Chelex-100), to remove Cr(III) from synthetic aqueous solutions. Both resins exhibited a large collective adsorption with Cr(III) ion, TP 207 showing stronger binding and a strong chelating ability. The sorption of Cr(III) ions on the resins is very dependent on the pH of the solution, the equilibrium being very favourable at pH 4–5. At pH below 2, the hydrogen ions are likely to compete with the metal ions for binding sites on the surface of the resins, reducing the Cr(III) uptake.

Fernández-Olmo et al. and Ortiz et al. [139, 140] investigated the regeneration of spent Cr(III) passivation baths used in the plating industry by IE. It is noteworthy that the passivation of zinc-plated surfaces is performed to protect them against corrosion, besides decorative purposes. The Cr(III) conversion coatings have been widely used in the plating industry to substitute Cr(VI), but the life of the new baths is generally reduced owing to the contamination of iron and zinc species. Hence, the selective removal of iron ions from actual passivation baths with different iron concentrations by means of IE was examined [139]. It was shown that the use of the chelating resin Purolite S-957 allows extending the operational life of Cr(III) passivation solutions.

5.3.7 Classical and membrane SX

5.3.7.1 Solvent extraction

5.3.7.1.1 Spent pickling solutions

The recovery of acids from spent pickling solutions by SX is reported in the literature by various authors [1, 3, 29, 37, 141]. Agrawal and co-authors [141] noticed that Alamine 336 (trioctyl/decyl amine) was a good extractant for sulphuric acid generated during the process of pickling of steel in an Indian tube industry. In fact, the loading capacity of 35% Alamine 336 (0.72 M) in kerosene/10% isodecanol was found to be 82 g/L of H_2SO_4. Nevertheless, acid could not be stripped completely from the loaded organic phase with water. In what concerns the iron(II) in the raffinate, it was oxidized to Fe(III) and then was extracted by a solvent mixture with DEHPA [di(2-ethylhexyl) phosphoric acid] and methyl isobutyl ketone. Benedetto and Morais [142] used 70% TBP (tri-*n*-butyl phosphate) in isoparaffin to recover nitric and hydrofluoric acids from industrial wastes. Although the extraction was very high, the issues related to the co-extraction of metallic ions and management of salts were not discussed.

Agrawal and Sahu [143] investigated the extraction of sulphuric acid from a waste pickle liquor using Cyanex 923 (mixture of trialkylphosphine oxides). The feed contained 91.9 g/L of H_2SO_4, 78–80 g/L of iron sulphate and traces of other impurities. The use of 1.85 M Cyanex 923 with an aqueous/organic ratio of 1/4 allowed extracting 74% of acid, which was stripped with hot water. The iron content in the raffinate was oxidized to Fe(III), FeO with high purity being produced.

Recently, Machado et al. [37] proposed the recovery of hydrochloric acid from galvanizing pickling baths by SX and distillation. Cyanex 923 was successfully used to extract hydrochloric acid from a pre-treated hot-dip galvanizing plant effluent. The stream processed contained ~50 g/L HCl, 160 g/L Fe(II), ~6 g/L Zn(II) and other minor elements. This extractant was selected due to the excellent results of the acid recovery from the loaded phase by distillation, which enabled obtaining HCl solutions with very high concentrations (i.e. 256 g/L). The distillate can be afterwards recycled to the pickling bath of the galvanizing industrial plant.

SX with undiluted Cyanex 923 has also been applied for recovering Fe(III), HNO_3 and HF from steel rinse waters [144]. The co-extraction of the target solutes from a solution containing ~1 g/L of Fe(III), ~2 g/L of HNO_3 and ~0.05 M of HF was in the range of 80–91% by using this solvating extractant. The selective stripping of iron(III) and acids was performed with water as an acid-stripping agent, the recovery of HNO_3 and HF and the iron concentration being within acceptable limits for the reuse of these acidic streams.

The Kawasaki Steel Process was developed in the 1980s for dealing with steel pickling waste liquors by combining SX technique with a ferrite formation technique [145]. In this process, DEHPA (30% in paraffins) is the extractant of Fe(III), which is stripped by NH_4HF_2 as $(NH_4)_3FeF_6$. The extraction of HNO_3 and HF is performed with TBP (70% in paraffins), the stripping being carried out with water. The recoveries of iron and HNO_3 attained 95% and the recovery of HF yielded 70% at industrial scale.

Agrawal and Sahu [146] proposed the extraction of iron from actual sulphate waste pickle liquor using Cyanex 923 diluted with kerosene. The composition of the feed processed was 60.9 g/L Fe(III), 53 g/L of acid and minor contaminants. Iron was successfully recovered with 40% Cyanex 923, by using oxalic acid as the strippant. Thus, Cyanex 923 is an alternative to the classical acidic extractant DEHPA for the selective extraction of Fe(III).

Various extractants have been examined for the recovery of zinc from hot-dip galvanizing effluents. The vast majority of workers have utilized TBP, which has been selected as one of the suitable extractants for the separation of zinc from iron in chloride medium. Mansur et al. [40] investigated the use of TBP, Cyanex 272 [bis(2,4,4-trimethylpentyl) phosphinic acid], Cyanex 301 [bis(2,4,4-trimethylpentyl) dithiophosphinic acid] and Cyanex 302 [bis(2,4,4-trimethylpentyl) monothiophosphinic acid] as extractants in HCl medium. The results obtained showed that TBP and Cyanex 301 are selective extractants for zinc over Fe(II) at very acidic conditions. It is worth mentioning that Fe(III) must be reduced to Fe(II) in a previous step otherwise the Zn/Fe separation factor decreases considerably. The main drawback of TBP is the volume of reagent required for the quantitative extraction of zinc. Additionally, TBP co-extracts HCl [147] and the water transfer to the organic phase might not be ignored. In what concerns the use of the dithiophosphinic acid, Cyanex 301, problems related to its chemical stability are reported elsewhere [40, 148].

Cook and co-authors [149] tested TBP and a range of secondary and tertiary amines (Hostarex A226, A324, and A327) for the selective recovery of Zn(II) from galvanizing effluent of high zinc concentrations. They identified TBP as the most favourable industrially available extractant for this purpose. The main attractive aspects of using this solvating extractant for recovering zinc from acidic chloride medium are as follows: (1) good separation of phases; (2) high selectivity over Fe(II); (3) easy stripping with water. The extraction reaction of zinc(II) in high concentration of HCl (>2.7 M) with TBP is expressed as [150]

$$2H^{+}_{aq} + ZnCl_{4\,aq}^{2-} + 2TBP_{org} \leftrightarrow H_2ZnCl_4 \cdot 2TBP_{org} \tag{5.11}$$

As acidity of the aqueous phase decreases, the following reactions can occur [149]:

$$H^{+}_{aq} + ZnCl_{3\,aq}^{-} + 3TBP_{org} \leftrightarrow HZnCl_3 \cdot 3TBP_{org} \tag{5.12}$$

$$ZnCl_{2\,aq} + 2TBP_{org} \leftrightarrow ZnCl_2 \cdot 2TBP_{org} \tag{5.13}$$

The extracted chlorometallate can be stripped into water, the reaction being described as

$$H_nZnCl_m \cdot 2TBP_{org} + H_2O_{aq} \leftrightarrow ZnCl_{2\,aq} + nH^{+}_{aq} + (m-2)Cl^{-}_{aq} + 2TBP_{org} \tag{5.14}$$

Recently, Wieszczycka proposed the recovery of zinc from spent pickling solution by using a pyridineketoxime extractant, namely the oxime of 1-(3-pyridyl) undecan-1-one [151]. In fact, this oxime showed to be an effective extractant of Zn(II) ions from the hydrochloric acid solutions and very selective in the presence of Fe(II) and Fe(III), which is very promising as an alternative for substituting the classical extractant TBP. Nevertheless, further studies are necessary to test the system at pilot plant with actual spent pickling solution. Table 5.6 lists the various extractants used in the recovery of acids, zinc and iron from spent pickling liquor.

Table 5.6: Acids and recovery of zinc and iron from pickling liquor by classical solvent extraction.

Extraction	Extractant	Remarks	Reference
HCl	Cyanex 923	After pretreatment of waste; with distillation of extractant	[37]
HF, HNO_3	TBP	Kawasaki Steel Process	[145]
H_2SO_4	Alamine 336	Poor stripping with water	[141]
Iron	DEHPA	Kawasaki Steel Process	[145]
	Cyanex 923	Stripping with oxalic acid	[146]
Zinc	TBP	Good stripping with water and selective over Fe(II); water transfer to the organic phase	[40, 81, 149, 152, 153]
	dibutyl butyl-phosphonate	Good stripping with water; water transfer to the organic phase	[154]
	DEHPA	In a second step (after TBP) to transfer Zn from chloride medium to sulphate	[155]
	Cyanex 921	Poor stripping with water, oxidation of Fe(II)	[153, 156]
	Cyanex 272, 301	Selective over Fe(II), Cyanex 301 stronger than Cyanex 272 but lack of chemical stability	[40, 81]
	Cyanex 272 salt	In an integrated process	[157]
	Cyanex 302	Poor extraction; third phase formation	[40]
	Hostarex A226	Selective over Fe(II)	[149]
	Hostarex A324, A327	Efficiency decreases with Zn concentration; needs modifier	[149]
	Alamine 336	Poor stripping with water	[156]
	Aliquat 336	Poor selectivity over Fe(II); emulsion formation	[152]
	3PC10[1]	Selective over Fe	[151]

Note: [1]1-(3-pyridyl)undecan-1-one oxime.

5.3.7.1.2 Electroplating solutions

The application of the SX process is considered as an attractive option for the removal and recovery of metals from spent electroplating baths and wasted rinsing bath solutions. During many plating operations, a substantial amount of bath solution adheres to the plated pieces. Consequently, valuable materials are lost as "drag-out" into the subsequent cascade rinse tanks. Since SX allows obtaining an aqueous solution with a high content in the target metal, it is a potential technique to be applied for recovering materials that can be recycled to the plating tank. Nevertheless, the studies on this topic are not very abundant.

Electroless nickel plating is one of the most important plating techniques, a large amount of its spent bath being generated with high concentrations of nickel and other components, as well as rinse waters. Several extractants have been employed by different investigators to extract nickel from this type of solutions (see Table 5.7). In fact, organophosphorus acids and chelating oximes are the reagents most used on the classical SX extraction of nickel [158]. The use of the well-known DEHPA needs a strict control of pH, the adequate values being within pH 3.0 and 5.0. The alkyl

Table 5.7: Recovery of metals from electroplating wastes by classical solvent extraction.

Extraction	Extractant	Remarks	Reference
Zinc	Aliquat 336	Separation from mix waste water	[167]
	DEHPA	Selective extraction over Cr(III)	[169]
Copper	LIX 984N-C	Efficient recovery from CESRBS[1]	[168]
		Separation from mix waste water	[167]
Iron	DEHPA	Applied to mix solution with Cr	[166, 167]
Nickel	Cyanex 301	Accumulation of Cu(II) in the organic phase	[170]
	DEHPA	Efficient recovery from NESRBS[2]	[46]
	LIX 84I	High extraction (>90% in 1 stage).	[160, 161]
	LIX860-IC	Ammonium sulphate feed	[159]
	LIX 984N-C	Emulsion formation	[46, 167]
Chromium(III)	DEHPA	Ammoniated DEHPA is necessary	[169]
	Cyanex 272	Separation from mix waste water	[167]
Chromium(VI)	TBP	Efficient recovery from electroplating effluent with Zn	[44]
	Cyanex 923	Co-extraction of Zn(II)	[164]
	Alamine 336	Co-extraction of Fe(III) and Zn(II)	[165]
	TBAB[3]	TBAB in dichloromethane	[171]

Notes: [1]CESRBS: copper electroplating second rinse bath solution; [2]NESRBS: nickel electroplating second rinse bath solution; [3]Tetra-*n*-butylammonium bromide.

phosphonic acidic extractants compared to alkyl phosphoric acids extract nickel at higher pH values with faster extraction and stripping kinetics. The alkylphosphinic acid Cyanex 272 exhibits a good performance of extraction under the pH range of 6.0–7.5, which requires high amount of alkali to maintain this operating condition. The alkylphosphinic acids, Cyanex 301 and Cyanex 302, extract nickel at very acidic medium, but are not stable and decompose [158]. Hydroxyoximes extractants exhibit slower kinetics of nickel extraction, but they have the ability of extraction under acidic medium in a broad range of pH values (i.e. 2.5–6.5) [158]. Sana et al. [159] investigated the extraction of nickel in ammonium sulphate solution with LIX860-IC (5-dodecylsalicylaldoxime) in kerosene using a vibro-mixer type extractor, but the acidic medium of the spent baths was not examined.

Tanaka et al. [160] studied the continuous extraction of nickel, in a mixer-settler, with 20% LIX 84-I in Shellsol D70 and 2% PC88A as an accelerator. A high nickel extraction efficiency (99.9%) was achieved using a countercurrent three-stage extraction process, while the most concentrated nickel sulphate solution (0.49 M) was obtained in the sulphuric stripping solution (two stages of stripping). The effects of the operating conditions on the extraction efficiency of this process were quantitatively examined on a subsequent study [161]. Kul and Cetinkaya [46] studied the use of DEHPA, Cyanex 272, LIX 84-I and LIX 984 N-C as the extractants of nickel. They found that DEHPA was the most convenient extractant to recover nickel from nickel electroplating second rinse bath solution. An electrolyte with 48 g/L of Ni, which can be recycled to the nickel electroplating bath, was generated by 0.5 M DEHPA in kerosene at the O/A ratio of 1/5 and pH 5.0 with two-stage countercurrent extraction. The stripping step was performed by using nickel electroplating first rinse bath solution (16 g/L of Ni) with 150 g/L of H_2SO_4. These authors noticed that LIX 984N-C can also be used for this purpose, but it is not recommended because stabilized emulsion may be formed [46].

SX is also an alternative method for the extraction and recovery of chromium from the solutions generated in the chrome-electroplating industries. Anion exchangers like high molecular weight amines have been used for the extraction of Cr(VI) from chloride and sulphate solutions of plating effluents [162]. Nevertheless, the solvating extractants, TBP, Cyanex 921 or Cyanex 923, the aldoxime LIX 84, besides the basic extractants, Alamine 336 or Aliquat 336, are important in chromium extraction because of their coordination ability and the stability of the complexes formed [163]. In what concerns Cr(III), the cation exchangers like DEHPA and carboxylic acids have been used to remove it from liquors [162]. Sahu et al. [44] noticed that Cr(IV) can be extracted quantitatively from chloride solution at pH 1.0 with TBP. The use of this extractant allowed recovering 93% of Cr(VI) and 76% of Zn(II) from a real electroplating effluent (50 ppm Cr, 1.0 g/L Zn), the selective stripping being accomplished with 0.1 M NaOH and 10% H_2SO_4, respectively [44]. Agrawal et al. [164] stated that Cyanex 923 is a potential extractant for recovering Cr(VI) from industrial effluents under highly acidic conditions (i.e. pH 0.5). This reagent, with a concentration as low as 2.5%, was found effective for the extraction of Cr (VI) with concentration varying from 54 to 1,100 ppm. The presence of several other metal ions such as Cr (III), Fe (II), Cu (II), Ni (II), Zn (II), Ca (II), Mg (II), which generally accompany Cr(VI) in the effluents of electroplating industries, was also examined. Zn (II) was co-extracted with Cr(VI), but the other ions were found to be non-interfering. Zn(II) was scrubbed with 0.25 M HCl solution, before stripping of Cr with 1 M NaOH.

Alamine 336 diluted with refined palm oil showed to be very efficient in removing Cr(VI) over a pH range of 1–4, the extraction being up to 95% in a single contact of phases [165]. The presence of contaminants like iron and zinc had a negative effect on the extraction of Cr(VI). The stability of the solvent system needs to be further investigated.

The removal of iron(III) from Cr-electroplating solution was checked by El-Nadi and El-Hefny [166] using DEHPA as the extractant. The organic phase was stripped with 4 M H_2SO_4 solution and the iron was precipitated using NaOH. A purity of chromium of 97.5% was obtained in the raffinate. Kul and Oskay [167] also used DEHPA for the separation and recovery of iron from real mix electroplating waste water containing Cu, Ni, Fe, Cr and Zn at pH 1.6. The extraction step of iron was carried out at pH 1.0 with 10% DEHPA, after separating zinc with 10% Aliquat 336 and copper with 5% LIX 984N-C. Regarding to nickel and chromium, they were extracted by 15% LIX 984N-C, at pH 5.25, and 10% Cyanex 272, at pH 6.0, respectively. The selection of LIX 984N-C for the recovery of copper was based on a previous work published elsewhere [168]. In fact, Kul and Çetinkaya [168] developed a complete hydrometallurgical process at the laboratory scale for recovering copper from the copper electroplating second rinse bath solution containing ~2.5 g/L Cu by SX using 30% LIX 984N-C in kerosene. An electrolyte with ~97 g/L Cu, which could be recycled to the copper electroplating bath, was generated after stripping be performed with copper electroplating first rinse bath solution (7.8 g/L Cu) with 550 g/L H_2SO_4.

Even though there is no relevant application of SX for recovering acid from electroplating solutions, many studies on the processing of waste streams/solutions containing acid are published in the literature. Agrawal and Sahu [1] reviewed all the possible hydrometallurgical options for the recovery of acid from spent acidic solutions from steel and electroplating industries. Major stress has been given to SX methods.

5.3.7.1.3 Etching solutions

As stated before, copper etching is considerably an important process in electronics industry, particularly in the fabrication of PCB. The most attractive method for the removal and recovery of copper from the ammoniacal etchants is the SX process and is presently in practice in industry. The use of hydroxyoximes (mainly LIX 84-I) and β-diketones (mainly LIX 54) for extracting copper(II) from ammoniacal solutions has been studied [172–180]. The equilibrium of copper extraction from ammoniacal media by β-diketones and oxime extractants may be described by the following equation [178]:

$$Cu(NH_3)^{2+}_{4,aq} + 2HR_{org} \leftrightarrow CuR_{2,org} + 2NH_{3,aq} + 2NH^{+}_{4,aq} \tag{5.15}$$

For instance, Kyuchoukov et al. [175] showed that LIX 54 (1.26 M) is able to extract copper from an aqueous feed containing 134 g/L Cu(II), 8 M NH_3 and ~4.8 M Cl^-; the loaded organic solution can be completely stripped with 1.58 M H_2SO_4. Actually, the β-diketone LIX 54 has demonstrated some advantages for copper recovery from ammoniacal medium, when compared with the hydroxyoxime LIX 84-I [174, 175, 181]. It is worth mentioning that LIX 54 provides low ammonia loading, ease of stripping and fast transfer kinetics. Nevertheless, various studies on the use of LIX 84-I for the recovery of Cu(II) from ammoniacal medium have revealed interesting results. Gameiro et al. [179] examined the extraction of Cu(II) from a solution containing 1.0 g/L Cu, 0.2 M $(NH_4)SO_4$ and 0.3 M NH_3 (pH 9.5) with LIX 84-I (0.09 M), in a pulsed sieve-plate column. The results of the pilot plant experiments demonstrated the feasibility of operating the extraction process in this type of column, with efficiencies of copper removal in the range of 90.5–99.5%. The same authors also investigated the extraction of Cu(II) by LIX 84-I, besides LIX 54, by applying liquid membrane techniques, as it will be referred later. Recently, Wieszczycka et al. [14] reported the extraction of copper(II) ions from ammonia/ammonium chloride solutions with hydrophobic pyridineketoximes. They compared the extraction properties of the studied pyridylketoximes with those obtained with commercial reagents (hydroxyoximes and β-diketones). The oxime of 1-(2-pyridyl)tridecan-1-one was found to be a better extractant for the extraction of Cu(II) from ammonia/ammonium chloride solutions but not from solutions with a high free ammonia content (above pH 9).

5.3.7.2 Membrane-based solvent extraction (MBSX)

MBSX has received deserved attention in recent years due to the low inventory of extractant, low energy consumption, process selectivity and the possibility of simultaneous extraction and stripping of the target compounds [2, 182]. Particularly, liquid membrane configurations in hollow fibre modules present the benefit of high surface area per unit volume. Non-dispersive solvent extraction (NDSX) using hollow fibres as membrane contactors is shown as an alternative to the classical extraction in mixer-settlers and columns. The aqueous and organic streams flow through the lumen and shell side of the fibres and come into contact at the pore mouth. One of the two phases wets the microporous fibre (membrane) and a slight overpressure should be applied to the other phase to stabilize the interface and to avoid the mixing of phases. The solute is thus transported from the feed to the organic phase, through the membrane, without phase dispersion. This technique overcomes the problems that often arise in operating traditional extraction equipment, such as the shortcomings related to loading, flooding and entrainment of organic phase in the aqueous effluent. The best-known commercialized hollow fibre module is the Liqui-Cel® Extra-Flow contactor with microporous polypropylene fibres offered for several contact areas by 3M Liqui-Cel^{TD} Membrane Contactors. Figure 5.2 shows a schema of a Liqui-Cel hollow fibre module with the flowing of phases. Since the fibres are

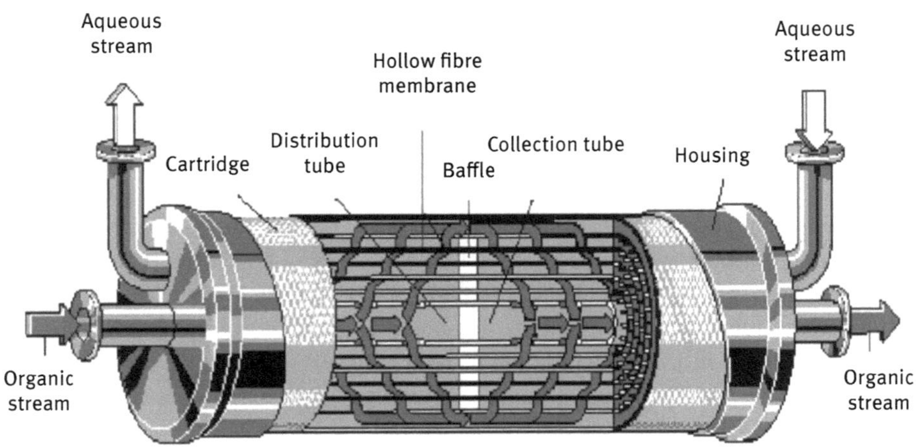

Figure 5.2: Schema of a Liqui-Cel® Extra-Flow membrane contactor with flow of the organic phase through the shell side of the hollow fibres.

hydrophobic, they are impregnated with the organic phase that commonly flows through the shell side of the fibres whereas the aqueous feed phase flows through the lumen side of the fibres. The simultaneous extraction and stripping steps are accomplished by using two hollow fibre modules.

The smart technique with strip/organic dispersion methodology like pseudo-emulsion hollow fibre strip dispersion (PEHFSD), also referred as emulsion pertraction technology, is an attractive alternative among the different liquid membrane configurations. PEHFSD couples the advantages of carrying out the extraction and stripping steps in a single contactor and overcomes the problems of membrane stability that are addressed to emulsion liquid membranes (ELM) [182]. In PEHFSD technique, the aqueous stripping solution is dispersed in the organic membrane solution containing the extractant reagent, a pseudo-emulsion being formed. This phase flows through the shell side of the fibres while the aqueous feed phase, which contains the target species to be extracted, flows through the inner side of the microporous fibres of the hollow fibre module.

In what concerns the recovery of zinc from spent pickling baths, several studies have focused the viability of applying NDSX and PEHFSD [2, 38, 183–187]. Actually, these techniques proved to be efficient alternatives to perform the separation of zinc from iron in HCl media employing TBP as the extractant and water as the stripping agent [2, 38, 185, 188]. The kinetics of zinc recovery was found to be promoted by PEHFSD in comparison with NDSX configuration and increasing the TBP concentration in the range of 20–50%; the transport of iron was enhanced by PEHFSD, high TBP concentration and increasing stripping volume [2]. The species involved in zinc transfer through the organic liquid membrane were confirmed to be $HZnCl_3{\cdot}3TBP$ and $ZnCl_2{\cdot}2TBP$ [2]. In the sequence of this work, Carrillo-Abad and co-authors [39] examined the electrowinning of the stripping solutions coming from the treatment of spent pickling baths by PEHFSD. The results showed that the pretreatment step based on MBSX technology is beneficial for the zinc electrodeposition process.

In recent years, several efforts have been put on the regeneration of spent chromium-based passivation baths by applying the PEHFSD liquid membrane technique [189–192]. It is worth emphasizing that in zinc electroplating, after zinc deposition, a chemical passivation layer is usually applied for inhibiting the corrosion [140, 188]. The new conversion coatings based on Cr(III) are more sensible to the presence of zinc and iron impurities, which reduces significantly the lifetime of the baths with respect to the previously employed Cr(V) [188]. Urtiaga et al. [189] and Diban et al. [190] examined the performance of the liquid membrane process at bench scale using Cyanex 272 as the carrier and H_2SO_4 as the strippant. The feed phases investigated were provided by a plating industry with contents of Cr(III) and Zn(II) attaining ~10 and 12–13 g/L, respectively. The iron concentration was in the range of 20–100 mg/L (65% Fe(III)). The phosphinic acid Cyanex 272 was selected as the carrier, since the extraction of chromium, at the working pH (~2.1) of the passivation bath, is not significant [189]. The results showed that zinc and iron were practically removed from the passivation bath, its reuse being thus possible; the zinc concentration in the stripping solution attained a very high value and iron transferred to the stripping solution was negligible; metallic zinc could be obtained by electrodeposition. Diban et al. [191] and García et al. [192] further intensified the study on PEHFSD for the selective separation of zinc and iron impurities from used Cr(III) passivation baths. The results obtained at industrial scale, with a membrane area of 40 m^2, demonstrated the easiness of the scale-up of the technique [192]. The mass transfer flux of Zn was 4.5 g/(m^2 h), while the flux of Fe was 0.025 g/(m^2 h) [192].

In respect to Cr(VI) present in waste water, several works on the applying of PEHFSD were published, being reviewed by Padilla et al. [193]. Nevertheless, more studies are necessary dealing with actual electroplating effluents.

The use of ELM for the removal and recovery of Ni(II) from electroplating solutions was reported by Ruey-Shin and Jiann-Der [194] and Eyupoglu and Kumbasar [195]. Table 5.8 summarizes the most relevant works carried out on liquid membrane techniques with potential to be applied on the processing of electroplating wastes.

Table 5.8: Summary of applications of liquid membrane techniques on electroplating and spent pickling effluents.

Extraction/Recovery	Technique	Extractant	Strippant	Reference
Recovery of zinc from spent pickling effluents	NDSX	TBP	Water	[155, 183, 184]
	PEHFSD			[38, 39, 185]
	NDSX, PEHFSD			[2]
Extraction of Zn and Fe from Cr(III) passivation baths	PEHFSD	Cyanex 272	H_2SO_4	[189–192]
Extraction of Cr(VI) from simulated industrial waste	PEHFSD	Cyanex 923	Hydrazine sulphate	[196]
Extraction of Cd(II) from acidic and neutral media	PEHFSD	Cyanex 923	Water	[197]
Extraction of Ni(II) from spent Cr–Ni electroplating bath	ELM	LIX 63, 2BDA	HCl	[195]
Recovery of nickel ions from a electroplating rinse solution	ELM	DEHPA	H_2SO_4	[194]

Several studies on MBSX dealing with etching solutions have been reported in the literature. Yang and Kocherginsky [13, 198, 199] published relevant studies concerning the use of the β-diketone LIX 54 in developing a hollow fibre supported liquid membrane (HFSLM) system to recover copper from ammoniacal solutions. In this arrangement, Cu(II) was transferred from the aqueous feed phase (lumen side) to the stripping phase (shell side) through the fibre membrane impregnated with LIX 54. Sengupta et al. [200] investigated the efficacy of LIX 84-I for recovering Cu(II) from ammoniacal medium by using ELM, the kinetics of extraction being found very fast. These authors also examined the same ELM system for dealing with Cu-Zn ammoniacal mixtures [201]. Gameiro et al. [202, 203] investigated the use of LIX 54 as the extractant of Cu(II) from ammoniacal solutions with the liquid membrane techniques ELM and NDSX. The same authors further examined the performance of the extractants, LIX 54 against LIX 84-I, in both liquid membrane configurations, i.e. ELM and NDSX [204]. NDSX allowed recovering practically all the copper from the feed solution by using LIX 54 or LIX 84-I as the carrier, but LIX 54 proved to be more advantageous than LIX 84-I in terms of transfer kinetics. The recovery of the target metal was found to be slightly lower, when the ELM system was used, which was due to the occurrence of non-idealities such as the leakage of the internal stripping phase to the external feed phase and water transport. It is noteworthy that LIX 84-I exhibited a very high extraction rate, but the accumulation of Cu(II) in the membrane was higher than in the case of LIX 54.

Table 5.9 lists the applications of liquid membrane techniques (HFSLM, NDSX and ELM) on the processing of Cu-ammoniacal effluents.

Table 5.9: Summary of applications of liquid membrane techniques on Cu-ammoniacal effluents.

Extraction/Recovery	Technique	Extractant	Strippant	Reference
Recovery of copper from ammoniacal medium	HFSLM	LIX 54	H_2SO_4	[13, 198, 199]
	NDSX	LIX 54		[203–205]
	ELM	LIX 54		[202, 204]
	NDSX, ELM	LIX 84-I		[204]
	ELM	LIX 84-I		[200]
Extraction of zinc and copper-zinc mixtures from ammoniacal media	ELM	LIX 84-I	H_2SO_4	[201]

5.4 Conclusions

Electroplating and other metal finishing operations generate wastes that are considered as hazardous materials, due to presence of acids and metals. The classical neutralization method of the acid content and precipitation of the metals from such waste streams produce a very large volume of sludge, which is an issue of environmental concern, besides the problem of its disposal. Therefore, the removal of acids and metals, the reclamation of the water and their reuse in the process has been the focus of the treatment methods under research.

Electrodialysis is being evaluated as a cleaner production strategy that can be applied to waste water treatment in electroplating industry. Electrodialysis should also be pointed out as one of the BAT for treating spent pickling solutions generated in steel industry. However, the spray roasting and IE/retardation process are widely applied. Currently, acid retardation is the most used process for the purification of mixed stainless steel pickle liquor. IE has also been examined as a potential technique for the removal of heavy metals from electroplating baths and rinse waters.

SX is the BAT for regenerating spent ammoniacal etching solutions and for treating rinse water produced in the manufacture of PCBs. This is the only application of SX currently in use in surface treatment industry. Nevertheless, this technique is well investigated and developed, being very attractive to be applied on the processing of several streams, like the spent pickling solutions.

In the context of surface treatment industry, liquid membranes separation processes using hollow fibres as contactors, such as NDSX and PEHFSD technique, have also been extensively studied. Very favourable results have been obtained regarding the recovery of copper from ammoniacal etchant, the regeneration of spent chromium-based passivation baths (zinc electroplating) and the treatment of spent pickling solutions from hot-dip galvanizing processes.

The environmental and economic benefits of the various treatment methods under inspection depend on the costs of baths regeneration and waste water treatment, cost of the new equipment/implementation of the new technique or cost associated with the changes/renewal on the existing technology, besides the legislation aspects.

References

[1] Agrawal A, Sahu KK. An overview of the recovery of acid from spent acidic solutions from steel and electroplating industries. J Hazard Mater. 2009;171:61–75.

[2] Laso J, García V, Bringas E, Urtiaga AM, Ortiz I. Selective recovery of zinc over iron from spent pickling wastes by different membrane-based solvent extraction process configurations. Ind Eng Chem Res. 2015;54:3218–24.

[3] Regel-Rosocka M. A review on methods of regeneration of spent pickling solutions from steel processing. J Hazard Mater. 2010;177:57–69.

[4] Fu F, Wang Q. Removal of heavy metal ions from wastewaters: a review. J Env Manag. 2011;92(3):407–18.

[5] European Commission. Reference document on best available techniques for the surface treatment of metals and plastics. Available at: http://eippcb.jrc.ec.europa.eu. Accessed: 30 Oct 2017.

[6] Loto CA. Electrodeposition of zinc from acid based solutions: a review and experimental study. Asian J Appl Sci. 2012;5:314–26.

[7] Biddulph C. Zinc electroplating. Choosing the best process for your operation. Prod Finish. 2015;106–15.

[8] Korzenowski C, Rodrigues MA, Bresciani L, Bernardes AM, Ferreira JZ. Purification of spent chromium bath by membrane electrolysis. J Hazard Mater. 2008;152:960–7.

[9] Tenório JA, Espinosa DC. Treatment of chromium plating process effluents with ion exchange resins. Waste Manage. 2001;21:637–42.

[10] DiBari G. Nickel plating. Metal Finish. 2001;99:270–88.

[11] Dydo P, Babilas D, Jakóbik-Kolon A, Franczak A, Nycz R. Study on the electrodialytic nickel concentration from electroplating industry waste. Sep Purif Technol. 2017;53(8):1241–8.

[12] Kyuchoukov G, Bogacki MB, Szymanowski J. Copper extraction from ammoniacal solutions with LIX 84 and LIX 54. Ind Eng Chem Res. 1998;37:4084–9.

[13] Yang Q, Kocherginsky NM. Copper recovery and spent ammoniacal etchant regeneration based on hollow fiber supported liquid membrane technology: from bench-scale to pilot-scale tests. J Membr Sci. 2006;286:301–9.

[14] Wieszczycka K, Kaczerewska M, Krupa M, Parus A, Olszanowski A. Solvent extraction of copper(II) from ammonium chloride and hydrochloric acid solutions with hydrophobic pyridineketoximes. Sep Purif Technol. 2012;95:157–64.

[15] Cakir O. Copper etching with cupric chloride and regeneration of waste etchant. J Mater Process Technol. 2006;175:63–8.

[16] Steel Statistical Yearbook 2016. World steel association 2016. Available at: https://www.world-steel.org/en/dam/jcr:37ad1117-fefc-4df3-b84f-6295478ae460/Steel+Statistical+Yearbook+2016.pdf. Accessed: 3 Nov 2017.

[17] World steel in figures 2017. World steel association Available at: https://www.worldsteel.org/en/dam/jcr:0474d208-9108-4927-ace8-4ac5445c5df8/World+Steel+in+Figures+2017.pdf. Accessed: 3 Nov 2017.

[18] Kladnig WF. New development of acid regeneration in steel pickling plants. J Iron Steel Res Int. 2008;15:1–6.

[19] Leonzio G. Recovery of metal sulphates and hydrochloric acid from spent pickling liquors. J Clean Prod. 2016;129:417–26.

[20] Stocks C, Wood J, Guy S. Minimisation and recycling of spent acid wastes from galvanising plants. Res Conserv Recycl. 2005;44:153–66.

[21] Brown CJ, Olsen DR. Regeneration of hydrochloric acid pickle liquors by crystallization. In: Dutrizac JE, Riveros PA, editors. Iron control technologies. Proc of the 3rd Intern Symp on Iron Control in Hydrom. Montreal, Canada, 2006:831–43.

[22] Brown CJ Recovery of stainless steel pickle liquors: purification vs. regeneration. CISA International Steel Congress, September 2002. Technical paper 158, Beijing China. Available at: http://www.ecotec.com/techpapers/TP%20158%20APU%20vs%20roaster.pdf. Accessed: 8 Aug 2017.

[23] Rogener F, Sartor M, Bán A, Buchloh D, Reichardt T. Metal recovery from spent stainless steel pickling solutions. Resour Conserv Recy. 2012;60:72–7.

[24] Galvez JL, Dufour J, Negro C, Lopez-Mateos F. Kinetics of K_2FeF5H_2O (s) and CrF_32H_2O (s) crystallization from stainless steel. Ind Eng Chem Res. 2007;46:5221–7.

[25] Forsberg KM, Rasmuson AC. Crystallization of metal fluoride hydrates from mixed hydrofluoric and nitric acid solutions, Part I, Iron(III) and Chromium(III). J Cryst Growth. 2010;312:2351–7.

[26] Agrawal A, Naman N, Dubey SK. A review on regeneration process of waste pickling acid at steel industries. Int J Eng Res Rev. 2014;2:70–3.

[27] Regel-Rosocka M, Cieszyńska A, Wiśniewski M. Methods of regeneration of spent pickling solutions from steel treatment plants. PolishJ Chem Techn. 2007;9:42–5.

[28] Devi A, Singhal A, Gupta R, Panzade P. A review on spent pickling liquor. Int J Env Sci. 2013;4:284–95.

[29] Devi A, Singhal A, Gupta R, Panzade P. A study on treatment methods of spent pickling liquor generated by pickling process of steel. Clean Techn Environ Policy. 2014;16:1515–27.

[30] Scarazzato T, Panossian Z, Tenorio JA, Perez-Herranz V, Espinosa DC. A review of cleaner production in electroplating industries using electrodialysis. J Clean Prod. 2017;168:1590–602.

[31] Paulino AT, Minasse FA, Guilherme MR, Reis AV, Muniz EC, Nozaki J. Novel adsorbent based on silkworm chrysalides for removal of heavy metals from wastewaters. J Colloid Interface Sci. 2006;301:479–87.

[32] Golder AK, Dhaneesh VS, Samantac AM, Rayc S. Electrotreatment of industrial copper plating rinse effluent using mild steel and aluminum electrodes. J Chem Technol Biotechnol. 2009;84:1803–10.
[33] Barakat MA. New trends in removing heavy metals from industrial wastewater. Arabian J Chem. 2011;4(4):361–77.
[34] Peng C, Jin R, Li G, Li F, Gu Q. Recovery of nickel and water from wastewater with electrochemical combination process. Sep Purif Technol. 2014;136:42–9.
[35] Naseem R, Tahir SS. Removal of Pb(II) from aqueous solution by using bentonite as an adsorbent. Water Res. 2001;35:3982–6.
[36] Schmidt B, Wolters R, Kaplin J, et al. Rinse water regeneration in stainless steel pickling. Desalination. 2007;211:64–71.
[37] Machado RM, Gameiro ML, Rodrigues JM, Ismael MR, Reis MT, Carvalho JM. Recovery of hydrochloric acid from galvanizing industrial effluents. Sep Sci Technol. 2017;52:1333–40.
[38] Carrera JA, Bringas E, San Román MF, Ortiz I. Selective membrane alternative to the recovery of zinc from hot-dip galvanizing effluents. J Membr Sci. 2009;326:672–80.
[39] Carrillo-Abad J, Garcia-Gabaldon M, Ortiz-Gandara I, et al. Selective recovery of zinc from spent pickling baths by the combination of membrane-based solvent extraction and electrowinning technologies. Sep Purif Technol. 2015;151:232–42.
[40] Mansur MB, Rocha SD, Magalhães FS, Benedetto JD. Selective extraction of zinc(II) over iron(II) from spent hydrochloric acid pickling effluents by liquid–liquid extraction. J Hazard Mater. 2008;150:669–78.
[41] Kumari S, Agrawal A, Bagchi D, Kumar V, Pandey BD. Synthesis of copper metal/salts from copper bleed solution of a copper plant. Min Proc Ext Met Rev. 2006;27:159–75.
[42] Dutra AJ, Rocha GP, Pombo FR. Copper recovery and cyanide oxidation by electrowinning from a spent copper-cyanide electroplating electrolyte. J Hazard Mater. 2008;152:648–55.
[43] Frenzel I, Holdik H, Stamatialis DF, Pourcelly G, Wessling M. Chromic acid recovery by electro-electrodialysis I. Evaluation of anion-exchange membrane. J Membr Sci. 2005;261:49–57.
[44] Sahu SK, Verma VK, Bagchi D, Kumar V, Pandey BD. Recovery of chromium(VI) from electroplating effluent by solvent extraction with tri-n-butyl phosphate. Indian J Chem Technol. 2008;15:397–402.
[45] Chang WS, Chen SS, Sie CH, Nguyen NC, Cheng HH, Hsu HT. Recovery of chromium from plastic plating wastewater by cetyltrimethylammonium bromide MEUF and electrodialysis. Desalin Water Treat. 2014;55:2408–15.
[46] Kul M, Cetinkaya U. Recovery of nickel by solvent extraction from electroplating rinse bath solution. Solvent Extr Ion Exch. 2010;28:225–43.
[47] Dermentzis K, Valsamidou E, Lazaridou A, Kokkinos NC. Nickel removal from wastewater by electrocoagulation with aluminum electrodes. J Eng Sci Tech Rev. 2011;4(2):188–92.
[48] Singh V, Ram C, Kumar A. Physico-chemical characterization of electroplating industrial effluents of Chandigarh and Haryana Region. J Civil Environ Eng. 2016;6:2–6.
[49] Dufour J, Negro C, Heras F, López-Mateos F. Recovery of the metals from pickling liquors of stainless steel by precipitation methods. ISIJ Int. 2001;41:801–6.
[50] Chen Q, Luo Z, Hill C, Xue G, Tyrer M. Precipitation of heavy metals from wastewater using simulated flue gas: sequent additions of fly ash, lime and carbon dioxide. Water Res. 2009;43:2605–14.
[51] Huisman JL, Schouten G, Schultz C. Biologically produced sulphide for purification of process streams, effluent treatment and recovery of metals in the metal and mining industry. Hydrometallurgy. 2006;83:106–13.

[52] Baltpurvins KA, Burns RC, Lawrance GA, Stuart AD. Use of the solubility domain approach for the modeling of the hydroxide precipitation of heavy metals from wastewater. Environ Sci Technol. 1996;30:1493–9.
[53] Dahlgren L. Treatment of spent pickling acid from stainless steel production. Master of Science Thesis, Stockholm, Sweden: Royal Institute of Technology, 2010.
[54] Brooks CS. Metal recovery from industrial wastes. Jom. 1986;38:50–7.
[55] Raymond LA, Stephen LJ. Acid recycle process with iron removal. US patent 20090145856, 2009.
[56] Mughal ST, Shafique T, Aftab T, Bashir F. Treatment of electroplating effluent. J Chem Soc Pak. 2008;30:29–32.
[57] Hoak DR, Lewis CJ. Treatment of spent pickling liquor with limestone and lime. Ind Eng Chem. 1945;37:553–9.
[58] Hoak RD, Lewis CJ, Sindlinger CJ, Klein B. Lime treatment of waste pickle liquor. Ind Eng Chem. 1947;39:131–5.
[59] Hoak DR, Lewis CJ. Pickle liquor neutralization: economic and technology factor. Ind Eng Chem. 1948;40:2062–7.
[60] Hoak RD, Sindlinger CJ. New technique for waste pickle liquor neutralization. Ind Eng Chem. 1949;41:65–70.
[61] Lewis CJ, Boynton RS. Acid neutralization with lime for environment control and manufacturing process. Bull no 16. National Lime Association. Available at: http://www.lime.org/documents/publications/free_downloads/Acid_Neutralization_with_Lime.pdf. Accessed: 30 Oct 2017.
[62] Devi A, Singhal A, Gupta R. Treatment of spent pickling liquor. 4th Int Conf Rec Inov Sci Eng Manag 2016:287–92.
[63] Forsell BA, Niklasson RJ Process for neutralization and regeneration of aqueous solution of acids and dissolved metals. US Patent 3800024 1974.
[64] Wang K, Li J, McDonald RG, Browner RE. The effect of iron precipitation upon nickel losses from synthetic atmospheric nickel laterite leach solutions: statistical analysis and modelling. Hydrometallurgy. 2011;109:140–52.
[65] Fortkamp U, Tjus K, Jansson Å. Factors influencing crystallisation from mixed acid pickling baths for stainless steel. Report of Swedish Environmental Research Institute B1519, Stockholm 2003 Available at: http://www.ivl.se/download/18.343dc99d14e8bb0f58b7439/1445515529160/B1519.pdf. Accessed: 18 Oct 2017.
[66] Forsberg KM, Rasmuson AC. Recycling of waste pickle acid by precipitation of metal fluoride hydrates. Miner Eng. 2007;20:950–5.
[67] Tjus K, Bergström R, Fortkamp U, Forsberg K, Rasmuson A Development of a recovery system for metals and acids from pickling baths using nanofiltration and crystallization. Report of Swedish Environmental Research Institute, 2006 Available at: http://www.ivl.se/download/18.343dc99d14e8bb0f58b74e0/1445515632292/B1692.pdf. Accessed: 18 Oct 2017.
[68] Olsen DR, Blumenschein CD System and method for converting the spent remnants of a first pickling acid solution into a usable second pickling acid solution. Patent US7351391 B1, 2005 Available at: http://www.google.co.id/patents/US7351391. Accessed: 18 Oct 2017
[69] Kehrmann A Method of producing ferrous sulfate heptahydrate. Patent US 7097816, 2006 Available at: https://www.google.ch/patents/US7097816. Accessed: 20 Oct 2017.
[70] Özdemir T, Öztin C, Kincal NS. Treatment of waste pickling liquors: process synthesis and economic analysis. Chem Eng Commun. 2006;193:548–63.
[71] Hermoso J, Dufour J, Galvez JL, Negro C, Lopez-Mateos F. Nickel hydroxide recovery from stainless steel pickling liquors by selective precipitation. Ind Eng Chem Res. 2005;44:5750–6.
[72] Sartor M, Buchloh D, Rögener F, Reichardt T. Removal of iron fluorides from spent mixed acid pickling solutions by cooling precipitation at extreme temperatures. Chem Eng J. 2009;153:50–5.

[73] Energy-saving regeneration of hydrochloric acid pickling liquor. 2000 Availabe at: https://www.nrel.gov/docs/fy00osti/28232.pdf. Accessed: 18 Oct 2017.

[74] Brown M. Sheedy, process and apparatus for regeneration of volatile acids. Patent WO1995004844 A1 1995 Feb 16 Accessed: 8 Oct 2017. Availabe at: https://www.google.com/patents/WO1995004844A1?hl=pt-PT&en.

[75] Brown CJ, Sheedy M. The Fluorex process for regeneration of nitric/hydrofluoric stainless steel pickle liquors. In: Dutrizac JE, Monhemius AJ, editors. Proceedings of the 2nd International Symposium on Iron Control in Hydrometallurgy. Ottawa, 1996:457–9.

[76] Lunner SE, Guozhu Y, Method in connection with steel production. Patent US8246716 B2, 21 Aug 2012.

[77] Blomquist SI, Markku TS, Pouru HJ, Nyman BG, Lindroos KH, Koivunen TT, et al. Method for regenerating pickling acids. Patent US4526650 A, 2 July 1985. Available at: http://www.google.ch/patents/US4526650. Accessed: 14 Oct 2017.

[78] Baerhold F, Mitterecker S. Acid recovery in the steel and metallurgical industry process characteristics and fuel saving options. In Canadian Institute of Mining, Metallurgy and Petroleum, ed., IMPC 2016: XXVIII International Mineral Processing Congress Proceedings. ISBN: 978-1-926872-29-2.

[79] Baerhold F, Lebl A, Stratrecevic J. Recycling of spent acids and iron via pyrohydrolisis. In: Dutrizac JE, Riveros PA, editors. Iron control technologies. Proc of the 3rd Intern Symp on Iron Control In Hydrom. Montreal, 2006:789–803.

[80] Adham K, Lee C, Small D. Energy consumption for iron chloride pyrohydrolisis: a comparison between fluidized beds and spay roasters. In: Dutrizac JE, Riveros PA, editors. Iron control technologies. Proc of the 3rd Intern Symp on Iron Control in Hydrom. Montreal, Canada, 2006:815–29.

[81] Ferreira AS, Mansur MB. Statistical analysis of the spray roasting operation for the production of high quality Fe2O3 from steel pickling liquors. Process Saf Environ Prot. 2011;89:172–8.

[82] Muthu N, Aziz FA, Yusuff RB. Minimization of spent acid waste from galvanizing plant in Malaysia. Global J Res Eng Civil Struct Eng. 2013;13(6). Online ISSN: 2249-4596 & Print ISSN: 0975-5861.

[83] Schiemann M, Wirtz S, Scherer V, Bärhold F. Spray roasting of iron chloride FeCl2: numerical modelling of industrial scale reactors. Powder Technol. 2013;245:70–9.

[84] Bascone D, Cipollina A, Morreale M, Randazzo S, Santoro F, Micale G. Simulation of a regeneration plant for spent pickling solutions via spray roasting. Desalin Water Treat. 2016;57:23405–19.

[85] Boricha AG, Murthy GV. Preparation, characterization and performance of nanofiltration membranes for the treatment of electroplating industry effluent. Sep Pur Techn. 2009;65:282–9.

[86] Hosseini SS, Bringas E, Tanc NR, Ortiz I, Ghahramania M, Shahmirzadia MA. Recent progress in development of high performance polymeric membranes and materials for metal plating wastewater treatment: a review. J Water Proc Eng. 2016;9:78–110.

[87] Bonner FJ, Donatelli AA. Diffusion dialysis and acid recovery in metal operations. Technical Report No. 49 of The Massachusetts Toxics Use Reduction Institute, University of Massachusetts Lowell, 2001 Available at: https://www.turi.org/content/download/8148/139602/file/2001%20Report%2049%20Bonner%20and%20Donatelli%20-%20Diff%20Dialysis%20and%20Acid%20Recov%20in%20Metal%20Ops.pdf. Accessed: 24 Oct 2017.

[88] Bailey DE Recycling of nitric-hydrofluoric acid with diffusion dialysis. Available at: http://www.mech-chem.com/images/stories/Acid-Brochure/Nitric-HF.pdf. Accessed: 24 Oct 2017.

[89] Bailey DE. Optimizing anodizing baths with diffusion dialysis. Metal Finishing. 1998;14–7.

[90] Negro C, Blanco M, Lopez-Mateos F, et al. Free acids and chemicals recovery from stainless steel pickling baths. Sep Sci Technol. 2001;36:1543–56.

[91] Jeong J, Kim MS, Kim BS, Kim SK, Kim WB, Lee JC. Recovery of H2SO4 from waste acid solution by a diffusion dialysis method. J Hazard Mater. 2005;124:230–5.

[92] Wei C, Li X, Deng Z, Fan G, Li M, Li C. Recovery of H2SO4 from an acid leach solution by diffusion dialysis. J Hazard Mater. 2010;176:226–30.

[93] Luo J, Wu C, Wu Y, Xu T. Diffusion dialysis of hydrochloric acid with their salts: effect of co-existence metal ions. Sep Purif Technol. 2013;118:716–22.

[94] Acid Recovery Systems. AP-300 to AP-3000 (Large Scale Series), Mech-Chem Associates, Inc. Available at: http://www.mech-chem.com/images/stories/brochures/PDF/Acid-Recovery-Units-Large%20Scale.pdf. Accessed: 24 Oct 2017.

[95] Luo J, Wu C, Wu Y, Xu T. Diffusion dialysis-concept, principle and applications. J Membr Sci. 2011;366:1–16.

[96] Baltazar V, Harris GB, White CW. The selective recovery and concentration of sulphuric acid by electrodialysis. Hydrometallurgy. 1991;30:463–81.

[97] Scarazzato T, Buzzi DC, Bernardes AM, Romano Espinosa DC. Treatment of wastewaters from cyanide-free plating process by electrodialysis. J Clean Prod. 2015;91:241–50.

[98] Dzyazko Y, Belyakov V. Purification of a diluted nickel solution containing nickel by a process combining ion exchange and electrodialysis. Desalination. 2004;162:179–89.

[99] Benvenuti T, Krapf RS, Rodrigues MA, Bernardes AM, Zoppas-Ferreira J. Recovery of nickel and water from nickel electroplating wastewater by electrodialysis. Sep Purif Technol. 2014;129:106–12.

[100] Benvenuti T, Siqueira Rodrigues MA, Bernardes AM, Zoppas-Ferreira J. Closing the loop in the electroplating industry by electrodialysis. J Clean Prod. 2017;155:130–8.

[101] Tran AT, Mondal P, Lin J, Meesschaert B, Pinoy L, Van Der Bruggen B. Simultaneous regeneration of inorganic acid and base from a metal washing step wastewater by bipolar membrane electrodialysis after pretreatment by crystallization in a fluidized pellet reactor. J Membr Sci. 2015;473:118–27.

[102] Lu H, Wang Y, Wang J. Removal and recovery of Ni^{2+} from electroplating rinse water using electrodeionization reversal. Desalination. 2014;348:74–81.

[103] Lu H, Wang Y, Wang J. Recovery of Ni^{2+} and pure water from electroplating rinse wastewater by an integrated two-stage electrodeionization process. J Clean Prod. 2015;92:257–66.

[104] Eberhard FS, Hamawand I. Selective electrodialysis for copper removal from brackish water and coal seam gas water. Int J Environ Res. 2017;11:1–11.

[105] Caprarescu S, Purcar V, Sârbu A, Radu A, Ghiureab M, Maiora I. The use of electrodialysis for Cu^{2+} removal from wastewater. Rev Roum Chim. 2014;59:639–44.

[106] Peng C, Liu Y, Bi J, Xu H, Ahmed AS. Recovery of copper and water from copper-electroplating wastewater by the combination process of electrolysis and electrodialysis. J Hazard Mat. 2011;189:814–20.

[107] Cifuentes L, Crisostomo G, Alvarez F, Casas J, Cifuentes G. The use of electrodialysis for separating and concentrating chemical species in acidic Cu,Fe,As,Sb electrolytes. In: Dutrizac JE, Ji J, Ramanachandran V, editors. Electrorefining and electrowinning of copper. 4th International Conference Copper 99, vol. III. Warrendale, Pennsylvania, 479–94.

[108] Cifuentes L, Mondaca C, Casas JM. The effectiveness of membrane systems for the separation of anolyte and catholyte in a lab-scale copper electrowinning cell based on reactive electrodialysis. Minerals Eng. 2004;17:803–9.

[109] Cifuentes L, Garcia I, Ortiz R, Casas JM. The use of electrohydrolysis for the recovery of sulphuric acid from copper-containing solutions. Sep Purif Technol. 2006;50(2):167–74.

[110] Cifuentes L, García I, Arriagada P, Casas JM. The use of electrodialysis for metal separation and water recovery from CuSO4–H2SO4–Fe solutions. Sep Purif Technol. 2009;65:105–8.

[111] Frenzel I, Stamatialis DF, Wessling M. Water recycling from mixed chromic acid waste effluents by membrane technology. Sep Purif Technol. 2006;49:76–83.

[112] Nataraj SK, Hosamani KM, Aminabhavi TM. Potential application of an electrodialysis pilot plant containing ion-exchange membranes in chromium removal. Desalination. 2007;217:181–90.

[113] Moura RC, Bertuol DA, Ferreira CA, Amado FD. Study of chromium removal by the electrodialysis of tannery and metal-finishing effluents. Int J Chem Eng. 2012. Article ID 79312. DOI: 10.1155/2012/179312

[114] Tor A, Buyukerkek T, Cengelolu Y, Ersoz M. Simultaneous recovery of Cr(III) and Cr(VI) from the aqueous phase with ion-exchange membranes. Desalination. 2005;171:233–41.

[115] Chaudhary AJ, Ganguli B, Grimes SM. The regeneration and recycle of chromium etching solutions using concentrator cell membrane technology. Chemosphere. 2006;62:841–6.

[116] Chen SS, Li CH, Hsu HD, Lee PC, Chang YM, Yang CH. Concentration and purification of chromate from electroplating wastewater by two-stage electrodialysis processes. J Hazard Mater. 2009;161:1075–80.

[117] Khan J, Tripathi BP, Saxena A, Shahi VK. Electrochemical membrane reactor: in situ separation and recovery of chromic acid and metal ions. Electrochim Acta. 2007;52:6719–27.

[118] Cherif AT, Elmidaoui A, Gavach C. Separation of Ag^{+}, Zn^{2+} and Cu^{2+} ions by electrodialysis with a monovalent cation specific membrane and EDTA. J Membr Sci. 1993;76:39–49.

[119] Kubal M, Machula T, Strnadová N. Separation of calcium and cadmium by electrodialysis in the presence of ethylenediaminetetraacetic acid. Sep Sci Technol. 1998;33:1969–80.

[120] Taghdirian HR, Moheb A, Mehdipourghazi M. Selective separation of Ni(II)/Co(II) ions from dilute aqueous solutions using continuous electrodeionization in the presence of EDTA. J Membr Sci. 2010;362:68–75.

[121] Frioui S, Oumeddour R, Lacour S. Highly selective extraction of metal ions from dilute solutions by hybrid electrodialysis technology. Sep Purif Technol. 2017;174:264–74.

[122] Babilas D, Dydo P. Selective zinc recovery from electroplating wastewaters by electrodialysis enhanced with complex formation. Sep Purif Technol. 2017;192:419–28.

[123] Oztekin E, Altin S. Wastewater treatment by electrodialysis system and fouling problems. The Online J of Sci Techn. 2016;6(1). Available at: https://www.tojsat.net/journals/tojsat/articles/v06i01/v06i01-14.pdf. Accessed: 25 Oct 2017.

[124] Nagarale RK, Gohil GS, Shahi VK. Recent developments on ion-exchange membranes and electro-membrane processes. Adv Colloid Interface Sci. 2006;119:97–130.

[125] Machado RM, Gameiro ML, Krupa M, et al. Selective separation and recovery of zinc and lead from galvanizing industrial effluents by anion exchange. Sep Sci Technol. 2015;50:2726–36.

[126] Dizge N, Keskinler B, Barlas H. Sorption of Ni(II) ions from aqueous solution by Lewatit cation-exchange resin. J Hazard Mater. 2009;167:915–26.

[127] Marañón E, Suárez F, Alonso F, Fernández Y, Sastre H. Preliminary study of iron removal from hydrochloric pickling liquor by ion exchange. Ind Eng Chem Res. 1999;38:2782–6.

[128] Marañón E, Fernández Y, Suárez FJ, Alonso FJ, Sastre H. Treatment of acid pickling baths by means of anionic resins. Ind Eng Chem Res. 2000;39:3370–6.

[129] Csicsovszki G, Kekesi T, Török TI. Selective recovery of Zn and Fe from spent pickling solutions by the combination of anion exchange and membrane electrowinning techniques. Hydrometallurgy. 2005;77:19–28.

[130] Miesiąc I. Removal of zinc(II) and iron(II) from spent hydrochloric acid by means of anionic resins. Ind Eng Chem Res. 2005;44:1004–11.

[131] Priya PG, Ahmed Basha C, Ramamurthi V, Begum SN. Recovery and reuse of Ni(II) from rinse-water of electroplating industries. J Hazard Mater. 2009;163:899–909.

[132] Karegar S, Bhargavi M, Divekar SV. Treatment of wastewater from chrome plating industry by ion exchange method. IJRET: Intern J Res Eng Techn. 2015;4:393–401.

[133] Revathi M, Saravanan M, Chiya AB, Velan M. Removal of copper, nickel, and zinc ions from electroplating rinse water. Clean Soil Air Water. 2012;40:66–79.
[134] Papadopoulos A, Fatta D, Parperis K, Mentzis A, Haralambous KJ, Loizidou M. Nickel uptake from a wastewater stream produced in a metal finishing industry by combination of ion-exchange and precipitation methods. Sep Purif Technol. 2004;39:181–8.
[135] Juang RS, Kao HC, Chen W. Column removal of Ni(II) from synthetic electroplating waste water using a strong-acid resin. Sep Purif Technol. 2006;49:36–42.
[136] Revathi M, Basha CA, Velan M. Removal of copper(II) ions from synthetic electroplating rinse water using polyethyleneimine modified ion-exchange resin. Desalination Water Treat. 2016;57:20350–67.
[137] Reynolds D Apparatus and process to regenerate trivalent chromium bath. US Patent US5269905, 1993.
[138] Gode F, Pehlivan E. A comparative study of two chelating ion exchange resins for the removal of chromium(III) from aqueous solution. J Hazard Mater. 2003;100:231–43.
[139] Fernández-Olmo I, Ortiz A, Urtiaga A, Ortiz I. Selective iron removal from spent passivation baths by ion exchange. J Chem Technol Biotechnol. 2008;83:1616–22.
[140] Ortiz A, Fernández-Olmo I, Urtiaga A, Ortiz I. Modeling of iron removal from spent passivation baths by ion exchange in fixed-bed operation. Ind Eng Chem Res. 2009;48:7448–52.
[141] Agrawal A, Kumari S, Ray BC, Sahu KK. Extraction of acid and iron values from sulphate waste pickle liquor of a steel industry by solvent extraction route. Hydrometallurgy. 2007;88:58–66.
[142] Benedetto J, Morais CA. In. Proceedings of VI Southern Hemisphere Meeting on Mineral Technology, Rio de Janeiro: CETEM/MCT, 2001:410–5.
[143] Agrawal A, Sahu KK. Value addition to sulfate waste pickle liquor of steel industry using hydrometallurgical processes. Metall Mater Trans B. 2009;40B:877–85.
[144] Lobo-Recio MA, Alguacil FJ, Lopez-Delgado A. Processing of steel rinse waters through coextraction and selective stripping. AIChE J. 2004;50:1150–5.
[145] Watanabe T, Hoshino M, Uchino K, Nakazato Y. A new recovery process for stainless steel pickling waste liquor with an iron removal process using solvent extraction. In: Dutrizac JE, Ed. Iron control in hydrometallurgy. Toronto: Ellis Howard Ltd; 1986. p. 549–70.
[146] Agrawal A, Sahu KK. Studies on solvent extraction of iron(III) as a step for conversion of a waste effluent to a value added product. J Environ Manage. 2011;92:3105–11.
[147] Lum KH, Stevens GW, Perera JM, Kentish SE. The modelling of $ZnCl_2$ extraction and HCl co-extraction by TBP diluted in ShellSol 2046. Hydrometallurgy. 2013;133:64–74.
[148] Flett DS. Solvent extraction in hydrometallurgy: the role of organophosphorus extractants. J Organomet Chem. 2005;690:2426–38.
[149] Cook SJ, Perera JM, Stevens GW, Kentish SE. The screening of extractants for the separation of Zn(II) from Australian hot-dip galvanizing effluent. Sep Sci Technol. 2011;46:2066–74.
[150] Morris DC, Short EL. Zinc chloride and zinc bromide complexes. Part II. solvent-extraction studies with zinc-65 as tracer. J Chem Soc. 1962;2662–71.
[151] Wieszczycka K. Recovery of Zn(II) from multielemental acidic chloride solution with hydrophobic 3-pyridineketoxime. Sep Purif Technol. 2013;114:17–23.
[152] Cierpiszewski R, Miesiąc I, Regel-Rosocka M, Sastre AM, Szymanowski J. Removal of zinc(II) from spent hydrochloric acid solutions from zinc hot galvanizing plants. Ind Eng Chem Res. 2002;41:598–603.
[153] Dessouky SI, El-Nadi YA, Ahmed IM, Saad EA, Daoud JA. Solvent extraction separation of Zn(II), Fe(II), Fe(III) and Cd(II) using tributylphosphate and CYANEX 921 in kerosene from chloride medium. Chem Eng Process. 2008;47:177–83.
[154] Grzeszczyk A, Regel-Rosocka M. Extraction of zinc(II), iron(II) and iron(III) from chloride media with dibutyl butylphosphonate. Hydrometallurgy. 2007;86:72–9.

[155] Lum KH, Stevens GW, Kentish SE. Development of a process for the recovery of zinc sulphate from hot-dip galvanizing spent pickling liquor via two solvent extraction steps. Hydrometallurgy. 2014;142:108–15.

[156] Regel-Rosocka M, Sastre AM, Szymanowski J. Recovery of zinc(II) from HCl spent pickling solutions by solvent extraction. Environ Sci Technol. 2001;35:630–5.

[157] Gameiro ML, Machado RM, Rodrigues JM, Ismael MRC, Reis MTA, Carvalho JMR. Recovery of zinc from HCl industrial effluents, Proc. of the 20th Int. Solv. Ext. Conf. (ISEC 2014), 934–940, Würzburg, Germany, 7–11 Sept, 2014.

[158] Chauhan S, Patel T. A review on solvent extraction of nickel. Inter J Eng Res Techn (IJERT). 2014;3(9):1315–22.

[159] Sana T, Shiomori K, Kawano Y. Extraction rate of nickel with 5-dodecylsalicylaldoxime in a vibro-mixer. Sep Purif Technol. 2005;44:160–5.

[160] Tanaka M, Huang Y, Yahagi T, Hossain MK, Sato Y, Narita H. Solvent extraction recovery of nickel from spent electroless nickel plating baths by a mixer-settler extractor. Sep Purif Technol. 2008;62:97–102.

[161] Huang Y, Tanaka M. Analysis of continuous solvent extraction of nickel from spent electroless nickel plating baths by a mixer-settler. J Hazard Mater. 2009;164:1228–35.

[162] Agrawal A, Kumar V, Pandey BD. Remediation options for the treatment of electroplating and leather tanning effluent containing chromium – a review. Mineral Proc Extract Metall Rev. 2006;27:99–130.

[163] El-Hussaini OM, Lasheen TA, Helmy EM, Hady MA, Mana AA. Liquid-liquid extraction/recovery of chromium(VI) from some industrial waste solutions using Alamine 336 in kerosene. J Dispers Sci Technol. 2012;33:1179–87.

[164] Agrawal A, Pal C, Sahu KK. Extractive removal of chromium(VI) from industrial waste solution. J Hazard Mater. 2008;159:458–464.

[165] Bachmann RT, Wiemken D, Tengkiat AB, Wilichowski M. Feasibility study on the recovery of hexavalent chromium from a simulated electroplating effluent using Alamine 336 and refined palm oil. Sep Purif Technol. 2010;75:303–9.

[166] El-Nadi YA, El-Hefny NE. Removal of iron from Cr-electroplating solution by extraction with di (2-ethylhexyl) phosphoric acid in kerosene. Chem Eng Process Process Intensif. 2010;49:159–64.

[167] Kul M, Oskay KO. Separation and recovery of valuable metals from real mix electroplating. Hydrometallurgy. 2015;155:153–60.

[168] Kul M, Cetinkaya U. Recovery of copper by LIX 984N-C from electroplating rinse bath solution. Hydrometallurgy. 2009;98:86–91.

[169] Sze YK, Xue L. Extraction of zinc and chromium(III) and its application to treatment of alloy electroplating wastewater. Sep Sci Technol. 2003;38(2):405–25.

[170] Singh R, Khwaja AR, Gupta B, Tandon SN. Extraction and separation of nickel(II) using bis (2,4,4-trimethyl pentyl) dithiophosphinic acid (CYANEX 301) and its recovery from spent catalyst and electroplating bath residue. Solvent Extr Ion Exch. 1999;17:367–90.

[171] Venkateswaran P, Palanivelu K. Solvent extraction of hexavalent chromium with tetrabutyl ammonium bromide from aqueous solution. Sep Purif Technol. 2004;40:279–84.

[172] Szymanowski J. Hydroxyoximes and copper hydrometallurgy. CRC Press, 1993.

[173] Sze YK, Wong JC. A study of a solvent extraction method for the regeneration of ammoniacal etching solutions of copper. Environ Technol. 1994;15:785–93.

[174] Schwab W, Kehl R. Copper recycling from spent ammoniacal etching solutions and their regeneration by solvent extraction. Proc XX Int Mineral Process Congr. 1997;4:285–92.

[175] Kyuchoukov G, Bogacki MB, Szymanowski J. Copper extraction from ammoniacal solutions with LIX 84 and LIX 54. Ind Eng Chem Res. 1998;37:4084–9.

[176] Alguacil FJ, Alonso M. Recovery of copper from ammoniacal/ammonium sulfate medium by LIX 54. J Chem Tech Biotechnol. 1999;74:1171–5.
[177] Alguacil FJ, Alonso M. The effect of ammonium sulfate and ammonia on the liquid–liquid extraction of zinc using LIX 54. Hydrometallurgy. 1999;53:203–9.
[178] Ismael MR, Gameiro ML, Carvalho JM. Extraction equilibrium of copper from ammoniacal media with LIX 54. Sep Sci Technol. 2004;39:3859–77.
[179] Gameiro ML, Machado RM, Ismael MR, Reis MT, Carvalho JM. Copper extraction from ammoniacal medium in a pulsed sieve-plate column with LIX 84-I. J Hazard Mater. 2010;183:165–75.
[180] Ochromowicz K, Jeziorek M, Wejman K. Copper(II) extraction from ammonia leach solution. Physicochem Probl Miner Process. 2014;50:327–35.
[181] Flett DS, Melling J. Extraction of ammonia by commercial copper chelating extractants. Hydrometallurgy. 1979;4:135–46.
[182] San Román MF, Bringas E, Ibañez R, Ortiz I. Liquid membrane technology: fundamentals and review of its applications. J Chem Technol Biotechnol. 2010;85:2–10.
[183] Ortiz I, Bringas E, San Román MF, Urtiaga AM. Selective separation of zinc and iron from spent pickling solutions by membrane-base solvent extraction: process viability. Sep Sci Technol. 2004;39:2441–55.
[184] Samaniego H, San Román MF, Ortiz I. Kinetics of zinc recovery from spent pickling effluents. Ind Eng Chem Res. 2007;46:907–12.
[185] Carrera JA, Muñoz E, Bringas E, San Román MF, Ortiz I. Influence of operation variables on the recovery of zinc from spent pickling effluents using the emulsion pertraction technology. Desalination. 2009;245:675–9.
[186] Bringas E, San Román MF, Urtiaga AM, Ortiz I. Membrane contactors (NDSX and EPT): an innovative alternative for the treatment of effluents containing metallic pollutants. Inter J Environ Waste Manage. 2012;9:201–20.
[187] Lum KH, Cook SJ, Stevens GW, Perera JM, Kentish SE. Zinc chloride and hydrochloric acid coextraction from galvanizing pickling waste in the presence of iron(II). Results with hollow fiber membrane contactors. Ind Eng Chem Res. 2014;53:4453–61.
[188] Bringas E, San Román MF, Urtiaga AM, Ortiz I. Integrated use of liquid membranes and membrane contactors: enhancing the efficiency of L–L reactive separations. Chem Eng Process. 2013;67:120–9.
[189] Urtiaga AM, Bringas E, Mediavilla R, Ortiz I. The role of liquid membranes in the selective separation and recovery of zinc for the regeneration of Cr(III) passivation baths. J Membr Sci. 2010;356:88–95.
[190] Diban N, Mediavilla R, Urtiaga A, Ortiz I. Zinc recovery and waste sludge minimization from chromium passivation baths. J Hazard Mater. 2011;192:801–7.
[191] Diban N, García V, Alguacil F, Ortiz I, Urtiaga A. Temperature enhancement of zinc and iron separation from chromium(III) passivation baths by emulsion pertraction technology. Ind Eng Chem Res. 2012;51:9867–74.
[192] García V, Steeghs W, Bouten M, Ortiz I, Urtiaga A. Implementation of an eco-innovative separation process for a cleaner chromium passivation in the galvanic industry. J Clean Prod. 2013;59:274–83.
[193] Padilla JI, Garcia-Diaz I, Urien A, Rodríguez O, Lopez F, Alguacil GJ. Membrane-based extraction with strip/organic dispersion methodologies for metals removal and recovery from wastewaters. Desalination Water Treat. 2012;40:282–97.
[194] Ruey-Shin J, Jiann-Der J. Recovery of nickel from a simulated electroplating rinse solution by solvent extraction and liquid surfactant membrane. J Membr Sci. 1995;100:163–70.

[195] Eyupoglu V, Kumbasar RA. Extraction of Ni(II) from spent Cr–Ni electroplating bath solutions using LIX 63 and 2BDA as carriers by emulsion liquid membrane technique. J Ind Eng Chem. 2015;21:303–10.
[196] Sonawane JV, Pabby AK, Sastre AM. Pseudo-emulsion based hollow fiber strip dispersion (PEHFSD) technique for permeation of Cr(VI) using Cyanex 923 as carrier. J Hazard Mater. 2010;174:541–7.
[197] Rathore NS, Leopold A, Pabby AK, Fortuny A, Coll MT, Sastre AM. Extraction and permeation studies of Cd(II) in acidic and neutral chloride media using Cyanex 923 on supported liquid membrane. Hydrometallurgy. 2009;96:81–7.
[198] Yang Q, Kocherginsky NM. Copper removal from ammoniacal wastewater through a hollow fiber supported liquid membrane system: modeling and experimental verification. J Membr Sci. 2004;234:1–10.
[199] Yang Q, Kocherginsky NM. Big Carrousel mechanism of copper removal from ammoniacal wastewater through supported liquid membrane. Sep Purif Technol. 2007;54:104–16.
[200] Sengupta B, Bhakhar MS, Sengupta R. Extraction of copper from ammoniacal solutions into emulsion liquid membranes using LIX 84 I. Hydrometallurgy. 2007;89:311–8.
[201] Sengupta B, Bhakhar MS, Sengupta R. Extraction of zinc and copper–zinc mixtures from ammoniacal solutions into emulsion liquid membranes using LIX 84I. Hydrometallurgy. 2009;99:25–32.
[202] Gameiro ML, Bento P, Ismael MR, Reis MT, Carvalho JM. Extraction of copper from ammoniacal medium by emulsion liquid membranes using LIX 54. J Membr Sci. 2007;293:151–60.
[203] Gameiro ML, Ismael MR, Reis MT, Carvalho JM. Recovery of copper from ammoniacal medium using liquid membranes with LIX 54. Sep Purif Technol. 2008;63:287–96.
[204] Gameiro ML, Ismael MR, Reis MT, Santos SM, Carvalho JM. Extraction of copper from aqueous solutions by liquid membrane processes. Solvent Extr Ion Exch. 2010;28:85–108.
[205] Gameiro ML, Ismael MR, Reis MT, Carvalho JM. Non-dispersive solvent extraction of copper from ammoniacal solutions. Can Metall Q. 2008;47:301–6.

Katarzyna Staszak

6 Chemical and petrochemical industry

Abstract: The potential sources of various metals in chemical and petrochemical processes are discussed. Special emphasis is put on the catalysts used in the industry. Their main applications, compositions, especially metal contents are presented both for fresh and spent ones. The focus is on the main types of metals used in catalysts: the platinum-group metals, the rare-earth elements, and the variety of transition metals. The analysis suggested that chemical and petrochemical sectors can be considered as the secondary source of metals. Because the utilization of spent refinery catalysts for metal recovery is potentially viable, different methods were applied. The conventional approaches used in metal reclamation as hydrometallurgy and pyrometallurgy, as well as new methods include bioleaching, were described. Some industrial solutions for metal recovery from spent solution were also presented.

Keywords: spent catalyst, metal content, metal recovery

6.1 Introduction

The main sources of metals in chemical and petrochemical industry are spent catalysts. The application of catalysts is widespread in the chemical industry in areas ranging from pharmaceuticals and specialty chemicals to polymers and petroleum processing. Over 60% of all chemicals are produced using catalysts in 90 % of all industries chemical processes [1, 2]. Eighty percent of catalysts are in the solid form, less common are homogeneous and biocatalysts – about 17% and 3 %, respectively [3]. Application of catalysts in large-scale chemical and petrochemical processes has started at the beginning of the twentieth century. Nowadays the researchers still look for the new catalysts with better activity, selectivity and durability. Consequently, compositions of catalysts are improved and changed.

The common utilization of catalytic processes in industry reflects the economic and environmental benefits achieved through catalysis. On the other hand industries generate thousands of tons per year of spent catalysts containing up to 35% valuable, but usually hazardous, metals such as Ni, Co, Mo, Cr, Cu, Zn, Fe, Ti, V, W, Pd, Ru, Rh [4]. The United State Environmental Protection Agency (EPA) classified these materials as "hazardous waste as one posing a substantial or potential hazard to human health and the environment" [5]. The spent catalyst may also be, due to

This article has previously been published in the journal *Physical Sciences Reviews*. Please cite as: Staszak, K. Chemical and petrochemical industry. *Physical Sciences Reviews* [Online] **2018**, *3* (4). DOI: 10.1515/psr-2018-0025

https://doi.org/10.1515/9783110547061-006

their high metal concentrations, considered as a secondary ore. To deal with the spent catalyst problem several methods are proposed: regeneration (to extend their operation life) and reuse, preparation of useful materials using spent catalysts as raw materials, recovery and recycling of valuable metals and disposal in landfills [6, 7]. These procedures are consistent with the waste management hierarchy and are important from the ecological and economical point of view. Their selection depends of the technical possibility and of course cost effectiveness. Recovery of metals from spent catalysts depends on several factors, such as the chemical composition, the price of metals, the environmental directories and the operational costs [6, 8]. Several methods are proposed, generally based on pyrometallurgy and hydrometallurgy (with chemical and microbial leaching) methods. The regeneration of spent catalysts can only be applied for a few times, and on a limited number of catalytic systems. Disposal in landfills is the least preferred due to both decreasing availability of landfill space, as well as the concern for pollutions arising from the possible leaching of heavy metals [9].

In this chapter, the potential sources of spent catalysts in chemical and petrochemical processes were presented. Special emphasis was put on the chemical composition of the catalysts, especially metal contents. The possibility of recovery of metals from spent catalysts by different methods was also described.

6.2 Basic information about catalysis and catalysts

Catalysis is one of the fundamental pillars of the chemical industry. It is also a basic feature of life processes and environmental chemistry. These technologies help to minimize the consumption of raw materials in the chemical, petroleum and other industries, to process renewable raw materials, such as biomass, into valuable chemicals and to reduce or eliminate the application and generation of hazardous substances. Moreover, new catalytic technologies help to reduce gas pollution (e.g. H_2S in gas and oil mining and by CH_4 in coal mining, NO_x from flue gases of power plants) and protect the ozone layer [10, 11]. Catalysts have been used for large-scale productions of chemicals since the beginning of the twentieth century. The oldest examples are Pt for SO_2 oxidation, Fe for NH_3 synthesis, and Zn–Cr oxide for methanol synthesis. The history of catalysis and its contribution to chemical industries has already been widely described in the literature [2, 12–17].

The International Union of Pure and Applied Chemistry (IUPAC) has defined a catalyst as “a substance that increases the rate of a reaction without modifying the overall standard Gibbs energy change in the reaction” [18]. The chemical process of increasing the reaction rate is called catalysis. Because the catalyst is restored after each catalytic reaction, it is both a reactant and a product of the reaction which doesn’t influence on the thermodynamical equilibrium composition of the reaction.

The most important characteristics of catalysts are as follows [2]:

- Activity – catalyst allows occurring of reaction. With a more active catalyst, the productivity is higher and/or production rate per volume of reactor becomes larger. Thus, the reactor volume can be made smaller and the milder reaction conditions can be used.
- Selectivity (chemical, stereo or region) – with using catalyst the selection of one (or more) desirable reactions among many reactions that would possibly occur is enabled. The catalyst allows choosing a reaction which produces a thermodynamically unfavorable but valuable product. Another function of catalyst is stereoselectivity that produces one of the two stereoisomers which have thermodynamically the same stability. Higher selectivity of reactions allows lowering the costs of separation, purification and waste treatments and the lower the amount of reagents necessary.
- Durability (catalyst life) – the performance of catalyst changes (usually it gradually deteriorates) during their use, and after a certain period, it has to be substituted by a fresh catalyst or reactivated by an appropriate method. Commercial catalyst usually has a lifespan from several months to 10 years.

These three functions are main properties of a catalyst which determine its choice in the appropriate process. Moreover, other very important properties include the ease of regeneration of spent catalysts (in order to increase the lifetime and to reduce the problems related to spent catalyst disposal), the toxicity (the easy and safety in handling, and the disposal of spent catalysts) and the price in all life cycle [19].

Catalysts can be classified in several ways. It is possible to classify the catalysts on the basis of the type of catalysis process which occur, the kind of main components or key-properties responsible for their catalytic behavior. The catalysis, as well as catalyst, can be classified according to the phases involved in the process. When the reactants, the products and the catalyst are in the same phase, usually the liquid one, it is homogeneous catalysis. The catalysts, mainly acids, bases, salts or organometallic compounds, are dissolved in a solvent (it can be also the reactant or product). In heterogeneous catalysis the catalyst, the reactants and the products are in different phases. Usually, the catalyst is a solid, mainly inorganic solids, such as metal oxides, sulfides or chlorides, or organic solids, such as modified polymers, and the reactants and products are in the liquid or vapor phase. This type of catalyst can be used in the following forms [19]:
- as a powder, in slurry reactors, in the presence of a liquid phase;
- in the form of pellets, in the presence of a liquid phase in trickle-bed reactors;
- as pellets in the presence of gaseous reactants in fixed-bed reactor and
- in small-sized particles in fluid- or transport-bed reactors.

The catalyst's material is characterized by the relative amount of several components, its shape, size, pore size distribution and surface area. Wherein, the catalytic performance of heterogeneous catalyst is affected by the type of main components (see below) and its loading of supported catalysts, type of catalyst (supported, powder, skeletal); type of support (active carbon, alumina, silica) [20]. Important parameters for the active metal are the surface area, the dispersion (typically only 10–60% of the metal atoms are exposed), the size of the crystallites (typically in the range 20–200 Å), the location in the pores of the support and oxidation state (reduced or unreduced). Important support parameters are the particle size (for slurry catalysts typically 1–100mm), the surface area (typically in the range of 100–1500m^2/g), the pore structure (pore volume, pore size distribution) and acid–base properties [21].

The heterogeneous catalysis is the most widely used in industry (about 80%). The main reason of this is the easier separation of catalysts from reactants and products after the separation, regeneration and recovery of catalysts is also easier, in comparison to the homogeneous catalysis. Moreover, there is no problem with the possibility of contamination of the products by the catalyst, like in homogeneous catalysis.

Classification of catalysts according to the kind of main component with examples is given [2]:
- metal with or without support – Pd, Pt, Pd-Re, Pd/carbon, Raney Ni
- metal oxides of transition metals – Mo, Co, V, Cr, Fe oxides
- metal coordination compounds (molecular) – Rh, Ru, Pd complexes
- metal sulfides – Mo, Co sulfides
- metal halides, sulfates, etc. – Cu, Pd chlorides
- typical elements – zeolites, Al_2O_3
- biocatalysts with or without support – enzyme, bioorganisms
- organic molecule – organic molecule without metals
- organic polymers – ion-exchange resins.

Four main classes of the catalysts, based on the key-properties responsible for their catalytic behavior can be presented as [19]:
- Redox catalysts – for oxidation, hydrogenation, dehydrogenation and halogenation reactions. These catalysts include the transition elements (in the form of salts, complexes, sulfides, chlorides, oxides or metals) as the main component. Exceptions are: Ag, Cu, ZnO, SbO_x, and $SbCl_x$.
- Acid–base catalysts – for alkylation, dehydration, hydration, oligomerization, cracking, H-transfer and isomerization reactions. The key feature of all the catalysts for these reactions is their acidity or basicity.
- Polyfunctional catalysts – for reforming, oligomerization and aromatization of paraffins. They are a mixture of redox and acid–base catalysts.
- Catalysts for the transformation of CO – It is the peculiar reactivity of CO that makes it possible to separate the catalysts for its transformation into a separate class.

As it was mentioned above commercial catalyst has limited durability, usually from several months to 10 years, depending on the process and type of catalyst. In the case of cracking catalysts, catalyst mortality may be in the order of seconds, while in ammonia synthesis the iron catalyst may last for 5–10 years [22]. The loss of catalytic activity is a big problem in industrial catalytic processes. It is assumed that costs of catalyst replacement are billions of dollars per year. The main reasons of this are several factors as deactivation by poisonous matters (poisons). These compounds may come from feedstock, products of reactions or from apparatus. They gradually cover active sites of catalyst. Moreover, the deposits of coke or carbonaceous matters on active sites of catalysts can occur. They are formed by the condensation/polymerization of reactants, intermediates, and/or products. Sintering of active components and support materials during the process, as well as mechanical destruction of catalyst due to compression, abrasion, etc. and chemical deterioration and vaporization (or dissolution) of active components and supports are also possible. The problem of catalysts durability has already been widely described in the literature [2, 23–29].

6.3 Chemical and petrochemical catalysts

Chemical and petrochemical processes are connected each other. In the petrochemical industry, the basic organic chemicals are produced as methanol, olefins (ethylene, propylene, butadiene), aromatics (benzene, toluene, xylenes). These chemicals are precursors to a variety of chemical products and are generally referred to as primary petrochemicals. Both chemical and petrochemical industries are a large-scale in the word. This sector produces thousands of products from polymers to pharmaceuticals and specialty chemicals. How great the scale of this industry is, say, the number of refineries – over 700 oil refineries all over the world today [30, 31]. Most of these refineries are located in Asia-Pacific and North America regions. About 50% of world crude distillation capacity (81.9 million b/d) is processed in these refineries. Refineries in Western Europe (105 refineries, about 18% of world crude distillation capacity) have the largest average refining capacity in the world (139,000 b/d) [32, 33]. Catalytic processes represent about 83% of the world crude distillation capacity, distributed as 45% hydrotreating (HDT), 17% fluid catalytic cracking (FCC), 14% naphtha reforming (N-REF), 5% hydrocracking (HCK) and 2% isomerization (ISOM) [33]. These proportions have remained practically steady during time. In chemical synthesis, catalysts are used both in organic and nonorganic chemical industries, from simple oxidation reactions, e.g. SO_2 or NO oxidations, synthesis of basic chemicals as nitric or sulfuric acid, ammonia or methanol, to fine compounds.

The variety of catalysts and their widespread use in industry makes spent catalysts a valuable source of metals. Due to the considerable differences in their composition, depending on the process, it is not possible to clearly indicate the methods of their recovery. Therefore, in order to make it easier to find potential sources of metals in the following sections, catalysts are divided into main components. If it was possible, the total compositions and their specific application in the process are given. In view of the fact that catalysts are largely protected by patentability, it is not always possible to accurately analyze the content of individual metals. In addition, the composition varies depending on the degree of durability of the catalyst. These factors can determine the different compositions in the tables presented below.

6.3.1 The platinum-group metals (PGMs)

The six PGMs consist of the late second- and third-row transition series elements – platinum (Pt), palladium (Pd), ruthenium (Ru), rhodium (Rh), iridium (Ir) and osmium (Os). The platinum metals and their complexes have many useful catalytic properties include resistance to chemical attack, excellent high-temperature characteristics, stable electrical properties, high stability in various oxidation states, functional-group tolerance and their ability to catalyze reactions under milder conditions with higher selectivity as compared to other metals [34]. The first industrial application of PGMs in chemical industry was platinum used in process of oxidation of ethanol to acetic acid or acetaldehyde, the oxidation of sulfur dioxide to sulfuric acid (the contact process) and catalytic combustion. Typical catalysts employed in the chemical processing industry incorporate PGMs which are deposited on catalyst supports such as soluble or insoluble alumina, silica/alumina or zeolites. The compositions of the catalysts with PGMs, for various processes are listed in Table 6.3. Using PGM complexes as a catalyst the large variety of reactions is proposed including hydrogenations, hydrosilylations, oxidations, dehydrogenations, hydrogenolysis, carbon-carbon and carbon-heteroatom coupling, carbonylations, and hydroxylations [35]. From PGMs the most commonly used in chemical industry is platinum, the least – osmium (see Figure 6.1). Such a distribution of metal consumption in processes is dictated by many factors. It's not just about their properties, but also their availability and the price.

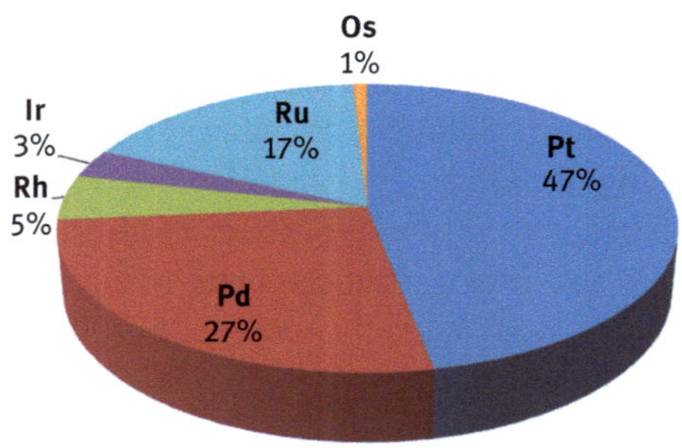

Figure 6.1: Distribution of PGMs applications in the chemical industry (mass percent) [34].

The chemical industry together with the electronics and automotive applications is the main recipient of PGMs, with distribution presented in Figure 6.2. PGMs with a wide variety of uses as a component of catalysts indicate the development of methods for their recover is necessary.

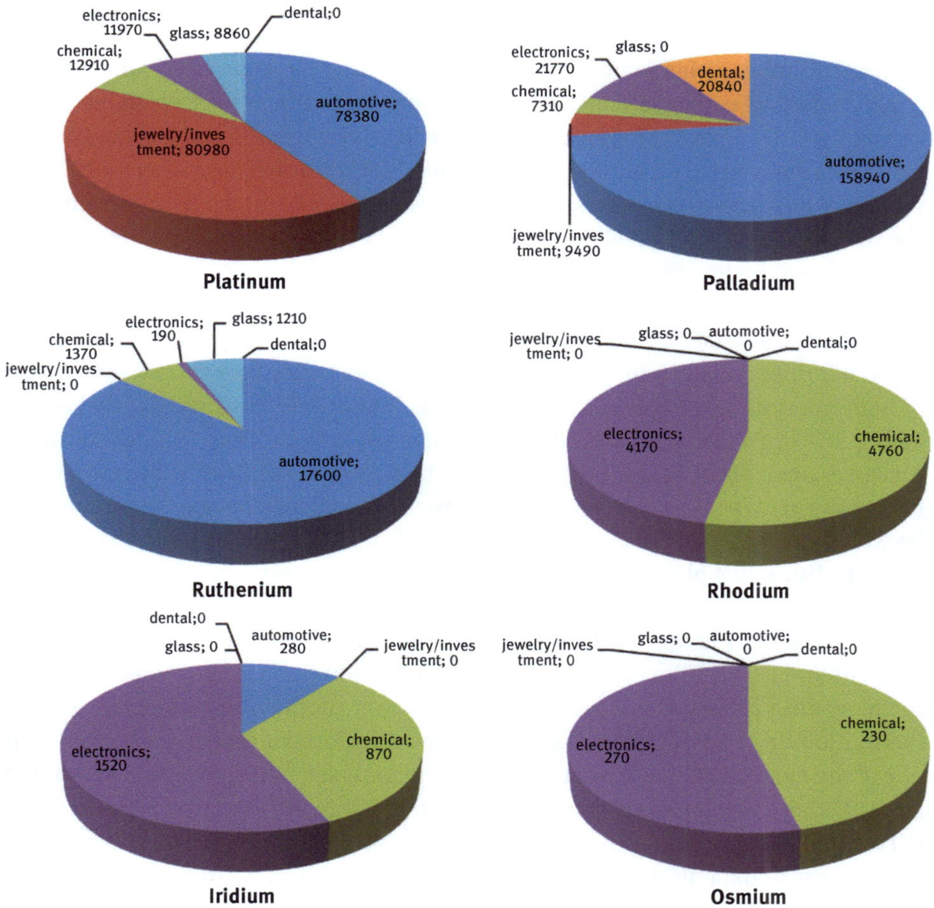

Figure 6.2: PGM demand by application, in kg [34].

Platinum black, the active Pt catalyst, is applied in syn-addition, forming cis-alkene reactions. Platinum (IV) in the form of dioxide PtO_2, also known as Adams' catalyst is used for the hydrogenation of various functional groups and dehydrogenation in organic synthesis [36]. Moreover platinum catalysts play important role in the hydrogenation of nitro compounds to amines and ketones to alcohols, reductions of alkenes, oxidative dehydrogenation of alcohols. Platinum is also used in the petroleum industry as a catalyst in catalytic reforming of straight-run naphthas into higher octane gasoline [37]. The most frequently used bimetallic catalysts today in this process are Pt–Re and Pt–Sn [13]. Furthermore platinum is proposed for photocatalytic organic synthesis. Excellent example of such process is the photocatalytic N-alkylation of ammonia (in alcohol) to amines or formation of secondary amines from primary amines in aqueous solutions using Pt–TiO_2 as photocatalyst [38].

Palladium complexes are commonly used in carbon–carbon bond-forming reactions, including carbonylations, carbon–carbon and carbon–heteroatom reactions (primarily C–C, C–O, C–N and C–F), and allylic alkylations. Suitable examples are $PdCl_2$ – carbonylation catalyst in commercially Wacker-Hoechst process for the conversion of ethylene to acetaldehyde and the production of acetone from propene and glycol derivatives; $PdCl_2(MeCN)_2$ – various cross-coupling reactions (Heck, Negishi, Grignard) and hydrocarboxylation catalyst [39–41]; palladium (II) acetate salt – catalyst precursor in a wide variety of different transformations [42]. Palladium deposited on calcium carbonate is applied for the hydrogenation of alkynes to alkenes (i.e. without further reduction into alkanes). Interesting solution of connection advantages of homogeneous and heterogonous catalysis is Pd EnCat™ Encapsulated Palladium Catalysts [43]. Pd EnCat™ is found in wide range applications from C–C bond forming processes, including Suzuki, Heck, Carbonylation, Sonogashira and Stille coupling, to reductions of carbonyls, alkenes, nitro groups and epoxides [44–46]. Palladium catalysts are, in details, described in the book of prof. Tsuji [47] and in works [48, 49].

Rhodium, like platinum, can be used for the hydrogenation of most functional groups, mainly carbonyl groups and carbocyclic aromatic and heteroaromatic system. Rhodium is one of the first metals used in homogeneous catalysis. This element and its complexes are proposed in several important processes including the Monsanto process developed in early 1970s for production of acetic acid from methyl iodide (carbonylation reaction), the hydroformylation process, which is one of the most important reaction in industry catalyzed by homogeneous catalysts [35, 50], and homogeneous hydrogenation with Wilkinson's catalyst [51]. Rhodium (III) chloride hydrate salt ($RhCl_3{\cdot}xH_2O$), like the osmium and ruthenium analogues, is a mixture of oxochloro and chlorohydroxo complexes in multiple (III/IV) oxidation states. It is used as a catalyst precursor in transformation reactions as the hydrosilylation of ketones and the carbonylation and isomerization of olefins [34]. The bis(1,5-cyclooctadiene)rhodium (I) and chloro(norbornadiene)-rhodium (I) salts are applied in chemical industries as catalyst precursors for the enantioselective hydrogenation of olefins and isomerization (H migration) of allylamines to enamines [52], the hydroboration catalyst to produce secondary alcohols from olefins [53], the hydrosilylation of ketones and in transfer hydrogenation and decarbonylation. Detailed description of rhodium with the most common oxidation states, namely Rh(I) and Rh(II), utilization, as catalysts and pre-catalysts for synthetic applications is presented, in details, in book *Modern Rhodium-Catalyzed Organic Reactions* edited by Prof. P. Andrew Evans [54]. Several types of application of rhodium for following asymmetric rhodium-catalyzed organic reactions are clearly written: hydrogenation, hydroboration, conjugate addition, olefin isomerization and hydroacylation, hydroformylation, hydrosilylation and silylformylation, cycloisomerization and cyclotrimerization, alderene, allylic substitution, carbocyclizations, cyclopropanation and carbon–hydrogen insertion, oxidative amination, ylide rearrangements, 1,3-dipolar cycloadditions.

Ruthenium and its complexes are utilized as catalysts in the hydrogenation of aromatic rings without the hydrogenolysis of any amino an hydroxyl group present on the ring, hydrogenation of ketones, aldehydes (especially in an aqueous solution) and carboxylic acids and olefin metathesis reactions [34]. Ruthenium (III) chloride hydrate ($RuCl_3 \cdot nH_2O$) is used in a variety of different manners for catalysis. Its solution in butanol is utilized in metathesis polymerization of norbornene to form Norsorex, an elastomer [55]. It finds application in alkyne functionalization (e.g. the formation of vinyl carbamates with secondary amines, acetylene and carbon dioxide) [56], oxidation of alcohols to aldehydes and ketones by amine-n-oxides [57], and in olefin cleavage to form diacids. Moreover ruthenium (III) chloride hydrate is a precursor for other ruthenium complexes utilized in catalysis, including dichlorotris(triphenylphosphine)ruthenium (II) ($RuCl_2(PPh_3)_3$) which is used as a catalyst for the oxidation of alcohols to aldehydes and ketones [57, 58], the reduction of diallyl α-oxalyl carboxylates with formic acid to form α-hydroxycarboxylic acids [34], the hydrogenation of nitro groups, imines and ketones [59], and as an isomerization catalyst.

Irydium and its complexes are utilized in industrial catalysis mainly in Cative process for the carbonylation of methanol to acetic acid. This process was developed in 1996 by BP Chemicals, and iridium catalyst package, named Cativa™ is based on iridium (III) acetate solution [60]. Cative process is alternative for the conventional rhodium-based Monsanto technology with several benefits as an inherently stable catalyst system, the production of byproduct propionic acid is reduced which leads to reduced purification costs. Moreover in the Cativa process the purer product is obtained because the level of acetaldehyde is lower than in the rhodium process. Hydrogenation of any unsaturated species present is catalyzed by the iridium species, resulting in almost complete elimination of unsaturated condensation products and iodide derivatives. The chlorocarbonylbis(triphenylphosphine)-iridium (I) is proposed in the decarbonylation of formate esters to form alcohols and CO [61]. Iridium (I) salts, as bis(1,5-cyclooctadiene)iridium (I) tetrafluoroborate or chloro-1,5-cyclooctadiene Iridium (I) are the catalysts in enantioselective hydrogenations reactions. The very important iridium catalysts of homogeneous hydrogenation is iridium (I) hexafluorophosphate (1,5-cyclooctadiene)-(pyridine)-(tricyclohexylphosphine) complex, known as Crabtree's catalyst (from the name of its inventor prof. Robert H. Crabtree) [62]. This compound indicates high level of reactivity, particularly for highly substituted alkenes, which contrasted with the then known rhodium and ruthenium catalysts. More details about iridium complexes in organic synthesis can be found in books [63, 64]. Chiral iridium catalysts, used in the asymmetric, stereoselective hydrogenation of largely unfunctionalized olefins are described, in details, in the Church and Andersson review [65].

Osmium and its complexes are catalysts, mainly in dihydroxylation transformation reaction [34]. The development of this process, especially with its enantioselective variant, was part of the focus of the 2001 Nobel Prize being awarded to K. Barry Sharpless. Osmium is a good catalyst for hydrogenations such as the conversion of unsaturated aldehydes to unsaturated alcohols or halonitrobenzenes to halo anilines. The oxochloro and chlorohydroxy complexes of osmium in various (III/IV) oxidation states ($OsCl_3 \cdot xH_2O$) are proposed as dihydroxylation and aminohydroxylation catalysts or as asymmetric catalysts in presence of chiral ligands [66]. Triflatopentaammineosmium (III) triflate is used in organic synthesis to promote dipolar cycloadditions with pyrroles [67].

The results presented in Table 6.3, as well as the short information presented in 6.3.1, indicate that PGMs are widely used in many applications in chemical and petrochemical processes. Therefore, PGMs are also called *Vitamin of modern industry* and *First and foremost high-technology metal* [68]. Moreover, the data presented in Table 6.3 indicate that the PGMs content in the catalysts is about 1% or higher, which means that PGMs content is about several kilograms per ton of catalyst. Thus, these spent catalysts can be treated as secondary ore of PGMs.

6.3.2 Rare-earth elements (REEs)

REEs are cerium (Ce), dysprosium (Dy), erbium (Er), europium (Eu), gadolinium (Gd), holmium (Ho), lanthanum (La), lutetium (Lu), neodymium (Nd), praseodymium (Pr), promethium (Pm), samarium (Sm), scandium (Sc), terbium (Tb), thulium (Tm), ytterbium (Yb) and yttrium (Y). They are widely used in fluorescent lamps, permanent magnets, batteries, catalysts and by the metallurgical and nuclear energy industries [69]. Moreover, REEs are used in the production of some explosives, special steels and alloys, and as getters. Wherein nearly 25% of all obtained REEs is used as catalysts (Figure 6.3) [70], mainly as cerium (IV) or

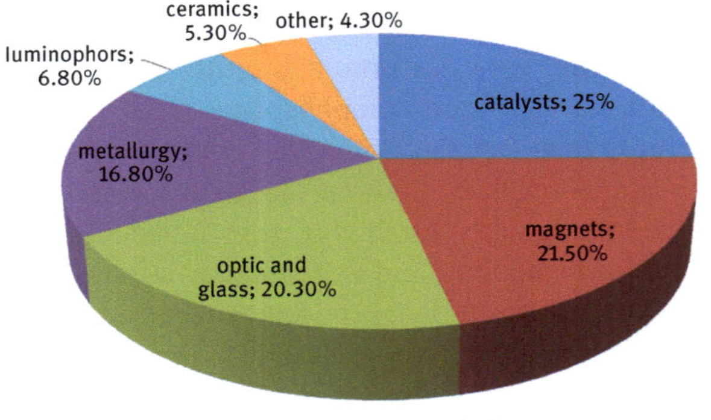

Figure 6.3: REEs demand by application [70].

lanthanum (III) oxides. The compositions of the fresh and spent catalysts with REEs and their applications are listed in Table 6.4.

Cerium in the form of cerium (IV) oxide is proposed in several chemical processes as: dehydrogentation of ethylbenzene into styrene [71], oxidative dehydrogenation of paraffins [72], dehydrogenation of n-butane [73], hydrogenation of toluene with the formation of crotonic aldehyde [74] and its hydrogenation to alcohols [75]. With lanthanum (III) oxide, cerium (IV) oxide play important role in the petroleum chemical industry in the FCC process, used to improve the yield and quality of gasoline produced from higher weight crude oil in refineries. Typically, FCC catalyst consists of a rare earth-exchanged USY-zeolite held in an amorphous silica-alumina matrix and several additives [76]. The composition of the spent FCC catalyst, usually known as the "equilibrium catalyst" (Ecat), is different in comparison to the fresh catalyst. It is possible to find in spent Ecat catalyst few new elements as nickel and vanadium, which do not exist in the fresh FCC catalyst [77]. It depends on both the composition of the original catalyst and the oil feed processed.

Lanthanum, as lanthanum (III) oxide has applications in environmental and catalysis fields. It is important component in catalyst for oxidative coupling of methane [78] and methanol [79]. In the mixture with CuO, La_2O_3 is proposed in the catalytic gaseous-phase oxidation of NH_3 [80]. Presence of La_2O_3 can enhance both catalyst performance and coking resistance [81] of catalysts due to improving the dispersion of the active phase [82].

Lanthanides such as europium and ytterbium are highly selective catalysts for Diels–Alder reaction in common organic solvents [83]. Lanthanides also catalyze selective acetalization of saturated aldehydes in the presence of ketones and conjugated aldehydes, aziridine synthesis in protic solvent or electrophilic substitution with aromatics, allylation, pericyclization, radical reaction [84]. In chemical reaction the mixture of lanthanides is also proposed as selective catalyst. For example five lanthanide oxides (La_2O_3, Sm_2O_3, CeO_2, Pr_6O_{11} and Tb_4O_7) are used in oxidation process of butane to methacrolein [85]. The lanthanide oxides CeO_2, La_2O_3, and Pr_6O_{11} are proposed for catalytic ozonation because of their chemical and biological inertness, strong oxidizing power, recyclability and extended stability against chemical corrosion [86]. Lanthanide complexes in multifunctional asymmetric catalysis are, in details, described in work [87].

6.3.3 Other metals

In catalysis processes, besides PGMs and REEs, the variety of transition metals and their complexes are used, such as vanadium, chromium, manganese, iron, cobalt, nickel, copper and mentioned above PGMs. The d orbitals of these metals are incompletely filled with electrons. Thus, they can easily give and take electrons, which make them good potential catalysts. Moreover, transitions metals are characterized by variability of oxidation state and coordination number or bonding ability [88].

Catalyst in the form of monometallic are used in homogenous (Al, Ti, Ni, Zr) or heterogeneous oligomerization (Ni). Moreover, nickel finds application in selective hydrogenation [89] (known as Raney nickel), as well as Cu or Fe (production of NH_3 and paraffins). Mo and W is proposed for metathesis of olefins, while Fe or Co for Fischer–Tropsch reaction (sometimes association with Ru) [13, 19]. In the hydroprocessing operations, widely used in refineries to purify a large number of different petroleum cuts, the sulfur metal-type catalysts that associate metals from groups VI (Mo, W) with metals from group VIII (Co, Ni) are used. To improve the catalytic properties addition of various adjuvant elements, such as F, B, and Si are proposed. The catalysts based on the sulfide Ni–Mo or Ni–W are proposed also for hydrogenating olefins. Besides metallic form, metal oxides of transitions metal are applied in industrial for oxidation (Co, Mo, V, Mo–V), hydrogenation (Cr, Fe, Cu–Cr, Cu–Zn–Al) and dehydrogenation (Zn, Fe–Cr–K, Cr–K–Al). The mixture of above metals is presented in hydrodesulphurization (HDS) catalysts. Spent HDS catalysts generally consist of 10–30% molybdenum, 1–12% vanadium, 0.5–6% nickel, 1–6% cobalt, 8–12% sulfur, 10–12% carbon and the balance is alumina. These compositions are different in comparison to the fresh catalysts because during operation, trace amounts of vanadium and nickel impurities in the crude oil gradually deposit onto the catalysts. In such conditions the formation of metal sulfides (molybdenum, cobalt, nickel and vanadium) are favorable. Catalysts can also be contaminated by carbonaceous deposits [90]. Moreover, catalysts based on metals are proposed for photocatalytic organic synthesis. ZnO and Fe-doped ZnO are applied to the atom transfer radical polymerization (ATRP) process for the production of poly(methyl methacrylate) (PMMA) [91].

The types and concentration of metals in the industrial catalysts are presented in Table 6.5. These compositions contain only the main metals. It should be noted that usually in such catalysts there is a whole range of elements. An interesting summary of detailed content ranges of some metals in spent FCC catalyst is shown in the work [92] and for spent refinery processing catalyst from naphtha treated unit in work [93]. The results from these works presented in Table 6.1, respectively, indicate that catalysts are a valuable source of various metals.

The inorganic elements composition in catalyst depends not only on the type of catalyst but also on its form – fresh, spent (and the time in line) or rejuvenated (and the method of this rejuvenated). Therefore, it is not possible to provide an unequivocal catalyst composition and the results from the Table 6.3–Table 6.5 should be considered as estimates. To confirm this statement, the Table 6.6 shows a comparison of compositions for the same hydrodesulphurization catalyst, in different stages of its "life" [7]. It is seen that the spent catalyst is fouled with carbon (20 wt%) and vanadium (6.5 wt%) deposition. Moreover, surface area and pore volume are 62% and 86% lower, respectively, in comparison to the fresh catalyst. The rejuvenated

Table 6.1: Metals content in spent FCC catalysts, unit: wt% [92].

	Minimum	Maximum		Minimum	Maximum
Ag	<0.0001	0.0059	**Pb**	<0.0001	0.0140
As	<0.0001	0.0185	**S**	0.0025	0.6700
Be	0.0003	0.0004	**Sb**	<0.0001	0.1600
Bi	<0.0001	0.1700	**Se**	<0.0001	<0.0100
B	<0.0001	0.7410	**Sn**	<0.0001	<0.0100
Cr	0.0005	0.0678	Sr	<0.0001	0.0505
Cu	0.0003	0.0066	Ti	0.0171	1.2500
Fe	0.2300	0.5000	V	0.0310	0.7000
Ni	0.0210	1.4000	**Zn**	<0.0001	0.0315
P	<0.0001	2.2000			

Table 6.2: Metals content in spent refinery processing catalyst from naphtha treated unit [93].

Element	Unit	conc.	Element	Unit	conc.	Element	Unit	conc.	Element	Unit	conc.
Ag	ppm	10	**Cu**	Ppm	88	**Ni**	ppm	598	**Th**	ppm	13
Al	%	39.4	**Fe**	%	0.48	**P**	%	<0.1	**Ti**	%	<0.1
As	ppm	41	**Hf**	Ppm	130	**Pb**	ppm	<20	**U**	ppm	<10
Ba	ppm	17	**K**	%	<0.1	**Rb**	ppm	3.8	**V**	ppm	<10
Be	ppm	0.8	**La**	Ppm	3.1	**S**	%	0.50	**W**	ppm	18
Bi	ppm	<5	**Li**	Ppm	<1	**Sc**	ppm	<2	**Y**	ppm	<1
Ca	%	0.02	**Mg**	%	<0.1	**Se**	ppm	22	**Yb**	ppm	<1
Cd	ppm	<2	**Mn**	%	<0.1	**Si**	%	0.15	**Zn**	ppm	110
Ce	ppm	9	**Mo**	ppm	80,000	**Sn**	ppm	4.1	**Zr**	ppm	41
Co	ppm	24,000	**Na**	%	<0.2	**Sr**	ppm	11			
Cr	ppm	160	**Nb**	ppm	399	**Te**	ppm	<5			

catalyst, independent of the method used for it, contains no coke, and its V content is 80% lower than that of the spent catalyst. The surface area increased from 98 (spent) to 240 and 221 m^2/g, for rejuvenated catalyst, for method I and II, respectively. Measurements showed that over 92% of the HDS activity of fresh catalyst was recovered by rejuvenation, with 28% for spent catalyst. The main difference between two methods of rejuvenated of catalyst is the leaching process – present only in method I.

6.4 Techniques of metal recovery from spent catalysts

Spent catalysts can be managed through the following approaches: (1) regeneration and reuse (2) thermal, chemical or microbial recovery and recycling of valuable metals, and (3) landfilling. Wherein number one is the best proposal, the last one – the ultimate when there is no other possibility. The regeneration has its limitations.

Table 6.3: PGMs content in catalysts, unit: wt%.

Name/Producer	Pt	Pd	Rh	Ru	Ir	Os	Al	Fe	Other metal	Others	Support	Application	Ref
CB-5;CB-8/ SINOPEC Corp., China	0.4; 0.15	–	–	–	–	–	NDA	NDA	Re (0.3)	Cl	γ-Al_2O_3	reforming	[94]
NDA	0.08	–	–	–	–	–	10.2	0.28	Si (0.85); Ni (0.005) K, Ca, Sr, Zr, Sn	S, Cl	γ-Al_2O_3	reforming	[95]
NDA	0.35	–	–	–	–	–	52.8	0.02	NDA	C (0)	γ-Al_2O_3	Reforming	[96]
NDA*	0.37	–	–	–	–	–	50.6	0.05	NDA	C (3.5)	γ-Al_2O_3	Reforming	[96]
NDA	0.39	–	–	–	–	–	50.8	0.03	Sn (0.42); In (0.28) Li (0.47)	C (0)	γ-Al_2O_3	n-Paraffin dehydrogenation	[96]
NDA**	0.37	–	–	–	–	–	48.3	0.15	Sn (0.41); In (0.26) Li (0.44)	NDA	γ-Al_2O_3	n-Paraffin dehydrogenation	[96]
NDA/Kima Company for Fertilizers, Aswan, Egypt	13.7	–	1.3	–	–	–	NDA	16.9	Ni (2.46); Ca (1.15); Mg (1.35); Cu (0.15); Zn (0.17); Pb (0.17); As (0.62); Re (0.84)	Si (8.25)	NDA	Oxidation of ammonia	[97]
NDA/Kima Company for Fertilizers, Aswan, Egypt	16.76	0.14	1.87	–	–	–	NDA	NDA	Ag (0.01); Au (0.0012)	Si	NDA	Oxidation of ammonia	[98]

(continued)

Table 6.3: (*continued*)

Name/Producer	Pt	Pd	Rh	Ru	Ir	Os	Al	Fe	Other metal	Others	Support	Application	Ref
AR-405/Axens; R-134/UOP, Korea	0.45; 0.24	–	–	–	–	–	NDA	NDA		NDA	Al_2O_3	Reforming	[99]
NDA	0.089	–	–	–	–	–	10.2	0.28	Ni (0.005); Si (42); K, Ca, Sr, Zr, Sn	S, Cl	NDA	Petroleum industry	[100]
NDA/Universal Oil Products	0.22	–	–	–	–	–	13.2	0.4	Zn (0.19); Re (0.44); Cu (0.15)	NDA	Al_2O_3	Aromatization of naphtha	[101]
OleMax®302/ Repsol Polímeros SA, Portugal	–	0.03 as PdO	–	–	–	–	NDA	NDA	Cr(III) oxide (0.03)	NDA	Al_2O_3	Hydrogenation	[102]
OleMax®302/ Repsol Polímeros SA, Portugal	–	0.013 as PdO	–	–	–	–	53	NDA	NDA	NDA	Al_2O_3	Hydrogenation	[102]
LD-265/Axens, Korea	–	0.25	–	–	–	–	NDA	NDA	NDA	NDA	Al_2O_3	Hydrogenation	[99]
NDA	–	–	–	4.3	–	–	–	NDA	NDA	NDA	SiO_2	Hydrogenation	[103]
NDA	–	–	–	–	20–40	–	NDA	NDA	NDA	NDA	γ-Al_2O_3	Decomposition of hydrazine	[104]
NDA	–	–	–	–	5	–	NDA	0.15	Cu (0.15)	NDA	NDA	Hydrogenation of nitroarene	[63]

Notes: NDA – No data available; white row – fresh catalyst; gray row – spent catalyst; *Time in line – about 3 months, with three times regeneration. ** Time in line – about 30 days, without regeneration.

Table 6.4: REE content in catalysts, unit: wt%.

Name/Producer	REE	V	Cr	Ni	Al	Fe	Other metal	Others	Support	Application	Ref
NDA/Valero, Houston, TX, USA	Rare earth elements (REE) contents, in µg/g: Y (14.0); La (13,617.3); Ce (651.8); Pr (14.7); Nd (32.1); Sm (24.2); Eu (0.7); Gd (3.8); Tb (0.3); Ho (1.4); Dy (0.3); Er (0.7); Tm (<0.1); Yb (0.4); Lu (0.1); Th (5.7)	NDA	NDA	NDA	NDA	NDA	NDA	NDA	55–60 % Al_2O_3/ 35–40 % SiO_2	Fluid catalytic cracking (FCC)	[105]
ORIM Spa/ Macerata, Italy	La (3.02); Ce (0.23)	–	–	0.02	17.35	0.33	Ti (0.43); Zr (0.16)	Si (12.75); S (0.013)	Al_2O_3/SiO_2	FCC	[106]
NDA/from an oil-refining factory in southern China	La (1.92); Ce(0.29)	<0.01	<0.01	<0.01	22.78	0.15	Na (0.18); K (0.1); Mg (<0.01); Ca (<0.01); Zn (<0.01); Ti (<0.01);	Si (20.48); S (0.1) P (0.3)	Al_2O_3/SiO_2	FCC	[107]
NDA	La (0.27); Ce (0.347)	<0.0033	–	0.0031	14.11	0.388	Ti (0.398); Ba (0.1684); Sb (<1.9 ppm);	Si(26); S (0.344)	NDA	FCC	[108]

(continued)

Table 6.4: (*continued*)

Name/Producer	REE	V	Cr	Ni	Al	Fe	Other metal	Others	Support	Application	Ref
NDA	La (0.3490); Ce (0.4077)	0.1455	–	0.393	16.15	0.65	Ti (0.643); Ba (0.0093); Sb (0.1326);	Si (33.2); S (0.11)	NDA	FCC	[108]
NDA/FCC catalyst factory in Hunan, China	La_2O_3 (2.06); CeO_2 (2.16)	–	–	–	33.46 as Al_2O_3	–	CaO (1.14); Na_2O (4.02)	NDA	NDA	FCC	[109]
NDA/from Yongping Refinery, China	La (1.69); Ce (1.57); sum of REE (3.27)	–	–	–	23.55	0.55	Na (0.11); Ca (0.28)	Silicon (69)	NDA	FCC	[110]
NDA	La_2O_3 (3.07)	1.14 as V_2O_5	–	1.31 as NiO	–	–	NDA	NDA	40.36 % Al_2O_3/ 349.22 % SiO_2	FCC	[76]
NDA	CeO_2 (5)	–	–	–	–	–	Fe_2O_3 (58); K_2CO_3 (23); MoO_3 (2.5)	NDA	NDA	Dehydrogentation of ethylbenzene into styrene	[70]

Notes: NDA – No data available; nd – not detected, white row – fresh catalyst; gray row – spent catalyst.

Table 6.5: Other metal content in catalysts, unit: wt%.

Name/Producer	V	Cr	Co	Ni	Mo	Al	Fe	Other metal, comments	Others	Support	Application	Ref
Mo-Fe catalyst/ NDA	–	–	–	–	5	NDA	48 as Fe_2O_3	Al_2O_3 (15); Cu (1)	C + S (30)	Al_2O_3	Hydrodesulphurization (HDS)	[90]
Mo-Ni catalyst/ NDA	–	–	–	4 as NiO	18.5	NDA	NDA	Al_2O_3 (52.5)	Si (4.7); P (2)	NDA	HDS	[90]
Mo-Co catalyst/ NDA	–	–	1.94	3.2	8.14	NDA	NDA	Al_2O_3 (24.3)	C + S (25.37)	Al_2O_3	HDS	[90]
Mo-V catalyst/ NDA	1–15 as V_2O_5	–	–	1–12	1–10	NDA	NDA	NDA	C + S (2–12)	NDA	HDS	[90]
NDA	0.247	0.01 as Cr_2O_3	–	0.086	–	34.06 as Al_2O_3	0.31	BaO (1.3); in ppm: Pb (12); Sb (<5); Ag, Hg, Cd, Se, As (<1); Cu (105)	SiO_2 (61.3)	Al_2O_3/SiO_2	Fluid catalytic cracking (FCC)	[92]
NDA	0.0552	0.02 as Cr_2O_3	–	0.074	–	37.03 as Al_2O_3	0.81	BaO (0.02); in ppm: Pb (350); Sb (290); Ag, Hg, Cd, Se, As (<1); Cu (22)	SiO_2 (61.3)	Al_2O_3/SiO_2	FCC	[92]
NDA/Singapore Refining Company (SRC)	0.39	–	–	0.26	–	17.5, 33.08 as Al_2O_3	0.56	Sb (0.03)	0 (54.44); S (nd); Si (22.46); C (0.44)	Al_2O_3/SiO_2	FCC	[9]

(continued)

Table 6.5: (*continued*)

Name/Producer	V	Cr	Co	Ni	Mo	Al	Fe	Other metal, comments	Others	Support	Application	Ref
NDA/Singapore Refining Company (SRC)	nd	–	–	nd	–	19.1	0.28	Sb (nd)	0 (77.53); S (0.47); Si (10.48); C (0.15)	Al_2O_3/SiO_2	FCC	[9]
Criterion 424/ Criterion Catalysts Company Limited, Singapore	–	–	–	6.1	13.7	33.3	NDA	NDA	Si (2.9)	NDA	Hydroprocessing	[111]
NDA/from Italian refinery LC-Fining units	9.4	–	–	4.5	4.4	NDA	NDA	Form of metal: NiV_2S_4, Mo_4O_{11}, Ni_3S_4 (polydymite)	NDA	Al_2O_3	Hydroprocessing	[112]
UOP: S-12 Co/ Mo/Al_2O_3/from Iranian refinery (Naphtha Treater unit).	–	–	2.4	0.0598	8	39.4	NDA	NDA	NDA	Al_2O_3	Hydrotreating	[113]
KUB-3/New Chemical Syntheses Institute, Poland	–	–	–	45 as NiO	–	NDA	NDA	CaO	NDA	Al_2O_3	Hydrogenation of benzene to cyclohexane	[114]

(*continued*)

Table 6.5: (*continued*)

Name/Producer	V	Cr	Co	Ni	Mo	Al	Fe	Other metal, comments	Others	Support	Application	Ref
PKH-3/New Chemical Syntheses Institute, Poland	–	–	–	4 as NiO	14 as MoO_3	NDA	NDA	Al_2O_3 (72)	NDA	Al_2O_3	HDS	[115]
PSMC/New Chemical Syntheses Institute, Poland	–	–	–	–	–	NDA	NDA	Al_2O_3 (max40); CuO (min. 30); ZnO (min. 30)	NDA	Al_2O_3	Deep desulfurization process of the hydrocarbon	[115]
NDA/from petroleum refinery company, South Korea	11.4/ 14.3*	–	–	3.31/ 3.97*	2.58/ 2.82*	20.56 /22.6*	0.21/ 1.1*	Form of metal: $AlFeS_3$, NiV_2O_6, Vanadium phosphorus sulfide, $MoVO_x$, Al_2O_3, V_2O_5, MoO_2, Nickel aluminum oxide, Aluminum vanadium oxide, $Fe_2Mo_2O_9$, $NiS{\cdot}xH_2O$	NDA	NDA	NDA	[116]

(*continued*)

Table 6.5: (*continued*)

Name/Producer	V	Cr	Co	Ni	Mo	Al	Fe	Other metal, comments	Others	Support	Application	Ref
V_2O_5 catalyst/ Projects and Development India Limited (PDIL), Sindri (India).	3.5	<1	–	–	–	2	1.3	Pb (<1)	Ti (0.8)	NDA	NDA	[117]
NDA/from SK Petroleum Corp., South Korea,	11.6	–	–	2	1.4	19.5	0.3	Cd (0.2); Cu (0.8)	S (11.5); Si (1.1)	Al_2O_3	NDA	[118]
NDA/National Iranian Oil Refining and Distribution Company (NIORDC), used in Tehran refinery	–	–	–	2.5	5	23	NDA	Al_2O_3 (47.12); SiO_2 (31.41); CaO (0.28); Fe_2O_3 (2.18); NiO (3.63); As_2O_3 (0.1); MoO_3 (9.49)	Cl (0.13)	Al_2O_3/SiO_2	Hydrocracking	[119]
lbemarle KF-1015/National Iranian Oil Refining and Distribution Company (NIORDC)	–	–	–	2.4	6.6	24	NDA	Al_2O_3 (50.47); SiO_2 (33.44); CaO (0.19); Fe_2O_3 (1.13); NiO (4.05); Na_2O (0.49); MoO_3 (10.24)	NDA	Al_2O_3/SiO_2	Hydrocracking	[120]

(*continued*)

Table 6.5: (*continued*)

Name/Producer	V	Cr	Co	Ni	Mo	Al	Fe	Other metal, comments	Others	Support	Application	Ref
Purelyst MD-101/ PureSphere Co., Ltd, Korea (South)	–	–	–	–	–	NDA	NDA	CuO> 23 mol%, γ-MnO_2 > 69 mol %, K, Al_2O_3	NDA	Al_2O_3	CO oxidation for gas mask and fire escape hood; formaldehyde destruction at room temperature	[121]
Phillips catalysts	–	1	–	–	–	NDA	NDA	NDA	SiO_2	NDA	Polymerization of ethylene (production of high-density polyethylene (HDPE))	[122]
TZC-3/1/Grupa Azoty S.A., Poland	–	min. 7.3 as Cr_2O_3	–	–	–	–	min. 71.5 as Fe_2O_3	CuO (min. 1.25); Na (max. 0.1)	S (max. 0.015)	NDA	High temperature conversion of CO with water vapor in processes of obtaining hydrogen, syngas and ammonia	[123]

Notes: NDA – No data available; nd- not detected, white row – fresh catalyst; gray row – spent catalyst; * acetone washed/roasted catalyst.

Table 6.6: Characteristic of spent, fresh and rejuvenated HDS catalysts, unit: wt% [7].

	V	Co, as CoO	Ni	Mo, as MoO_3	Fe	Na	Others
Fresh	–	4.20	–	12.00	0.10	0.20	–
Spent	6.50	3.20	2.80	11.20	0.19	0.50	C (20.00); S (6.30)
Rejuvenated I	1.30	3.05	–	10.10	0.06	0.35	–
Rejuvenated II	1.60	4.01	–	11.80	0.05	0.27	–

It is available not for all type of catalysts and is only available one to several times for one catalyst. Therefore, it is important to develop methods of metal recovery from spent catalysts. Two approaches are used in metal reclamation: hydrometallurgy (metals are leached) and pyrometallurgy (using a heat treatment).

6.4.1 Pyrometallurgical recovery methods

There are two main approaches for metal recovery in pyrometallurgical processes. In the most used, the spent catalysts are melted with the addition of flux components and metal collectors (or related additives) at high temperatures to form a metal-containing alloy. Then metals are recovered through refining techniques. The catalysts may also be vaporized for recovery of metals or reduced *in situ* as a reforming means for their recycling [124].

Pyrometallurgical techniques are a crucial method to recover PGMs from spent catalyst [25]. These methods can be divided into three groups: smelting, chlorination and sintering processes. First step in smelting process is mixing spent catalyst with flux, collector and reducing agent. The prepared feed is smelted in a high-temperature plasma furnace, electric arc furnace or inductive furnace at a high temperature (usually above 1000°C) [124]. Usually the pretreatment steps are necessary before smelting such as dismantling/incineration of nonmetallic components, calcinations or reductions. The problem of the catalyst carriers, such as alumina is solved by their melting in the presence of suitable fluxes to obtain a low-viscosity liquid slag. After separation of metal phase enriched in PGMs from slag, the PGM-containing alloy is purified.

In gas-phase volatilization, metals from spent catalysts are volatilized by selective chlorination (chlorinated with chlorine or chlorides) and condensed in a cooler zone. This process occurs in high temperature between 250°C and 1200°C [124, 125]. PGM chlorides that are obtained in the process are separated by appropriate washing, adsorption on an activated carbon, or by distillation using their difference in vapor pressures between the metal chlorides. In this process high purity metals are obtained. Selective chlorination could be used not only for PGMs but also for other metals. This method is proposed for Co, Mo, Ti and W recover from spent hydrodesulfurization catalysts. Depending on the experimental conditions, it is possible to recover up to 98% of the Ni and Co, as chlorides from the chlorination residue and about 98% of the Mo, Ti and W and 80% of vanadium compounds in the condensates [4]. In chlorination process high purity metals are obtained. Disadvantages of such solution are corrosion of furnace and associated facilities and the utilization of hazardous gases as Cl_2 which causes health and environmental risks. Therefore this process doesn't find many industrial applications.

In the sintering process PGMs are recovered in the presence of plasma by *in situ* reduction of its oxidized PGMs components [126]. The processing conditions are following: temperature above 1200°C, N_2 as the central plasma gas allows decomposing of organic tar on the surface of spent catalyst and convert it to syngas which reduced oxide of PGM to its metallic form. This method has been successfully applied for spent platinum-based aluminum oxide catalyst for petrochemical industry.

Pyrometallurgy is appropriate only for relatively high metal concentration; otherwise large amounts of energy will be required to melt associated compounds. Moreover, this process generates the emission of SO_2 gas which is a severe environmental pollutant. Therefore, the researchers in their works resign from pyrometallurgy and focus on hydrometallurgical methods. However, pyrometallurgy is the most common industrial solution. Sometimes the pyrometallurgical and hydrometallurgical methods are used together, when spent catalysts are roasted before chemical leaching. This process is called thermal pretreatment, and it is used to improve the leaching efficiency for metal recovery [25].

6.4.2 Hydrometallurgical recovery methods

In hydrometallurgical processing, metals are leached directly or after pretreatment using suitable solutions, sometimes in the presence of oxygen, iodine, bromine, chlorine, hydrogen peroxide, etc. Generally, the following approaches are proposed as follows: (1) Acid leaching with H_2SO_4, HNO_3, HCl or $(COOH)_2$, often after roasting; (2) caustic leaching with NaOH, sometimes after roasting; (3) salt roasting with Na_2CO_3, NaCl or NaOH followed by leaching with water or Na_2CO_3; (4) chelation using ethylene diamine tetra acetic acid (EDTA) or another similar agent to complex the metal ions [77].

The leaching of a metal is mainly governed by the formation of a complex under the potential and pH of the solution in the presence of ionic species. Thus, the choice of appropriate leaching agent depends on the possibility to form metal-complexes. For example, Pt forms stable chloro-complexes in chloride solution at high acid concentration, thus Pt could be leached at a higher potential using an oxidizing agent. Pd and Rh form chloro-complexes at lower potential. They are leached, but the Rh leaching rate is relatively low [25]. The compilation of chemical leaching processes, described in the literature, is presented in Table 6.7. As there are a lot of results on this topic in literature, only exemplary data is outlined in the table to

Table 6.7: Leaching process with different leaching agents.

Catalyst	Leaching agent	Metal recovery	Comments	Ref.
Acid leaching				
Oxidation of ammonia	3 HCl:1 HNO_3	**Pt** (98 %)	– The optimal conditions: time 1.5 h, temp. 109°C, liquid/solid ratio of 10:1. – Platinum was separated from the leach liquor as ammonium hexachloroplatinate by two alternative methods, precipitation with ammonium chloride and solvent extraction with TOA (trioctylamine, basic extractant).	[97]
Petroleum catalysts	H_2SO_4	AR-405: **Al** (92%), **Pt** (52%) R-134: **Al** (98%), **Pt** (83%) LD-265: **Al** (90%), **Pt** (45%), **Pd** (45%)	– Three commercial catalysts were used: AR-405, R-134 and LD-265. – The optimal conditions for AR-405 and R-134: 6 M H_2SO_4, time 2–4 h, temp. 100°C, pulp density 220 g/L – The optimum conditions for LD-265: 8 M H_2SO_4, time 18 h, temp. 100°C, pulp density 220 g/L.	[99]
Hydro-processing	H_2SO_4, citric acid	**10%** H_2SO_4 with/without ultrasonic: **Mo** (74.1/67.9%), **V** (98.3/88.7), **Ni** (96.4/89.2), **Al** (45.4/37.2%) **10%** citric acid with/without ultrasonic: **Mo** (95.7/91.5%), **V** (94.8/83.7), **Ni** (95.7/88.2), **Al** (18.1/<20%)	– Metal leaching was more effective and selective for citric acid in comparison to sulfuric acid. – Al extraction was <20% in the case of citric acid – the separation of Al from Mo, V and Ni was possible. – Ultrasonic vibration increased the leaching effect. – The recovery of metals increased with temperature (from 30°C to 60°C).	[128]
From fertilizer industry	H_2SO_4	**Ni** (99%)	– The optimal conditions: 50% H_2SO_4, solid/liquid ratio 1:12, <500 micron particle size, contact time higher than 5 h, 800 rpm stirring rate, temp. 100°C. – Nickel recovery as nickel sulfate.	[129]
From ammonia plant	H_2SO_4	**Ni** (99%)	– The optimal conditions: 80% H_2SO_4, particle size of 0.09 mm, contact time 50 min, temp. 70°C. – Nickel recovery as nickel sulfate.	[130]

(continued)

Table 6.7: (*continued*)

Catalyst	Leaching agent	Metal recovery	Comments	Ref.
Petrochemical industry	H_2SO_4	**Ni** (98%)	– The optimal conditions: 8% H_2SO_4, contact time 2 h, pulp density 10%, temp. 90°C – Nickel in the form of nickel sulfide or oxalate was finally separated with overall 90% and 80% recovery, respectively.	[131]
Catalytic reforming	$HF+H_2O_2$	**Pt** (5–10%), **Al** (99%)	– Hydrogen peroxide improved the leaching kinetics to less than 1 h (4 h without H_2O_2)	[96]
Basic leaching				
Hydro-processing	NH_4OH, conc. 10%; $(NH_4)_2CO_3$, conc. 10%; $(NH_4)_2S_2O_3$, conc. 10%	**Mo** (80%), **V** (70%), **Ni** (60%), **Al** (10%)	– Ratio spent to solvent 1:40. – Temp. 50°C and 6 h leaching time.	[132]
Hydro-processing	NaOH	**Mo** (98%), **V** (95%), **Al** (92%), **Ni** (0%)	– Process under nitrogen pressure 30 bar. – Temp. was varied from 150°C to 250°C.The best results was obtained for 250°C.	[132]
Sulfuric acid manufacture	NaOH, conc. 50 g/L	**V** (55.61%); with pre-treatment by calcinations **V** (71.88%)	– Vanadium recovery was increased when the catalyst is pretreated by calcining at 450°C for 4 h to remove coal and sulfur waste in the form of gas, as CO_2 and SO_2. – The results for acid leaching are better than basic one.	[133]
Salts leaching				
Hydro-processing	Na_2CO_3	**Mo** (95.7%), **V** (95%), **Ni** (95%)	– The separation of Ni from mixture of V and Mo was obtained. – Al in the form $Al(OH)_3$ was formed. – Ratio spent to solvent 1:1.15, pH>10. – Temp. 50°C and 6 h leaching time.	[132]
HDR	$KHSO_4$	**Co, Ni, Mo, Al** (>85%)	– The procedure was based on fusion of samples with $KHSO_4$. – After fusion, the solid was dissolved in water (90–100°C). – The optimum temperature range (450–500°C): insoluble matter after fusion reaches a minimum amount. – Co, Ni, Mo and Al were recovered by precipitation techniques or selective solvent-extraction procedures.	[134]

demonstrate the efficiency of metals recovery from spent catalysts. Recovery hydrometallurgical techniques of PGMs from spent catalysts are described in detail in reviews [25, 68], for Mo, V, Ni and Co in [127].

The recovery of metal from spent catalyst by chelation process occurs at relatively lower temperatures, and the chelating agents can be reused in the extraction process itself [135]. The whole process contains two steps: chelation and dechelation of metal complex. Various chelating agents are proposed in the literature. Their compilation is presented in Table 6.8. Chelating agents could be treated as effective extractants in solid/liquid extraction process [136]. They are characterized by high efficiency of metal extraction, high thermodynamic stability and good solubility of the metal complexes formed and low adsorption of the chelating agents on soils.

An interesting alternative to the chemical leaching is the microbial leaching, known as bioleaching [77]. The microorganisms used in this process have ability to

Table 6.8: Chelation process with different spent catalysts.

Catalyst	Chelating agent	Metal recovery	Comments	Ref.
Hydrodesulfurization	Ethylene diamine tetraacetic acid (EDTA)	**Co** (80.4%), **Mo** (84.9%) with 91% purity	– Significant metal extraction efficiency (72.7% Co and 76.5% Mo) was observed with recovered EDTA even after the fourth cycle of operation.	[135]
Hydroprocessing	EDTA, conc. 10%	**Mo** (97%), **V** (94%), **Ni** (95%), **Al** (14%)	– The EDTA recovery was about 95%.	[135]
Reforming	[*S*,*S*]-ethylenediamine disuccinic acid (EDDS)	**Ni** (84%)	– Biodegradable chelating agents were used. – The optimal conditions: particle size 100 μm, temperature 90°C, reaction time 6 h, MR 3.6, solid/liquid ratio 1:20 (g/mL), stirring speed 500 rpm, and pH 7. – Dechelation of Ni-EDDS complex was performed at pH 5 where more than 96% EDDS was recovered.	[137]
NiO/Al_2O_3 catalyst	EDTA	**Ni** (95%)	– The optimal conditions: 0.8 M EDTA, solid/liquid ratio 1:50 (g/ml), particle size 100 μm, pH 10, 10 h of chelation time, 700 rpm and 100°C. – Up to 95% of the EDTA was recovered without losing significant activity.	[138]
Steam reformer catalyst from an ammonia plant	EDTA	**Ni** (95%)	– In dechelation of Ni-EDTA complex mineral acids H_2SO_4 and HNO_3 were used (precipitation of EDTA about 97%). – Efficiency of chelation decreased with every successive recovery of EDTA (in the case of dechelation using H_2SO_4).	[139]

mobilize and leach metals from solid materials, based on following principles: (1) the transformation of organic or inorganic acids (protons); (2) oxidation and reduction reactions and (3) the excretion of complexing agents. This process has got high efficiency of metal recovery (63–99 %), similar to the chemical leaching. Wherein, this application is supported by the following advantages: low cost, low-energy consumption, mild operation conditions, without strict requirements of raw material composition. In bioleaching process the three groups of microorganisms are commonly used: autotrophic bacteria, heterotrophic bacteria and fungi. The reason for using of autotrophic bacteria is no necessity of an organic carbon source for their growth. Unfortunately, they are sensitive to high pH. In this operation conditions it is better to use heterotrophic bacteria and fungi. Some examples of bioleaching process using in recovery of metals from spent catalyst are presented in Table 6.9.

As is presented in Table 6.9 bioleaching is proposed in the scientific literature to recovery metals from spent catalysts but up to now they did not have any industrial application [147].

The results presented in Table 6.7–Table 6.9 indicate high hydrometallurgical potential in recovery of metals from spent catalysts.

6.4.3 Industrial processes

Because the problem of spent catalysts is high, and the need for metal recovery becomes very important issue, there has been an increase in the number of companies which collect, recycle, regenerate or recover spent catalysts in the world.

Moxba-Metrex (www.moxba.com) is a Dutch company, specialized in the responsible and environmentally friendly processing of spent catalyst and metal-containing residues, since 1974. They have created a complete recycling chain, turning spent catalyst and metal-containing residues back into first class products and base materials. In the first step the spent catalysts are separated, sieved and breaked. Then such material, depending on its type, is thermally treated in a fluidized bed kiln or batch furnace. The combustible part of the waste, containing hydrocarbons and sulfur compounds, is treated in the fluid bed process. Gasses are primarily cleaned in a cyclone and dust-collecting filter. The sulfur oxide compounds of the flue gas are treated with magnesium hydroxide in a series of scrubbing towers and converted to magnesium sulfate, which is refined to a quality suitable for use as an agricultural fertilizer product. The materials which contain metals are treated by pyrolysis or calcination with temperatures starting from 100°C up to 1200°C in the rotary batch. The generated heat is converted into electricity or used in the process. In the last step, the intermediate metal products are directed to the hydrometallurgical leaching process or pyrometallurgical electric arc melting process. The leaching process provides pure metal oxides and concentrates where the melting process produces special or custom made metal alloys. The schema of this process is presented in Figure 6.4.

Table 6.9: Bioleaching processes with different spent catalysts.

Catalyst	Microorganisms	Metal recovery	Comments	Ref.
Fluid catalytic cracking (FCC)	Fungi: *Aspergillus niger*	**Ni** (9%), **Fe** (23%), **Al** (30%), **V** (36%), **Sb** (64%)	– Bioleaching gave 2.7–20% higher metal extraction efficiency than chemical leaching with commercial organic acids (citric, oxalic and gluconic acids) at the same concentration. – Extraction efficiency generally decreased with increased pulp density.	[9]
Refinery processing catalyst from naphtha treater unit	Fungi: *Aspergillus niger*	**Co** (71%), **Mo** (69%), **Ni** (46%)	– The optimal conditions: pH of 5.0, a temperature of 31°C, a pulp density of 2 g/L, a rotation speed of 115 rpm, and using a 12% inoculums.	[93]
Hydrocracking	Fungi: *Aspergillus niger*	**W** (100%), **Fe** (77.8%), **Mo** (90.9%), **Ni** (65.8%), **Al** (14.2%)	– The main lixiviant in bioleaching using *Aspergillus niger* was gluconic acid. – The main agent in spent medium leaching was citric acid.	[140]
Petrochemical process	Bacteria: *Acidithiobacillus thiooxidans*	15 g/L spent catalyst conc.: **Ni** (88.3%), **Mo** (58. %), **V** (32.3%) 50 g/L spent catalyst conc.: **Ni** (88.3%), **Mo** (46.3%), **V** (94.8%)	– Bioleaching gave slightly higher metal recovery (8–18%) than chemical leaching with sulfuric acid. – A significant amount of molybdenum was found to be lost to the solid phase as MoO_3.	[118]
Hydrocracking	Bacteria: *Acidithiobacillus ferrooxidans*	**Mo** (87%), **Ni** (37%), **Al** (15%)	– The optimal conditions: pulp density of 0.9% (w/v), particle size of 60.7 µm, and aeration rate of 209 ml/min.	[119]
Hydrocracking	Fungi: *Penicillium simplicissimum*	**Mo** (98.8 ± 0.9%), **Ni** (46.5 ± 0.6%), **Al** (13.7 ± 0.4%)	– The four variables (yeast extract conc., $NaNO_3$ conc., sucrose conc., and pulp density) were considered. – Higher pulp density more than 4% w/v resulted in decreasing specific growth rate and biomass concentration.	[141]

(*continued*)

Table 6.9: (*continued*)

Catalyst	Microorganisms	Metal recovery	Comments	Ref.
Hydrocracking	Fungi: *Aspergillus niger*	**Mo** (99.5 ± 0.4%), **Ni** (45.8 ± 1.2%), **Al** (13.9 ± 0.1%)	– The optimal conditions: particle size 150–212 μm, sucrose 93.8 g/L, pulp density 3% w/v, and pH 7.	[142]
Petrochemical process	Bacteria: *Acidithiobacillus ferrooxidans (AF)* and *Acidithiobacillus thiooxidans (AT)*	*AF*: **Al** (91.1%), **Mo** (90.4%), **V** (92.5%), **Ni** (95.4%) *AT*: **Al** (85.3%), **Mo** (82.4%), **V** (97.3%), **Ni** (96.9%)	– With increase pulp densities leaching rate decreased for *AF* and increase for *AT*. – The optimal condition: pH 2.5 for *AF* and 2.0 for *AT*.	[143]
Petrochemical process	Thermophilic bacteria	**Ni** (92–97%), **V** (81–91%), **Al** (23–38%)	– At initial hours of process: lower particle size of spent catalyst results in increase yields of metals. – The final metals recovery was almost independent of particle size of catalyst.	[144]
Petrochemical process	Similar to bacteria: *Acidithiobacillus ferrooxidans*	**Ni** (95%), **V** (95%)	– The comparison between adapted and unadapted bacteria cultures was done. – Ni and V recovery decreased for unadapted bacteria to 85%.	[145]
Hydrocracking	Fungi: *Penicillium simplicissimum*	**W** (100%), **Fe** (100%), **Mo** (92.7%), **Ni** (66.43%), **Al** (25%)	– The optimal conditions for all of the metals: two-step bioleaching, pulp density (3% w/v).	[146]

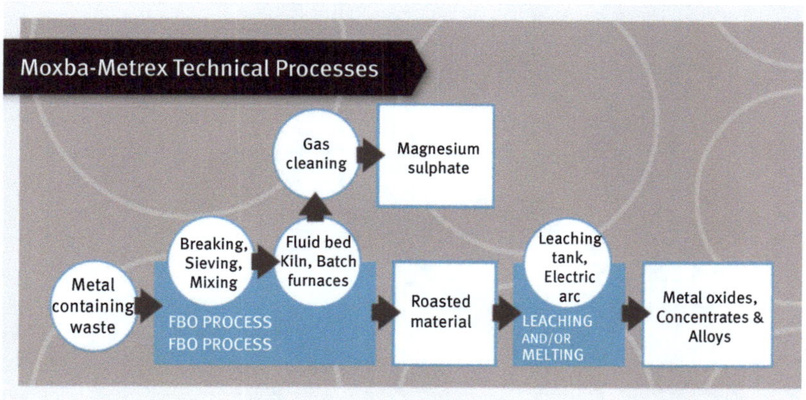

Figure 6.4: Moxba-Metrex technical processes, the materials are reproduced with the permission of Moxba-Metrex B.V [148].

Nickelhütte Aue GmbH (www.nickelhuette-aue.de) (Germany) recycles spent catalysts from petrochemical industries, chemical industries as well as from edible oil hydrogenation, cosmetic, fertilizer and pharmaceutical industries. The company is focused mainly on waste materials containing nickel, cobalt, copper, molybdenum, vanadium and tungsten. In smelting works they produce nonferrous metal concentrates and by following metallurgical processes, these concentrates are refined into nonferrous metals and Ni-, Cu-, Co-, V-chemicals. Moreover, in 2001, Nickelhütte Aue GmbH started to treat precious metals containing waste, for example spent catalysts [149].

Belgian company Sadaci N.V. (www.sadaci.be) is specialized in recycles spent catalysts containing molybdenum, i.e. hydroprocessing catalysts from oil-refineries, through thermal and hydrometallurgical treatments. The plant is located in Belgium where HDS are recycled with the recovery of molybdenum only, sold as technical-grade oxide and ferro–molybdenum (FeMo 65 %) alloy [147, 150].

Treibacher Industrie AG (www.treibacher.com), an international company with over 100 years of experience in chemistry and metallurgy located in Austria, recycles catalysts from the chemical, petrochemical, pharmaceuticals and foodstuffs industries. They are focused on materials that contain vanadium and/or molybdenum and/or nickel. Similar to the earlier described companies the pyrometallurgical process is applied [151].

The French company ERAMET (www.eramet.com) proposed the solution named Valdi for battery, catalyst and metal oxide recycling through pyrometallurgical processes with a strong commitment: “zero landfill and 100% recovery”. From 2017, Valdi will have annual processing capacity of 20,000 tons of batteries, 10,000 tons of spent oil catalysts and 8,000 tons of metal oxides, making it Europe’s #1 recycling site in its field [152]. As it is presented in Figure 6.5 two pyrometallurgical operations are used in this process: roasting in a specific furnace to remove the sulfur

Figure 6.5: Valdi process, the materials are reproduced with the permission of the company Valdi [152]. Copyright Valdi.

from spent catalyst and melting in a submerged electrode furnace to separate metal from mineral elements. The SO_2 from the roasting process is removed by calcinations operation.

Canadian company Globe Metal (globemetal.com) recycles spent nickel, molybdenum, tungsten, copper, platinum and cobalt catalysts, as well as Raney nickel catalyst scrap from chemical, petrochemical, food and oil industry. They use ISO-certified, advanced thermal chemical reactor (ATCR) technology [153].

In USA there are several companies specialized in spent catalyst recycling. ACI Industries, LTD (www.aci-industries.com) using pyro- and hydrometallurgy processes recovery metals such as precious metals, nickel, copper, zinc, cobalt, iron, molybdenum, vanadium and others [154]. Metal & Catalyst Resources (www.metalandcatalyst.com) has more than 40 years of experience in the business of catalyst and spent catalyst recycling and reclamation. The website presents the information that several metals contained in spent catalysts could be reclaimed by this company: nickel, cobalt, molybdenum, vanadium, copper, zinc, platinum, palladium, rhodium, rhenium, ruthenium, gold, or silver, as well as less common chromium, selenium, tellurium, yttrium, europium, niobium, samarium, neodymium, gallium, neodymium, gallium, gadolinium and others [155].

Very interesting solution in recovery of metals from spent catalyst is Tetronics (tetronics.com) plasma recovery for precious metal, especially PGMs. The main advantages of Tetronics patented plasma technology are connection of the highest recovery/operational flexibility with the lowest environmental impacts and cost base. In a single processing step the metals are separated from the less valuable catalyst material, which is vitrified into an approved, inert, reusable building product called Plasmarok®. In the same technology any hazardous organic material, such as dioxins etc. are destroyed [156].

The spent catalyst recycling is a large business and it is impossible to list all companies. These examples demonstrate that technical solutions already exist in the industry and are successfully used. These projects are based mainly on pyrometallurgical processes.

6.5 Conclusion

The analysis suggests that the secondary source of metals in the chemical and petrochemical sector is high, which can be attributed to the catalysts application. Among the metals present in the catalysts it can be mentioned the PGMs, the REEs and the variety of transition metals as vanadium, cobalt, nickel, copper and molybdenum.

Despite the fact that in the industry there are solutions for recovery of metals from spent catalyst, based mainly on pyrometallurgical processes, the researchers are still looking for new ones to improve the metal recovery efficiency based on hydrometallurgical and bioleaching.

References

[1] Shiju NR, Guliants VV. Recent developments in catalysis using nanostructured materials. Appl Catal A. 2009;356:1–17.

[2] Bartholomew CH, Farrauto RJ. Fundamentals of industrial catalytic processes. 2nd ed. Hoboken, NJ: John Wiley & Sons, Inc.; 2006.

[3] Misono M. Basis of heterogeneous catalysis. In: Misono M, editor. Studies in surface science and catalysis. vol. 176. Amsterdam, The Netherland: Elsevier B.V.; 2013:1–23.

[4] Gaballah I, Djona M, Mugica JC, Solozobal R. Valuable metals recovery from spent catalysts by selective chlorination. Resour Conserv Recy. 1994;10:87–96.

[5] US EPA. Hazardous waste management system: identification and listing of hazardous waste-spent catalysts from dual-purpose petroleum hydroprocessing reactors. Federal Register 2002. 2002;67(89):30811.

[6] Marafi M, Stanislaus A. Options and processes for spent catalyst handling and utilization. J Hazard Mater. 2002;101:123–32.

[7] Marafi M, Stanislaus A, Kam E. A preliminary process design and economic assessment of a catalyst rejuvenation process for waste disposal of refinery spent catalysts. J Environ Manage. 2008;86:665–81.

[8] Trimm DL. The regeneration or disposal of deactivated heterogeneous catalysts. Appl Catal A. 2001;212:153–60.

[9] Aung KMM, Ting Y-P. Bioleaching of spent fluid catalytic cracking catalyst using *Aspergillus niger*. J Biotechnol. 2005;116:159–70.

[10] Anastas PT, Kirchhoff MM, Williamson TC. Catalysis as a foundational pillar of green chemistry. Appl Catal A. 2001;221:3–13.

[11] Zamaraev KI. Catalytic science and technology for environmental issues. Catal Today. 1997;35:3–13.
[12] Lindström B, Petterson L. A brief history of catalysis. CatTech. 2003;7:130–38.
[13] Marcilly C. Present status and future trends in catalysis for refining and petrochemicals. J Catal. 2003;216:47–62.
[14] Smith JK. History of catalysis. In: Horvath I, editor(s). Encyclopedia of catalysis Vol. 3. Weinheim, Germany: John Wiley & Sons, 2003:447–79.
[15] Wisniak J. The history of catalysis. From the beginning to nobel prizes. Educ Quím. 2010;21:60–69.
[16] Armor JN. A history of industrial catalysis. Catal Today. 2010;163:3–9.
[17] Kieboom A, Moulijn JA, van Leeuwen PWNM, Van Santen RA. History of Catalysis. In: Averill BA, Moulijn JA, van Santen RA, van Leeuwen PWNM, editors. Catalysis: an integrated approach 2nd edition, studies in surface science and catalysis. vol. 123. Amsterdam, The Netherland: Elsevier B.V., 1999:4–17.
[18] McNaught AD, Wilkinson A. IUPAC compendium of chemical terminology. 2nd ed. Cambridge, UK: British Royal Society of Chemistry, 1997.
[19] Cavani F, Trifirò F. Classification of industrial catalysts and catalysis for the petrochemical industry. Catal Today. 1997;34:269–79.
[20] Blaser HU. Heterogeneous catalysis for fine chemicals production. Cataly Today. 2000;60:161–65.
[21] Blaser HU, Steiner H, Studer M. Heterogeneous hydrogenation: a valuable tool for the synthetic chemist. In: Beller M, Bolm C, editors. Transition metals for organic synthesis: building blocks and fine chemicals. vol. 2. 2nd ed. Weinheim, Germany: Wiley-VCH, Verlag GmbH & Co. KGaA, 2004:125–41.
[22] Bartholomew CH. Mechanisms of catalyst deactivation. Appl Catal A. 2001;212:17–60.
[23] Boskovic G, Baerns M. Catalyst Deactivation. In: Baerns M, editor. Basic principles in applied catalysis, springer series in chemical physics. Berlin, Germany: Springer-Verlag, 2004:477–553.
[24] Argyle MD, Bartholomew CH. Heterogeneous catalyst deactivation and regeneration: a review. Catalysts. 2015;5:145–269.
[25] Jha MK, Lee J-C, Kim M-S, Jeong J, Kim B-S, Kumar V. Hydrometallurgical recovery/recycling of platinum by the leaching of spent catalysts: A review. Hydrometallurgy. 2013;133:23–32.
[26] van Leeuwen PWNM. Decomposition pathways of homogeneous catalysts. Appl Catal A. 2001;212:61–81.
[27] Bleken FL, Barbera K, Bonino F, Olsbye U, Lillerud KP, Bordiga S, et al. Catalyst deactivation by coke formation in microporous and desilicated zeolite H-ZSM-5 during the conversion of methanol to hydrocarbons. J Catal. 2013;307:62–73
[28] Moulijn JA, van Diepen AE, Kapteijn F. Catalyst deactivation: is it predictable? What to do? Appl Catal A. 2001;212:3–16.
[29] Montero C, Ochoa A, Castaño P, Bilbao J, Gayubo AG. Monitoring NiO and coke evolution during the deactivation of a $Ni/La_2O_3–\alpha Al_2O_3$ catalyst in ethanol steam reforming in a fluidized bed. J Catal. 2015;331:181–92.
[30] Billege I. 700 Refineries supply oil products to the world. Nafta. 2009;60:401–03.
[31] U.S. Energy Information Administration (EIA) Report. Independent Statistics & Analysis, Available at: https://www.eia.gov/dnav/pet/pet_pnp_cap1_a_(na)_8CDHOR_BpSD_a.htm, Accessed: 9 Sept 2017.
[32] Nakamura D, Stell J. Worldwide refining survey. Oil Gas J. 2002;100:68–111.
[33] Silvy RP. Future trends in the refining catalyst market. Appl Catal A. 2004;261:247–52.
[34] Kettler PB. Platinum group metals in catalysis: fabrication of catalysts and catalyst precursors. Org Process Res Dev. 2003;7:342–54.
[35] Cornils B, Herrmann WA. Applied homogeneous catalysis with organometallic compounds. Weinheim, Germany: Wiley-VCH Verlag GmbH & Co. KGaA, 2002.

[36] King AO, Shinkai I. Platinum(IV) Oxide, in e-EROS encyclopedia of reagents for organic synthesis. Hoboken, New Jersey, USA: John Wiley & Sons, 2001.

[37] Augustine RL. Heterogeneous catalysis for the synthetic chemist. New York, USA: Marcel Dekker, 1996.

[38] Friedmann D, Hakki A, Kim H, Choi W, Bahnemann D. Heterogeneous photocatalytic organic synthesis: state-of-the-art and future perspectives. Green Chem. 2016;18:5391–411.

[39] Clerici MG, Ricci M, Strukul G. Formation of C–O bonds by oxidation. In: Chiusoli GP, Maitlis PM, editors. Metal-catalysis in industrial organic processes. Cambridge, UK: RSC, 2006: 23–78.

[40] Miller JA. Palladium-catalyzed cross coupling of Grignard reagents with in situ-derived enol phosphates. Tetrahedron Lett. 2002;43:7111–14.

[41] Ji Y, Jain S, Davis RJ. Investigation of Pd leaching from supported Pd catalysts during the Heck reaction. J Phys Chem B. 2005;109:17232–38.

[42] Ley SV, Ramarao C, Gordon RS, Holmes AB, Morrsion AJ, McConvey IF, et al. Polyurea-encapsulated palladium(II) acetate: a robust and recyclable catalyst for use in conventional and supercritical media. Chem Commun. 2002;1134–35

[43] Barros JC. Microencapsulated palladium catalysts: pd EnCat™. Aust J Chem. 2010;63:850–51.

[44] Yu J-Q, Wu H-C, Ramarao C, Spencer JB, Ley SV. Transfer hydrogenation using recyclable olyureaencapsulated palladium: efficient and chemoselective reduction of aryl ketones. Chem Comm. 2003;678–97.

[45] Bremeyer N, Ley SV, Ramarao C, Shirley IM, Smith SC. Palladium acetate in polyurea microcapsules: a recoverable and reusable catalyst for hydrogenations. Synlett. 2002;11:1843–44.

[46] Ramarao C, Ley SV, Smith SC, Shirley IM, DeAlmeida N. Encapsulation of palladium in polyurea microcapsules. Chem Comm. 2002;10:1132–33.

[47] Tsuji J. Palladium reagents and catalysts: new perspectives for the twenty-first century. West Sussex, UK: John Wiley & Sons, Ltd, 2004.

[48] Yin L, Liebscher Y. Carbon-carbon coupling reactions catalyzed by heterogeneous palladium catalysts. Chem Rev. 2007;107:133–73.

[49] Farina V. High-turnover palladium catalysts in cross-coupling and heck chemistry: a critical overview. Adv Synth Catal. 2004;346:1553–82.

[50] van Leeuwen PWNM, Claver C. Rhodium catalyzed hydroformylation. Dordrecht, The Netherlands: Kluwer Academic Press, 2000.

[51] Osborn JA, Jardine FH, Young JF, Wilkinson G. The preparation and properties of tris(triphenylphosphine)halogenorhodium(I) and some reactions thereof including catalytic homogeneous hydrogenation of olefins and acetylenes and their derivatives. J Chem Soc A. 1966;0:1711–32.

[52] Wu Y, Singh RP, Deng L. Asymmetric olefin isomerization of butenolides via proton transfer catalysis by an organic molecule. J Am Chem Soc. 2011;133:12458–61.

[53] Evans DA, Fu GC. The rhodium-catalyzed hydroboration of olefins: a mechanistic investigation. J Org Chem. 1990;55:2280–82.

[54] Evan PA, editor. Modern rhodium-catalyzed organic reactions.Weinheim, Germany: Wiley-VCH Verlag GmbH & Co. KGaA, 2005.

[55] Lefebvre F. Applications of the olefin metathesis reaction to industrial processes. In: Khosravi E, Szymanska-Buzar T, editor(s). Ring opening metathesis polymerisation and related chemistry. New York, NY, USA: Kluwer Academi Publishers, 2002:247–61.

[56] Mitsudo T, Hori Y, Yamakawa Y, Watanabe Y. Ruthenium catalyzed selective synthesis of enol carbamates by fixation of carbon dioxide. Tetrahedron Lett. 1987;28:4417–18.

[57] Sharpless KB, Akashi K, Oshima K. Ruthenium catalyzed oxidation of alcohols to aldehydes and ketones by amine-n-oxides. Tetrahedron Lett. 1966;17:2503–06.

[58] Hwang YH, Cheng CP. Dioxygen activation by [Co_2(μ-OH)L]{H_3L=2,6-Bis[(2-hydroxyphenyl)iminomethyl]-4-methylphenol}, catalysed cooxygenation of PPh_3 and alkenes, alcohols, a sulfide, a sulfoxide and diphenylacetylene. J Chem Soc Chem Commun. 1992;4:317–19.
[59] Plummer JS, Murahashi S, Zhao C, Murahashi SI. Dichlorotris(triphenylphosphine)ruthenium(II), in e-EROS Encyclopedia of Reagents for Organic Synthesis. Hoboken, New Jersey, USA: John Wiley & Sons, 2001.
[60] Jones JH. The Cativa™ process for the manufacture of acetic acid iridium catalyst improves productivity in an established industrial process. Platinum Metals Rev. 2000;44:94–105.
[61] Westcott SA. Trans-Carbonyl(chloro)bis(triphenylphosphine)iridium(I). e-EROS encyclopedia of reagents for organic synthesis. Hoboken, New Jersey, USA: John Wiley & Sons, 2001.
[62] Xu Y, Mingos DMP, Brown JM. Crabtree's catalyst revisited; Ligand effects on stability and durability. Chem Commun. 2008;199–201.
[63] Oro LA, Claver C. Iridium complexes in organic synthesis. Weinheim, Germany: Wiley-VCH, Verlag GmbH & Co. KGaA, 2009.
[64] Andersson PG. Iridium catalysis. Berlin, Germany: Springer-Verlag, 2011.
[65] Church TL, Andersson PG. Iridium catalysts for the asymmetric hydrogenation of olefins with nontraditional functional substituents. Coord Chem Rev. 2008;252:513–31.
[66] Kolb HC, VanNieuwenhze MS, Sharpless KB. Catalytic asymmetric dihydroxylation. Chem Rev. 1994;94:2483–547.
[67] Gonzalez J, Koontz JI, Hodges LM, Nillson KR, Neely LK, Myers WH, et al. Osmium-promoted dipolar cycloadditions with pyrroles: an efficient stereoselective synthesis of 7-Azanorbornanes. J Am Chem Soc. 1995;117:3405–21.
[68] Dong H, Zhao J, Chen J, Wu Y, Li B. Recovery of platinum group metals from spent catalysts: a review. Int J Min Proc. 2015;145:108–13.
[69] Alonso E, Sherman AM, Wallington TJ, Everson MP, Field FR, Roth R, et al. Evaluating rare earth element availability: a case with revolutionary demand from clean technologies. Environ Sci Technol. 2012;46:3406–14.
[70] Brykin AV, Artemov AV, Kolegov KA. Analysis of the market of rare earth elements (REEs) and REE catalysts. Catal Ind. 2014;6:1–7.
[71] Kovacevic M, Agarwal S, Mojet BL, van Ommen JG, Lefferts L. The effects of morphology of cerium oxide catalysts for dehydrogenation of ethylbenzene to styrene. Appl Catal A. 2015;505:354–64.
[72] Gärtner CA, van Veen AC, Lercher JA. Oxidative dehydrogenation of ethane: common principles and mechanistic aspects. ChemCatChem. 2013;5:3196–217.
[73] Madeira LM, Portela MF. Catalytic oxidative dehydrogenation of n-butane. Cat Rev. 2002;44:247–86.
[74] Sepúlveda-Escribano A, Silvestre-Albero J, Coloma F, Rodríguez-Reinoso F. Preparation and characterization of carbon-supported Pt-CeO2 catalysts. In: Corma A, Melo FV, Mendioroz SG, Fierro JLG, editor(s). Studies in surface science and catalysis Vol. 130. Amsterdam: Elsevier, 2000:1013–18.
[75] Abid M, Paul-Boncour V, Touroude R. Pt/CeO_2 catalysts in crotonaldehyde hydrogenation: selectivity, metal particle size and SMSI states. Appl Catal A. 2006;297:48–59.
[76] Silva JS, de Jesus-Neto RM, Fiuza RA, Gonçalves JP, Mascarenhas AJS, Andrade HMC. Alkali-activation of spent fluid cracking catalysts for CO_2 capture. Microporous Mesoporous Mater. 2016;232:1–12.
[77] Asghari I, Mousavi SM, Amiri F, Tavassoli S. Bioleaching of spent refinery catalysts: a review. J Ind Eng Chem. 2013;19:1069–81.
[78] Dixit M, Menon A, Baruah R, Bhargav A, Sharma S. Oxidative activation of methane on lanthanum oxide and nickel-lanthanum oxide catalysts. React Kinet Mech Cat. 2015;115:611–24.

[79] Hereijgers BPC, Eggenhuisen TM, De Jong KP, Talsma H, van der Eerden AMJ, Beale AM, et al. Understanding the promotion effect of lanthanum oxide on gold-based catalysts in the partial oxidation of methanol by in situ XAFS and DSC studies. J Phys Chem C. 2011;115:5545–54.
[80] Hung CM. Development of a copper-lanthanum oxide catalyst for the removal of gaseous ammonia by catalytic oxidation: physicochemical and electrochemical characterization of catalyst materials. Int J Energy Res. 2013;37:2001–08.
[81] La N, Al OO. Characterization and analysis of carbon deposited during the dry reforming of methane. Chin J Catal. 2009;30:1076–84.
[82] Gao J, Hou Z, Guo J, Zhu Y, Zheng X. Catalytic conversion of methane and CO_2 to synthesis gas over a La_2O_3-modified SiO_2 supported Ni catalyst in fluidized-bed reactor. Catal Today. 2008;131:278–84.
[83] Breslow R. Hydrophobic effects on simple organic reactions in water. Acc Chem Res. 1991;24:159–64.
[84] Sanzhong L, Baolian Z, Ming X, Jańczuk A, Wenhua X, Jinpei C, et al. Recent progress in lanthanide-catalyzed organic reactions in protic media. Chinese Sci Bulletin. 2001;46:1673–81.
[85] De Smet F, Ruiz P, Delmon B, Devillers M. Catalytic behaviour of multiphasic oxide catalysts containing lanthanides (La, Ce, Pr, Sm, Tb) in the selective oxidation of isobutene to methacrolein. Catal Lett. 1996;41:203–07.
[86] Wang J, Lou Y, Xu C, Song S, Liu W. Magnetic lanthanide oxide catalysts: an application and comparison in the heterogeneous catalytic ozonation of diethyl phthalate in aqueous solution. Sep Purif Technol. 2016;159:57–7.
[87] Shibasaki M, Yoshikawa N. Lanthanide complexes in multifunctional asymmetric catalysis. Chem Rev. 2002;102:2187–209.
[88] Wieszczycka K, Staszak K. Artificial metalloenzymes as catalysts in non-natural compounds synthesis. Coord Chem Rev. 2017;351:160–71.
[89] Zeifert B, Blásquez JS, Moreno JGC, Calderón HA. Raney-nickel catalysts produced by mechanical alloying. Rev Adv Mater Sci. 2008;18:632–38.
[90] Zeng L, Cheng CY. A literature review of the recovery of molybdenum and vanadium from spent hydrodesulphurisation catalysts Part I: metallurgical processes. Hydrometallurgy. 2009;98:1–9.
[91] Dadashi-Silab S, Atilla Tasdelen M, Mohamed Asiri A, Bahadar Khan S, Yagci Y. Photoinduced atom transfer radical polymerization using semiconductor nanoparticles. Macromol Rapid Commun. 2014;35:454–59.
[92] Furimsky E. Spent refinery catalysts: environment, safety and utilization. Catal Today. 1996;30:223–86.
[93] Gholami RM, Mousavi SM, Borghei SM. Process optimization and modeling of heavy metals extraction from a molybdenum rich spent catalyst by *Aspergillus niger* using response surface methodology. J Ind Eng Chem. 2012;18:218–24.
[94] Wu D. Advances in China's refining and petrochemical catalysts. Catal Today. 1996;30:21–31.
[95] Sun PP, Lee MS. Separation of Pt from hydrochloric acid leaching solution of spent catalysts by solvent extraction and ion exchange. Hydrometallurgy. 2011;110:91–98.
[96] de Sá Pinheiro AA, de Lima TS, Campos PC, Afonso JC. Recovery of platinum from spent catalysts in a fluoride-containing medium. Hydrometallurgy. 2004;74:77–84.
[97] Barakat MA, Mahmoud MHH. Recovery of platinum from spent catalyst. Hydrometallurgy. 2004;72:179–84.
[98] Mahmoud MHH. Leaching platinum-group metals in a sulfuric acid/chloride solution. Jom. 2003;55:37–40.
[99] Kim MS, Kim EY, Jeong J, Lee JC, Kim W. Recovery of platinum and palladium from the spent petroleum catalysts by substrate dissolution in sulfuric acid. Mater Trans. 2010;51:1927–33.

[100] Sun PP, Lee MS. Recovery of platinum from chloride leaching solution of spent catalysts by solvent extraction. Mater Trans. 2013;54:74–80.
[101] Motaghed M, Mousavi SM, Rastegar SO, Shojaosadati SA. Platinum and rhenium extraction from a spent refinery catalyst using *Bacillus megaterium* as a cyanogenic bacterium: statistical modeling and process optimization. Bioresour Technol. 2014;171:401–09.
[102] Paiva AP, Ortet O, Carvalho GI, Nogueira CA. Recovery of palladium from a spent industrial catalyst through leaching and solvent extraction. Hydrometallurgy. 2017;171:394–401.
[103] Duncan TM, Winslow P, Bell AT. The characterization of carbonaceous species on ruthenium catalysts with ^{13}C nuclear magnetic resonance spectroscopy. J Catal. 1985;93:1–22.
[104] Conrour JP, Pannetier G. Hydrazine decomposition over a supported iridium catalysts. J Catal. 1972;24:434–45.
[105] Reed DW, Fujita Y, Daubaras DL, Jiao Y, Thompson VS. Bioleaching of rare earth elements from waste phosphors and cracking catalysts. Hydrometallurgy. 2016;166:34–40.
[106] Innocenzi V, Farella F, De Michelis I, Vegliò F. Treatment of fluid catalytic cracking spent catalysts to recover lanthanum and cerium: comparison between selective precipitation and solvent extraction. J Ind Eng Chem. 2015;24:92–97.
[107] Zhao Z, Qiu Z, Yang J, Lu S, Cao L, Zhang W, et al. Recovery of rare earth elements from spent fluid catalytic cracking catalysts using leaching and solvent extraction techniques. Hydrometallurgy. 2017;167:183–86.
[108] Ferella F, Innocenzi V, Maggiore F. Oil refining spent catalysts: a review of possible recycling technologies. Resour Conserv Recy. 2016;108:10–20.
[109] Wang J, Huang X, Wang L, Wang Q, Yan Y, Zhao N, et al. Kinetics study on the leaching of rare earth and aluminum from FCC catalyst waste slag using hydrochloric acid. Hydrometallurgy. 2017;171:312–19.
[110] Ye S, Jing Y, Wang Y, Fei W. Recovery of rare earths from spent FCC catalysts by solvent extraction using saponified 2-ethylhexyl phosphoric acid-2-ethylhexyl ester (EHEHPA). J Rare Earths. 2017;35:716–22.
[111] Santhiya D, Ting Y-P. Use of adapted *Aspergillus niger* in the bioleaching of spent refinery processing catalyst. J Biotechnol. 2006;121:62–74.
[112] Beolchini F, Fonti V, Ferella F, Veglio F. Metal recovery from spent refinery catalysts by means of biotechnological strategies. J Hazard Mater. 2010;178:529–34.
[113] Mafi Gholami R, Borghei SM, Mousavi SM. Heavy metals recovery from spent catalyst using *Acidithiobacillus ferrooxidans* and *Acidithiobacillus thiooxidans*. International Conference on Chemistry and Chemical Engineering (ICCCE), 2010.
[114] Stołecki K, Gołębiowski A, Borowiecki T. The kinetics of hydrogenation of benzene over a nickel catalyst under industrial conditions. Przem Chem. 2004;83:564–66.
[115] http://www.ins.pulawy.pl, Accessed: 14 Sept. 2017.
[116] Srichandan H, Singh S, Kim DJ, Lee SW. A comparative study of metal extraction from spent catalyst using *Acidithiobacillus ferrooxidans*. Eng Technol. 2013;7:112–16.
[117] Painuly AS. Separation and recovery of vanadium from spent vanadium pentaoxide catalyst by Cyanex 272. ESR. 2005;4:7–13.
[118] Mishra D, Kim DJ, Ralph DE, Ahn JG, Rhee YH. Bioleaching of vanadium rich spent refinery catalysts using sulfur oxidizing lithotrophs. Hydrometallurgy. 2007;88:202–09.
[119] Shahrabi-Farahani M, Yaghmaei S, Mousavi SM, Amiri F. Bioleaching of heavy metals from a petroleum spent catalyst using *Acidithiobacillus thiooxidans* in a slurry bubble column bioreactor. Sep Purif Technol. 2014;132:41–49.
[120] Gerayeli F, Ghojavand F, Mousavi SM, Yaghmaei S, Amiri F. Screening and optimization of effective parameters in biological extraction of heavy metals from refinery spent catalysts using a thermophilic bacterium. Sep Purif Technol. 2013;118:151–61.

[121] http://www.ecotrade.org/homepage/puresphere/offerdetail.asp?cat=s&OfferID=114, Accessed: 14 Sept. 2017.

[122] Brown C, Krzystek J, Achey R, Lita A, Fu R, Meulenberg RW, et al. Mechanism of initiation in the Phillips ethylene polymerization catalyst: redox processes leading to the active site. ACS Catal. 2015;5:5574–83.

[123] Śpiewak Z, Gołębiowski A, Kruk J, Stołecki K, Baran P. The TZC-3/1 catalyst for high-temperature shift reaction. Przem Chem. 2007;86:437–39.

[124] Peng Z, Li Z, Lin X, Tang H, Ye L, Ma Y, et al. Pyrometallurgical recovery of platinum group metals from spent catalysts. JOM. 2017;69:1553–62.

[125] Kim C-H, Woo SI, Jeon SH. Recovery of platinum-group metals from recycled automotive catalytic converters by carbochlorination. Ind Eng Chem Res. 2000;39:1185–92.

[126] Chiang KC, Chen KL, Chen CY, Huang JJ, Shen YH, Yeh MY, et al. Recovery of spent alumina-supported platinum catalyst and reduction of platinum oxide via plasma sintering technique. J Taiwan Inst Chem Eng. 2011;42:158–65.

[127] Marafi M, Stanislaus A. Spent hydroprocessing catalyst management: a review: Part II. Advances in metal recovery and safe disposal methods. Resour Conserv Recy. 2008; 53:1–26.

[128] Marafi M, Stanislaus A. Waste catalyst utilization: extraction of valuable metals from spent hydroprocessing catalysts by ultrasonic-assisted leaching with acids. Int Eng Chem Res. 2011;50:9495–501.

[129] Al-Mansi NM, Abdel Monem NM. Recovery of nickel oxide from spent catalyst. Waste Manag. 2012;22:85–90.

[130] Ivascanu S, Roman O. Nickel recovery from spent catalysts. I Solvation process. Bul Inst Politeh Iasi Sect 2. 1975;2:47–52.

[131] Sahu KK, Agarwal A, Pandey BD. Nickel recovery from spent nickel catalyst. Waste Manage Res. 2005;23:148–54.

[132] Al-Sheeha H, Marafi M, Raghavan V, Rana MS. Recycling and recovery routes for spent hydroprocessing catalyst waste. Ind Eng Chem Res. 2013;52:12794–801.

[133] Calderón H, Endara D. Recovery of vanadium from acid and basic leach solutions of spent vanadium pentoxide catalysts. J Geolog Res Eng. 2015;4:213–18.

[134] Busnardo RG, Busnardo NG, Salvato GN, Afonso JC. Processing of spent NiMo and CoMo/Al_2O_3 catalysts via fusion with $KHSO_4$. J Hazard Mater. 2007;B139:391–98.

[135] Chauhan G, Pant KK, Nigam KDP. Metal recovery from hydroprocessing spent catalyst: a green chemical engineering approach. Ind Eng Chem Res. 2013;52:16724–36.

[136] Peters RW. Chelant extraction of heavy metals from contaminated soils. J Hazard Mater. 1999;66:151–210.

[137] Chauhan G, Pant KK, Nigam KDP. Extraction of nickel from spent catalyst using biodegradable chelating agent EDDS. Ind Eng Chem Res. 2012;51:10354–63.

[138] Goel S, Pant KK, Nigam KDP. Extraction of nickel from spent catalyst using fresh and recovered EDTA. J Hazard Mat. 2009;171:253–61.

[139] Vuyyuru KR, Pant KK, Krishnan VV, Nigam KDP. Recovery of nickel from spent industrial catalysts using chelating agents. Ind Eng Chem Res. 2010;49:2014–24.

[140] Amiri F, Yaghmaei S, Mousavi SM, Sheibani S. Recovery of metals from spent refinery hydrocracking catalyst using adapted *Aspergillus niger*. Hydrometallurgy. 2011;109: 65–71.

[141] Amiri F, Mousavi SM, Yaghmaei S. Enhancement of bioleaching of a spent Ni/Mo hydroprocessing catalyst by *Penicillium simplicissimum*. Sep Purif Technol. 2011;80:566–76.

[142] Amiri F, Mousavi SM, Yaghmaei S, Barati M. Bioleaching kinetics of a spent refinery catalyst using *Aspergillus niger* at optimal conditions. Biochem Eng J. 2012;67:208–17.

[143] Pradhan D, Mishra D, Kim DJ, Ahn JG, Chaudhury GR, Lee SW. Bioleaching kinetics and multivariate analysis of spent petroleum catalyst dissolution using two acidophiles. J Hazard Mat. 2010;175:267–73.
[144] Srichandan H, Singh S, Pathak A, Kim D-J, Lee S-W, Heyes G. Bioleaching of metals from spent refinery petroleum catalyst using moderately thermophilic bacteria: effect of particle size. J Environ Sci Health A. 2014;49:807–18.
[145] Kim D-J, Pradhan D, Ahn J-G, Lee S-W. Enhancement of metals dissolution from spent refinery catalysts using adapted bacteria culture – Effects of pH and Fe(II). Hydrometallurgy. 2010;103:136–43.
[146] Amiri F, Yaghmaei S, Mousavi S. Bioleaching of tungsten-rich spent hydrocracking catalyst using *Penicillium simplicissimum*. Bioresour Technol. 2011;102:1567–73.
[147] Akcil A, Vegliò F, Ferella F, Okudan MD, Tuncuk A. A review of metal recovery from spent petroleum catalysts and ash. Waste Manag. 2015;45:420–33.
[148] www.moxba.com, Accessed: 16 Oct 2017.
[149] www.nickelhuette-aue.de, Accessed: 16 Oct 2017.
[150] www.sadaci.be, Accessed: 16 Oct 2017.
[151] www.treibacher.com, Accessed: 16 Oct 2017.
[152] http://www.eramet.com, Accessed: 16 Oct 2017.
[153] globemetal.com, Accessed: 16 Oct 2017.
[154] www.aci-industries.com, Accessed: 16 Oct 2017.
[155] www.metalandcatalyst.com, Accessed: 16 Oct 2017.
[156] tetronics.com, Accessed: 16 Oct 2017.

Karolina Wieszczycka

7 Wastes generated by mineral extraction industries

Abstract: Mineral extraction industries at each mining step generate large volume of waste, and most of them represent potentially sources of crucial metals. This chapter describes extraction and main mining processes, principal classifications of the generated wastes, as well as the chemical properties of the waste material, their disposal and further treatment to recover metals. Most of the presented technological solutions are or have been tested in real processes or using real waste materials.

Keywords: metal recovery, mineral commodities, mineral extraction

7.1 Introduction

For the last years, global economic growth has resulted in a rapid increase in demand for a range of mineral commodities, as witnessed by the fact that up to 2014 world mineral extraction increased by a factor of 1.7 [1–3]. Technological evolution improving mining techniques influences on different types of mining waste and their management. Each mining step is able to generate large quantities of unmarketable or uneconomic materials, but their volumes, compositions and perspective of future utilization depend on the type of mining processing procedure and the type of the deposit. Numerous mining techniques have been developed to address the problem of extracting ores and materials, and the selection of the techniques depends on the variety of morphologies of natural deposits and the large variety of useful mineral substances.

Wastes are generated at all levels of the recovery process, mostly employed to meet demand. For a given mineral, it will have different physicochemical properties according to composition of ore, type of deposit and the conditions in which wastes have been generated. Waste is a permanently unused extractive product that is usually stored in the immediate vicinity of mining. The amount of mining waste that can be deposited in a mining center varies considerably and depends mainly on the mining processes and local regulations, e. g. U.S laws and regulatory agencies provide a framework and baseline for increased sustainability focus for mining activities. Within European Community, the EU directive on the management of waste from extractive industries (2006/21/EC) provides for measures, procedures and guidance of

This article has previously been published in the journal *Physical Sciences Reviews*. Please cite as: Wieszczycka, K. Wastes generated by mineral extraction industries. *Physical Sciences Reviews* [Online] **2018**, *3*. DOI: 10.1515/psr-2018-0026

https://doi.org/10.1515/9783110547061-007

waste management. It also provides a framework defined areas of intervention and associated measurable parameters and indicators. In Asia, the problem with mineral waste is much visible, e. g. China ranks the first place among gold production countries in the world. Meanwhile, an increase of hazardous industrial wastes like cyanide tailings generated by gold industry is becoming a serious problem. In China, probably each year, more than 20 million tons of cyanide tailings are produced [4–6].

7.2 Mineral commodities

Mining is a main source of mineral commodities, which are necessary for increasing and maintaining technological manufacturing. Some mineral deposits are formed, improved or preserved by geologic processes, and natural leaching process concentrates these minerals as ore of different grade. For example, in the last years, Brazil has been estimated to be the world's leading producer of niobium (90 % of world production), main producer of iron ore and one of the top five producers of mined bauxite and tin [7] (Figure 7.1). Canada is the leading producer of mined potash, the third-ranked producer of mined bauxite and nickel and one of the top five producers of gem-quality diamond, palladium and platinum. Australia is the world's leading producer of bauxite (29 %); however, it has also large resources of cobalt, copper, diamond, gold, iron ore, lead, lithium, manganese, tantalum and uranium. China even though has large resources of many minerals, but it is the world's leading producer of aluminum (Bayer process, 48 % of the total world production). Russia accounted for 28 % of total natural diamond production in the world, while Kazakhstan in 2014 was responsible for 41 % of uranium production [8]. The United State uses more mineral commodities than it produces. Some industrial minerals are concentrated in certain parts of the country. Phosphate mining is confined to Florida, North Carolina, Idaho, Utah and Wyoming. Sulfur is mined in the offshore Gulf of Mexico and Texas. Another source is sulfur recovered from power plants, smelters and petroleum refineries. The Carolinas and Georgia are the only sources of high-grade kaolin, alumina and bauxite. The United States has also a rare-earth element mine located in southeast California (e. g. deposit at Mountain Pass). Smaller nations may benefit from localized deposits of individual minerals. For example, Chile is the leading producer of mined copper and the second-ranked producer of refined copper in the world. Peru was one of the top five producers of mined copper, gold, lead, tin and zinc in the world and was among the top five producers of tin metal. Mexico was among the top five producers of mined lead in the world. Europe currently retains a modest position in world mining activity in terms of scale of production and mineral reserves [9], however, together with Central Eurasia accounting for 57 % of the world's production of uranium, 54 % of lignite coal and 42 % of potash.

Other mineral production is also without clear leadership. Titanium sponge production and refined nickel metal accounted for 39 % of world production, slightly less refined palladium (31 %), chromite (26 %), ammonia (26 %), zinc metal (23 %),

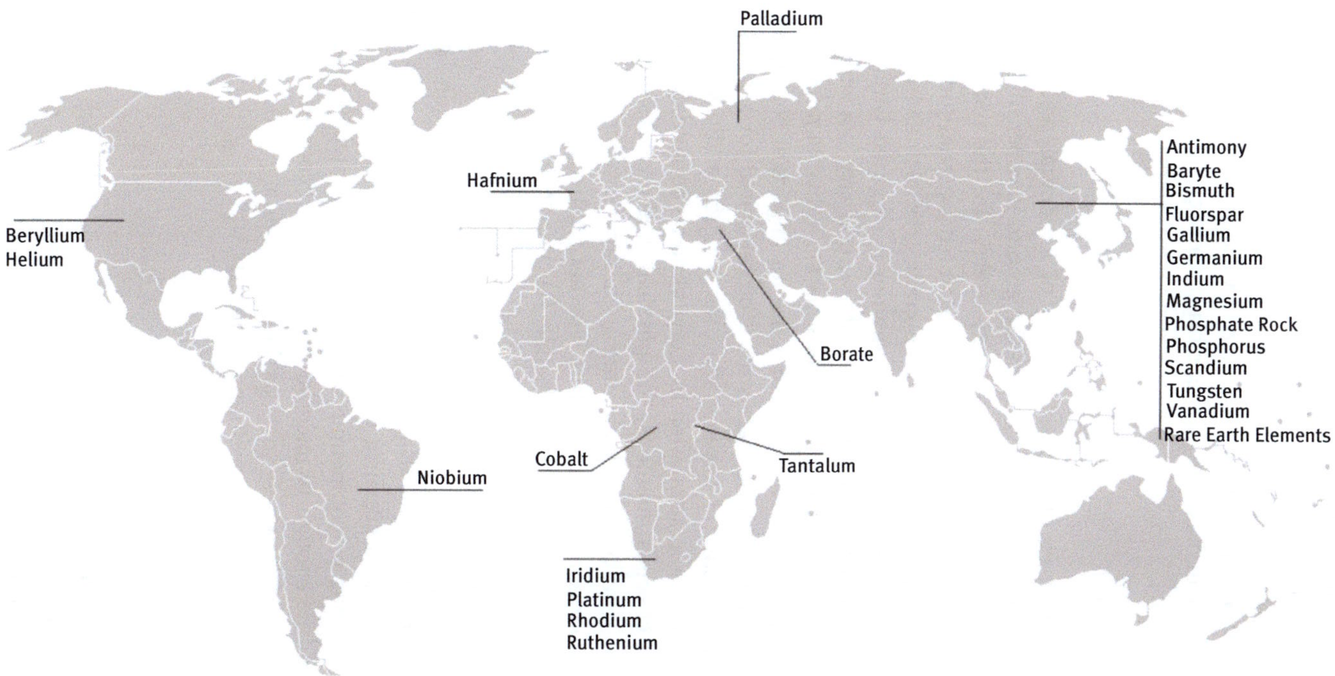

Figure 7.1: Global map of critical metals (according to data from [10]).

copper from secondary sources (22 %), aluminum (35 %), lead (22 %) and refined platinum (20 %). Most ores are mixtures of extractable minerals and useless or worth less rocky material known as gangue. Metallic minerals are often found in certain associations within which they may occur as mixtures of a wide range of particle sizes or as single-phase solid solutions or compounds. In order to produce metals, the ore minerals must be broken down by the action of various process, either alone or in combination, which can make the mining most economic.

7.3 Mining and enrichment

Mining is the first stage of the exploitation of a mineral resource. It can be defined as the extraction of material from the ground in order to recover the ore. The waste generated in this stage is mostly useless for the industry. The main order of the stage is to provide a more concentrated material for the procedures of extractive metallurgy. In the case of high-grade ores such as iron and bauxite, with mineral contents higher than 35 %, the ore is exported without processing stage [11]. Unfortunately, in many cases, it is necessary not only to separate valuable mineral from gangue, but also to separate valuable minerals from each other. For example, porphyry copper ores are an important source of molybdenum and the minerals of these metals must be separated for separate smelting. Therefore, the processing of lower grade ores required additional chemical (or biological), physical (e. g. separation by gravity) and a combination of processes. Comparing with chemical methods, the physical methods used in mineral processing consume relatively small amounts of energy. However, this not a general rule, and the important factor is deciding whether the treatment of the deposit is profitable, especially if ore grades decline [11–14]. Mineral processing makes lower smelter energy costs and lower slag production, but the treatment of low-grade ores is uneconomical and technologically difficult, because of the need to produce high-grade metals. Not only poor content ore threatens the economy of the mining, but also a complexity of the rocky material. Mineral ores commonly occur as mixtures of a wide range of particle sizes or as single-phase solid solutions or compounds [15]. The metal content varies for various ores, but should be enough to make metal extraction economic. Also the selections of additional processes such as beneficiation methods depend on the mineralogy and chemical composition of the deposit.

7.3.1 Physical enrichment

Broadly speaking, separation processes can be divided into two main categories: sizing separation and concentration separation. In some cases, sizing separation is carried out with a view to obtaining a product which must meet a size specification.

More importantly, however, sizing separation is applied to control the size of material to be fed to other equipment. Such control is necessary because equipment has an optimum size of material that it can handle in an efficient manner. The two basic operations in a sizing separation process are classification and screening. The chief objective of concentration separation is to obtain valuable minerals in more concentrated forms. The processes employed for accomplishing this objective rely on both physical and surface chemical properties of the minerals. The main concentration separation processes are gravity concentration, magnetic separation, electrostatic separation and flotation.

Physical separation is the mainstay of industrial mineral processing. It is a separation of the valuable mineral by exploiting differences in its physical characteristics. Some mineral properties that may be exploited are size, shape, density, magnetic susceptibility, electrical conductivity, or surface properties. For example, the separation of minerals based on the difference in the specific gravity of the valuable mineral to gangue and the fluid medium is gravity separation. Another method based on natural or induced differences in magnetic susceptibility or conductivity of the minerals is magnetic and electrostatic separation but the most frequently utilized method is froth flotation which is based on the surface chemistry properties of a mineral (Figure 7.2). The natural or modified surface property of a mineral determines its ability to attach to an air bubble and float to the surface. Physical separation methods are generally cost-effective, with regard to both capital and operating costs; therefore, the ever-varying grade and mineralogy are two major challenges faced by physical separation of major ores.

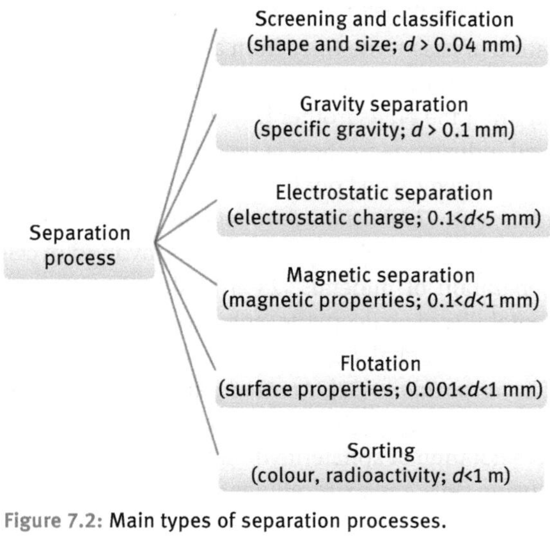

Figure 7.2: Main types of separation processes.

7.3.1.1 Flotation

Flotation is a selective process and can be used to achieve specific separations from complex ores. Initially developed to treat the sulfides of copper, lead and zinc, the field has been expanded to include platinum, nickel and gold-hosting sulfides and oxidized minerals, such as malachite [16]. There are numerous factors affecting flotation, from which the most important are particle size (separation of minerals sized from 5 μm to 3.3 mm [17]), pulp density, air bubble size, agitation intensity, conditioning time, particle surface charge, pH of slurry and type and amount of reagent. Chemistry-based solutions play a pivotal role in flotation process. They provide optimal metallurgical performance, especially in the scope of variability in operational factors, comminution and mineralogy [18]. Flotation has also nowadays found wide applications, particularly, in industrial wastewater treatment containing mainly metal ions [19–21]. Especially, the development of froth flotation allowed the exploitation of vast low-grade deposits which were previously uneconomical to treat. The reagents used in flotation are usually divided into two categories: surfactants, also known as collectors, and regulating or modifying agents. Surfactants are that class of reagents which, thanks to their amphiphilic structure, create a hydrophobic film on the mineral surface and thereby bring about the attachment of mineral particles with the air bubbles produced in the froth. The surfactants of particular importance in flotation may be classified as (i) monopolar and (ii) multipolar, but each group is subdivided into (i) thio-compounds, which act mainly as collectors for metallic sulfides; (ii) nonthio, ionizable compounds, which may act as both collectors and frothers; and (iii) nonionic compounds, some of which act mainly as frothers, while others act as depressants, flocculating agents and even as activators [22]. The main collector groups are categorized as is shown in Figure 7.3.

The frother should not in itself be a strong collector and should work in the presence of the other reagents necessary for flotation. An example of flotation with innovative collector is the beneficiation of cassiterite fines from a tailing slime in the Datun concentration plant using froth flotation process with benzohydroxamic acid as collector, lead nitrate as activator and pine oil as frother [24]. The classification of the reagents employed by the mineral industry as frothers is presented in Figure 7.4.

In recent years, several studies have been carried out on the use of biosurfactants as flotation reagents [25–27]. Flotation with *Thiobacillus ferrooxidans* may provide a novel approach to mineral processing in which the biological functions involved in cell adhesion play a key role in the separation of minerals [28–31]. Also the effect of *Leptospirillum ferrooxidans* on the floatability of chalcopyrite, sphalerite and pyrrhotite was studied. It was found that in the case of chalcopyrite, the flotation rate significantly increased in the presence of *L. ferrooxidans* [32]. Another paper indicated that the higher rate of leaching with bacteria is due to the bacteria increasing the pH at the surface of the pyrite [33]. The selective flotation of sphalerite from a sphalerite–galena mineral mixture has been achieved using cells and extracellular secretions of *Bacillus*

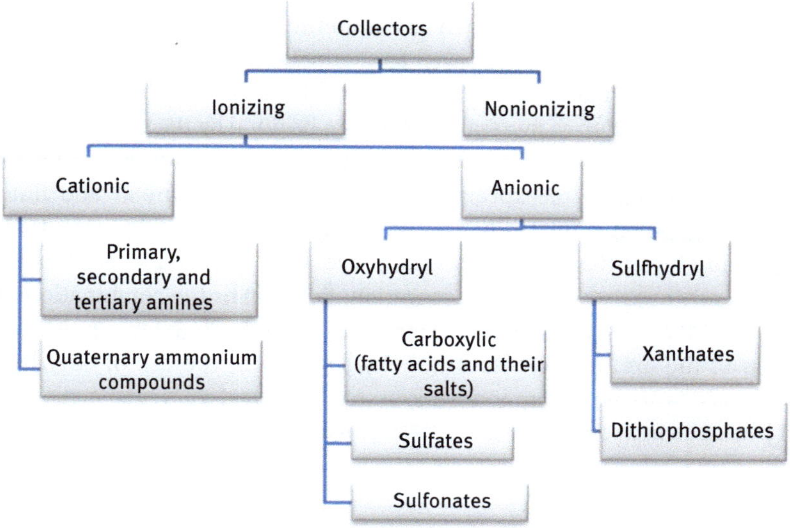

Figure 7.3: Classification of collectors [22, 23].

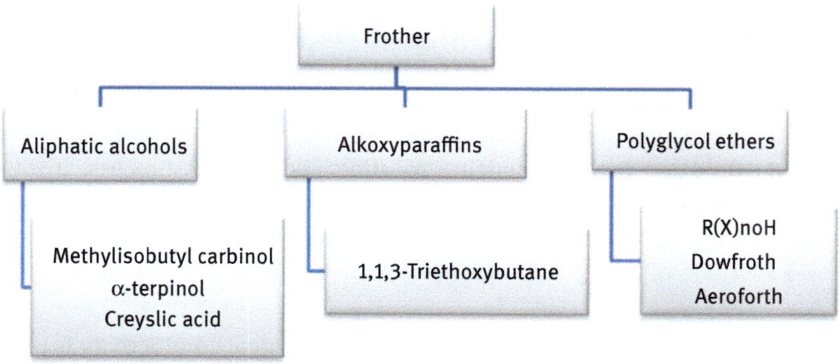

Figure 7.4: Classification of frothers [22].

megaterium [34]. After flotation, the obtained concentrates required to be dried for final processing. Dewatering methods are sedimentation, filtration and thermal drying.

7.3.1.2 Gravity separation

Gravity separation is based on differences in density between the minerals. This technique uses the differential movement of mineral particles in water due to their different hydraulic properties. It is widely used in processes of coal beneficiation and in the preconcentration of metalliferous ores [15, 35, 36]. The introduction of the flotation process has caused the gravity concentration methods to lose much of their importance. However, nonsulfide ores such as cassiterite, chromite and wolframite and certain nonmetallics do not respond well to economic flotation, and so gravity methods are still used for their treatment. Prediction of the applicability of gravity concentration to the separation of mineral mixture can be done by the concentration criterion, which is the ratio of differences between specific gravity of the heavy mineral and fluid medium to the specific gravity of the light mineral and fluid medium. Gravity separation is relatively easy when value of the ratio is larger than 2.5; however, a ratio lower than 1.25 makes the process not feasible. Most of metal concentration usually consists of gravity separation, flotation or the combination of both processes, and the choice among those methods depends on the mineralogical characterization of the ore. In the case of gravity separation, the use of only this technique was very rare; however, innovations over the last 25 years make the gravitational preconcentration of low-grade ores possible [37].

7.3.1.3 Magnetic separation

The most economic and environmental friendly method to recover minerals is magnetic separation. This is a separation and capture of fine magnetic particles by the magnetic force acting on the particles in a gradient magnetic field [38]. The principle of the process is the selective adsorption of the magnetite material onto a mineral surface, which makes it amenable to recover by the magnetic separation techniques [39, 40]. An example of this is cassiterite (tin-based mineral), which is often associated with traces of magnetite or wolframite that can be removed by magnetic separators [41]. The separability of low-grade hematite from Tangshan was also investigated via fluidized bed magnetizing roasting [42]. In this process, a low-grade hematite is converted to strongly magnetic magnetite or maghemite, which can then be recovered by low-intensity magnetic separation.

7.3.1.4 Electrostatic separation

Electrostatic separation is the method where property electrical conductivity is used to separate out desirable mineral from mixture. Electrostatic forces are generated by the action of an electric field on a charged particle. After contact with a conductor, both conducting and nonconducting particles become polarized, but only conducting particles achieve an equipotential surface [43]. Electrostatic separation can be applied for only a small number of minerals, e. g. rutile and ilmenite deposits. The major application of electrostatic separation is in the processing of beach sands, coal and alluvial deposits containing titanium minerals.

Although each of the methods is efficient individually, the recovery of metals such as gold, tin, titanium, zirconium and certain rare-earth element minerals from low-grade ore is investigated by combination of froth flotation process and magnetic separation [44] or gravity, magnetic and electrical processes [45]. For example, the separation of xenotime uses a combination of gravity separation, electrostatic separation and strong magnetic separation. However, flotation is always used for fine grain xenotime [46].

7.3.2 Waste produced at physical enrichment stage

The exploitation of a mineral resource and it beneficiation (if required) produces various types of waste, which composition depends on the used mining process. Mostly, there are materials that must be removed to gain access to the mineral resource. Waste rock is a permanently unused extractive product that is usually stored for an indefinite period in a landfill that is usually located in the immediate vicinity of mining. The amount of mining waste that can be deposited in a mining center varies considerably and depends mainly on the selectivity of the mining method. As a rule, opencast pits and quarries generate much more mining waste than an underground mine. Most of the waste rocks are inert and therefore cannot pose a significant threat to the environment, even if they are stored in large quantities. However, waste generated in nonferrous metal extraction industries may contain large quantities of heavy metals. Useless wastes from the extractive industries have therefore to be properly managed and additionally stabilized to prevent leaching of hazardous metals. Table 7.1 shows the content of main component of the exemplary mine waste rocks.

Waste from physical processing of mineral resources is one of the largest waste streams. The above-described enrichment processes considerably increase the value of the ore to allow economic smelting. However, the provision of clean concentrates, with little or no contamination with associated metals, is not always economically feasible, and this leads to loss as tailing. On the other hand, obtaining a very high-grade concentrate from low-grade minerals is also combined with large amount of waste.

Tailings are generated by the mineral processing of the ore (gravity, magnetic or electrostatic separation, flotation). The structure and composition of waste vary greatly from one process to another, e. g. as a result of milling and flotation they are generally finely grained rocks in suspension in water [52]. Enrichment processes are not fully efficient in performing the separation between minerals and worthless material. The mineral concentrate produced still contains a significant percentage of gangue as well as desirable metals remaining in the waste. The mineral processing efficiency depends highly on the technology chosen and the specificity of the ore deposits. However, it should not be forgotten that metal recovery often needs to be adjusted to the desired metal grade and economy of production [16]. The stored tailings still contain valuable metals and can become material of economic value. Solid tailings may also contain fines and slimes which may affect the stability of the landed waste and thus constitute a potential environmental problem. The long-term

Table 7.1: Example composition of waste rocks.

Mine, country	Component (wt. %)	Ref.
Aitik Cu Mine, Boliden Mineral AB, Sweden	SiO_2 (60–65), Al_2O_3 (15), CaO (3–5), MgO (2–3), Na_2O (1–2), K_2O (5–6), Fe (5–7), S (0.7–2), Cu (< 0.2), Zn (< 0.1), TiO_2 (< 0.7), MnO (< 0.4), BaO (< 0.6)	[47]
Antamina Copper–Zn–Mo Mine, Peru	Marble S (1.17), Fe, (3.48), Ca (18.4) and in ppm Cu (680), Pb (840), Zn (479), Sb (3.84), As (44), Mo (43.8); Hornfels S (0.24), Fe, (0.29), Ca (35.3) and in ppm Cu (162), Pb (160), Zn (401), Sb (5.95), As (20), Mo (5.5).	[48]
Libiola Fe–Cu Sulfide Mine, Eastern Liguria, Italy	SiO_2 (37.1), Fe_2O_3 (36), Al_2O_3 (9.1), CaO (0.9), MgO (14.4), TiO_2 (1.9), S (0.8), Na_2O (0.3), P_2O_5 (0.2), Cu (<0.2), Co (<0.01), Zn (<0.03), MnO (<0.1), Cr (<0.1), V (<0.03), Cd (<0.002), Ni (<0.1)	[49]
La Mine Doyon, Quebec, Canada	Al (3.3), Ca(1.2), Fe (3.2), K (< 1) Mg (3.1) Mn (0.05), Na (2.7), Si (34.2), S (36), Ti (5.4), Cr (0.1), Co (< 0.01), Zn (< 0.01), Pb (< 0.002), Cu (< 0.001)	[50]
Complexo Mínero-Industrial de Poços de Caldas (CIPC), Minas Gerais, Brazil	Al_2O_3 (29.5), K_2O (8.3) SiO_2 (47.9), Fe_2O_3 (5.5), CaO (0.03), MgO (0.1), TiO_2 (0.6), ZrO_2 (0.3), P_2O_5 (0.2), Na_2O (< 0.1), Ba (>0.1), As (<0.1), Mo (<0.1), Pb (<0.03), V (<0.1), Cr (<0.01), Zn(<0.01), Sr (<0.1), Y (<0.01).	[51]

management of tailings is also a concern because expensive long-term maintenance is not desirable. For example, wastes generated in Black Swan Nickel Mine and Silver Swan tailings (deposited in the Silver Swan Tailings Dam Storage Facility) are produced in 58,000 t/a and Cygnet tailings (deposited in the Cygnet Tailings Dam Storage Facility) are produced approximately in 325,000 t/a [53, 54]. The annual tailing deposits at "Żelazny Most" ("Iron Bridge") (Europe's largest tailings pond, Figure 7.5) are estimated at 20 to 26 million tons and in 2012, the total volume of waste was 522 million m^3. Due to content of Cu and Zn (0.55 % and 0.25 %, respectively) "Żelazny Most" may be considered as a source of further processing.

The residues from the metal recovery process are tailings consisting mostly of silicate minerals but also contain sulfide minerals, since a metal recovery of 100 % is not possible. The Silver Swan tailings compose 40 % of pyrite and 60 % of gersdorffite, which suggests that the waste material may contain economically viable concentration of cobalt. Molybdenum mining is also a source of valuable tailings. The main commercial source of molybdenum is molybdenite, together with other metals, e. g. Pb, Ni, W, Cu and Re, which is present as low-grade ores. Generally, molybdenum concentrates are obtained by selective flotation but the generated tailings still contain Mo, Pb and Cu [55, 56]. Iron ore tailings have been of great importance as secondary resource, especially as cement replacement, filling goaf and using as magnetization fertilizer [5, 57, 58]. Moreover, utilizing iron ore tailings to produce

Figure 7.5: Flotation tailings stored in the tank "Iron Bridge" (Copyright Paweł Litwin).

the building materials is a more effective solution for their resource recovery and management. The major chemical composition of the example tailings is presented in Table 7.2.

Cyanide tailings usually have high content of hazardous materials, which could cause environmental pollution. Iron content of cyanide tailings is high, and the processing of cyanide tailings as a secondary resource rather than waste has become a more attractive route for sustainable production on account of the depletion of iron ores and rapid rise of metal prices [64–67].

The next stage of metal production involved the smelting of ore concentrate. This is a melting operation in which the constituents, in the molten condition, partition into two or more phases, which may be slag, matte, speiss or metal [68]. In the smelting process, the valuable components of a resource material enrich one molten phase, whereas the gangue and other impurities are discarded in another, called the slag phase. The different types of smelting are reduction smelting, matte smelting and flash smelting. The principal objective of reduction smelting is to separate nonferrous base metals into a highly concentrated liquid phase and iron in the form of a discard slag containing low concentrations of nonferrous metals. Matte smelting is a heating of a concentrate to a desired temperature and oxidizing contained Fe to generate a molten matte and slag (mostly obtained matte is a molten mixture of sulfides). Flash smelting is oxygen-autogenous process, in which the

Table 7.2: Composition of tailings generated in mining process.

Mine, country	Components (wt.%)	Ref.
Boliden Mineral AB, Sweden	Fe (19.8), Si (15.7), Mn (6.4), Al (4.5), Mg (4.1), Ca (2.2), K (0.7), Na (0.3), P (< 0.1), Ti (0.2), elements in ppm: S (178,000), Zn (5,290), As (2,960), Pb (1850), Cu (640), Zr (94.8), Co (57.8),Cr (36.2), Sr (43 the,7, V (40.7) and Y, Sc, Ni, Mo, La, Hg, Be below 15 ppm.	[59]
Molybdenum Ore Plant, Henan, China	Pb (24.8), Mo (2.24), S (23.6), Fe (12.6), Si (12.9), Ca (3.45), Cu (0.77), Na (0.23).	[36]
Anshan Iron Ore Processing Plant, Liaoning, China	SiO_2 (47.4), Fe_2O_3 (24.2), Al_2O_3 (7.4), CaO (8.9), MgO (0.1), Na_2O (0.3), K_2O (0.7) or SiO_2 (73.3), Fe_2O_3 (11.6), Al_2O_3 (4.1), CaO (3.1), MgO (4.2), Na_2O (0.4), K_2O (0.95), TiO_2 (0.2), SO_2 (0.3), P_2O_5 (0.2), MnO (0.3)	[6]
Handan-Xintai Metallurgy Bureau, Handon, China	SiO_2 (82.3), Fe_2O_3 (14.7), Al_2O_3 (0.8), CaO (0.6), MnO (0.03) TiO_2 (0.02)	[5]
Musselwhite Gold Mine Ontario, Canada	SiO_2 (50.8), Al_2O_3(8.9), Fe_2O_3 (29), MnO (0.4), MgO (3.4), CaO (3.2), K_2O (0.8),Na_2O (0.02), P_2O_5 (0.2), Cr_2O_3(0.02), TiO_2 (0.5)	[60]
Copper Concentrator Plant, Katanga, Congo	Mn (12.3), Pb (5.7), Fe (3.4), Cu (1.7), Ni (1.3), Ca (1.3), Cd (0.01), Co(0.6)	[61]
Hcnan Zhongyua Gold Smelter, Sanmenxia, China	Au (5.6), Ag (50), Fe (27.7), SiO_2 (23.9), Al_2O_3(6.35), B_2O_3(3.96), SO_3 (5.37), CO_2 (23.5), CaO (2.6), Na_2O (1.8), K_2O (1.4), MgO (0.9), PbO (0.5), TiO_2 (0.5), ZnO (0.5)	[62]
Tailings from flotation operations, Sudbury basin in Ontario, Canada	Ni (0.27), Zn (0.57), Cu (0.14), Fe (52.9), S (28.7), Mg (0.8), Al (1.5), Si(4.8), Ca (1.1)	[63]

oxidized sulfide enables to reach reaction temperature and melt the feed material. This type of smelting enables to raise the matte grade up to white metal and finally reach blister copper instead of matte.

Metal slags generated from metallurgical processes are a group of promising recyclable sources of metal. The main example is copper slag which is produced during matte smelting and converting steps of copper production (2.2 t of slag is generated with 1 t of copper). Slags containing copper are either treated as waste or sold as value products, e. g. copper slag with a low content of CaO exhibits pozzolanic properties, but with much higher content it exhibits cementation properties and can be used as a replacement for Portland cement. Copper slag usually contains a certain quantity of valuable metals, such as copper, nickel, cobalt and iron [69]. Another example is slag generated during smelting of laterite concentrate to ferronickel [70]. The feed material for ferronickel production is moist saprolite, a type of laterite, which contains 1.5–3 % Ni, while the refined ferronickel, being suitable for making stainless steel and other ferrous alloys, contains 20–40 % Ni and 80–60 % Fe.

As it has been known, ore deposits usually contain several types of metallic minerals and valuable nonmetallic minerals are also associated (e. g. fluorites and barytes). However, they may not be present in sufficient amounts for their extraction to be economic. When these minor minerals are found in the concentrate, they are considered as impurities by metallurgists and are rejected in the slag. The same situation is when polymetallic ore mine is considered, when several elements are present at economic levels. In this case, the metal production requires separation of each concentrate, while other metals are treated as impurities [52]. In many old copper, iron and tin mines after pyrometallurgial procedure the ore or concentrate was also burned or smelted nearby to produce high-pure product simultaneously producing slag heaps which can be currently valuable source of metals. Most of the copper produced in the world is coming from sulfide ores that require oxidation at high temperatures. However, high-temperature metallurgical processes produce large quantities of slags, and this constitutes one of the main by-products of the metal production. The copper concentration in the discarded slag ranges from 0.2 and 15 % and depends upon many factors such as the type of ores processed, the type of furnaces and the cooling methods (Table 7.3). Also other valuable metals, such as nickel and cobalt, are present in slag partially due to entrapment of matte in sulfides. Some of them are dissolved in slag as oxides chemically bound with silica in fayalite or magnetite as lattice substituent [71]. However, with the increasing of oxygen enrichment during matte smelting, the proportion of metal oxides entrapped in smelter slag continuously increased. Especially cobalt, which resembles iron in terms of chemical properties, exists in slag predominantly in the oxide form [72]. In smelting and converting stages, the concentration of silica in the slag close to saturation is conducive. Avoidance of excessive oxidizing conditions is an additional conducive point. This inhibits the formation of magnetite, which makes the slag viscous, but when copper slag is crystalline, the major phases are usually fayalite along with other silicates. The conversion process removes Fe, S and other impurities from copper matte, thereby yielding Cu of high purity (99 %) and the slags containing significant amounts of the metal (2–15 %). The slag formed as a by-product of the copper matte smelting in a water-jacket furnace is essentially a mixed oxide phase and contains also copper and iron sulfides drawn mechanically but not dissolved. In some plants, flash smelting furnace slag (1.6–2 % Cu) and converter slag (5–7 % Cu) are cleaned in an electric arc furnace, where coke is used as a reductant and copper content of waste slag decreases to 0.5–0.8 % [73–76]. A similar procedure was used for copper smelting slag lying in the Küre region in the northern part of Turkey [77]. This ancient slag was subjected to carbothermic reduction to produce a Fe–Co–Cu alloy [78] or was reduced with coke/graphite in a DC arc furnace to yield a liquid alloy that was rich in iron, containing about 3.8 % Cu, 3.3 % Co and 2.15 % S [79]. Table 7.3 presents the major chemical composition of slags generated in mining process.

Table 7.3: Composition of slags generated in copper mining process.

Type of slag	Mine, country (technology)	Components (wt.%)	Ref.
Copper smelter slag	Lubumbashi plant, Congo (water-jacket furnace)	SiO_2 (31), FeO (29), ZnO (10), Al_2O_3 (7), CaO (10), MgO (5), Co (2), Cu (1.4),S (0.6), PbO (0.5)	[80]
Smelter slag	Khetri copper plant, Rajasthan, India	Cu (1.8), Ni (0.2), Co (0.2), Fe (46.5), Ca(2.1), Mg (0.97), Si (33.6), Al (1.6)	[81]
Converter slag	Ghatsila copper plant, Bihar, India	Cu (4.0), Ni (2.0), Co (0.5), Fe (38.3), Ca(4.1), Mg (2.7), Si (34.3), Al (0.1)	[82]
Copper smelting Slag	Küre, Kastamonu Turkey	Fe (51.5), SiO_2 (21.3), S (1.7), Cu (0.98), Co (0.5), Zn (0.2), Ni (0.004)	[83]
Copper–cobalt smelting slag	Yanggu Xiangguang Copper Co., Shandong, China	Fe (41.5), SiO_2 (24.7), CaO (5.3), Al_2O_3 (2.9), Cu (1.4), Co (0.7), S (0.7), Mg (0.6)	[84]
Converter slag	Ergani Mining Co., Turkey	Cu (2.6), Ni (0.05), Co (0.4), Fe (52), Zn (0.4), S (4.9)	[85]
Smelter slag	INCO's flash smelting furnace in Sudbury, Ontario, Canada	Cu (0.6), Ni (0.6), Co (0.2), Fe (38.6), Ca(1.1), Mg (1.7), Si (17.4), Al (2.5), Mn (0.03), Ti (0.15), S (0.9)	[70]
Smelter slag	Mount Isa Copper Smelter, Australia	Cu (3.7), Co (0.3), Zn (0.44), Fe (50), SiO_2 (22.5), Al_2O_3 (1.14), S (1.6), As (0.05), Pb (0.2) Ag (2 ppm)	[86]
Reverberatory slag	Refimet Division, Chile	Cu (1.2),Fe (32.2) *	[87]
Reverberatory slag	Sarcheshmeh, Kerman-Iran	Cu (0.66), Fe(38.3), S (1.3), Zn (2.4), Ti (0.77), Mg (0.16), CaO (8.4), SiO_2 (39), Al_2O_3 (6.1)	[88]
Smelter slag	El–Maady Co., Cairo, Egypt	Cu (20.7), Zn (1.3), Fe (11.4)*	[89]

*Other components have been not reported.

7.3.3 Chemical enrichment of low-grade ores

Low-grade ores having uneconomical grades mostly remain in the "waste" category, then are stored or used for blending with profitable ore, or are chemically leached by using selected lixiviants. The leaching involves the use of aqueous chemistry also assisted by oxidizing agents. Unlike physical separations, this technique is capable of yielding solutions of relatively pure metal ions, which usually can be further recovered using other procedures e. g. precipitation, solvent extraction, electrowinning and adsorption. Chemical treatment of ore or mineral waste is a heterogeneous reaction that takes place at the interface between a solid and liquid phase. The leaching of the mineral generates immense quantities of unless waste. That is because the amount of recoverable metal is generally just a small fraction of their total mass. The leaching process can be divided into two distinct types: *In situ* leaching (removal of the valuable components of a mineral deposit without physical extraction), dump leaching (leach solution flows by gravity through the dump pile), heap leaching (leaching of low-grade ore that has been crushed and deposited on a specially designed pad), pressure leaching (combination of elevated operating temperature and oxygen overpressure), leaching in vats and the most aggressive leaching agitation leaching. The optimum result in leaching is not only combined with leaching process conditions but also with pretreatment stage, mostly mineral concentration and thermal pretreatments. Mineral concentration is carried out by using any one or combinations of physical beneficiation processes such as magnetic and electrostatic separations and flotation, while the thermal pretreatment also includes a sulfatizing, chloridazing, alkali or reducing roasting [90].

In situ leaching has many environmental advantages over conventional mining because it generates less waste material and causes less surface disturbance [91]. The major environmental concern is postmining water quality. For example, *in situ* uranium leaching process selectively removes uranium and leaves most of the dangerous radioactive by-products in the ground [16]. The leaching may also take place in backfilled stopes or caved areas (in place leaching); however, this method is used where the permeability of the material has been enhanced through low-intensity blasting, hydraulic fracturing and chemically induced dissolution of gangue material [92]. So far, both *in situ* leaching and in-place leaching have only been commercially used for copper and uranium, but in case of the uranium mining the *in situ* leaching states 20 % of the total global uranium production [93]. Although lixiviants are available to leach various copper oxide and copper sulfide minerals, attempts at *in situ* leaching of copper in pristine formations have been very difficult because the leaching solution has not been able to adequately contact the ore minerals in the rock. The most successful *in situ* copper leaching has been in ore bodies that had been previously mined or broken by detonation [94–96]. In this method, an aqueous leach liquid is pumped into the area along with oxygen. In those instances wherein a substantial amount of pyrite is contained in the ore which is to be leached, the oxygen reacts with the pyrite to form sulfuric acid and ferric sulfate. Usually underground deposits of copper contain insufficient amounts of pyrite to generate sufficient sulfuric acid and ferric sulfate necessary to leach the copper from the deposits; therefore, this technique requires the increase the permeability of underground formations and mine waste dump formations. The increase in permeability can be achieved by using rubblize rock during conventional mining or by using hydraulic fracturing techniques [97].

The development of heap leaching and dump leaching technologies for low-grade ore has extended the world's ore resource base considerably. The processes were developed by the copper industry and have been extended to uranium and gold. Mining and milling operations, in general, leave behind numerous waste rock dumps, and these may bear significant amounts of metal value. One method of recovering such values is dump leaching, and this concept has been very useful in the recovery of copper from the huge dumps made of mined waste. In leaching dumps of copper sulfide using water and air copper passes into solution as copper sulfate. Leaching of low-grade or other waste material from open pit mines is already an important process. Especially, copper minerals such as chalcopyrite, chalcocite and, to a lesser extent, covellite and bomite are leached in dump. A site suitable for dump leaching must have surfaces impermeable to leach liquor and topography that makes it convenient to collect pregnant liquor at low points. In heap leaching, mined low-grade ores are broken and piled into relatively small heaps on impervious ground or on a concrete or asphalted surface with drainage channels and pipes to carry away the pregnant solution to a collection pond. A major advantage of heap leaching is the elimination of crushing process.

Pressure leaching is a well-established hydrometallurgical processing alternative for gold [98], copper [99], nickel, cobalt [100] and platinum group elements [101]. Pressure leaching, high-pressure leaching and pressure oxidation are conducted in an autoclave at carefully controlled temperatures and pressures. This type of leaching requires handling and grinding of all run-of-mine ores and disposal of tailings in impoundments. Pressure leaching involves a large capital investment and higher operating costs, but yield high recovery rates from difficult, complex or refractory ores. Mineral ore can be also chemically treated in open tanks under atmospheric pressure conditions to extract metal salts from the ore at an accelerated rate. This technique, called as tank leaching, requires handling and grinding of all run-of-mine ores and disposal of tailings. Tank leaching technique has been widely used in mining for the recovery of gold and silver (Merrill-Crowe Process [102]).

7.3.3.1 Leaching of low-grade Ni-Co ores

Production of nickel from saprolite or laterite through pyrometallurgial methods is highly energy intensive, e. g. the production of ferronickel from laterite requires high temperatures due to the high melting temperature of both the ferronickel and the slag [103]. Therefore, for this type of ores, the wet methods are recommended. Laterite hydrometallurgy is based mainly on leaching of limonite or smectite ores [104]. High-pressure acid leaching (HPAL) is usually used to treat limonite with the advantage of high Ni recovery. At these temperatures, the iron precipitates from solution as hematite, leaving a solution containing nickel and cobalt. Next, the cooled slurry is then preneutralized and sent to a countercurrent decantation circuit, where the solids and liquids are separated (removal of impurities such as Al, Cr, Cu, Fe, Mg, Mn and Zn). The impure leach solution is further neutralized to remove more contaminants. Atmospheric tank leaching as one of hydrometallurgical process for laterite is used to treat saprolite, even though the Ni recovery is lower than that of HPAL. Until now, there are only two commercial plants in mineral industry operating full-scale heap leaching of nickel laterite ores, the Murrin Murrin [105] and Çaldağ operations (Çaldağ Nickel laterites) [106]. The Murrin Murrin operation in Western Australia includes a satellite heap leaching circuit producing 2,000 t/a nickel from scats). Piaui Project is under development in Brazil by Brazilian Nickel based in the UK [107, 108]. In this process, the ore containing 1.0 % Ni and 0.05 % Co is amenable to sulfuric acid heap leaching which produces a pregnant leach solution containing nickel, cobalt (extraction above 80 %) and other metal species. The leach solution is then treated in a simple precipitation circuit where limestone is added to remove iron and other impurities. The remaining solution is then passed through an ion-exchange process which separates and concentrates the nickel and cobalt solutions which are then precipitated as high-purity nickel and cobalt hydroxide. Starved acid leach technology (SALT) [109] is another process developed to recover nickel and cobalt from low-grade mineral materials. SALT was developed and patented by Search Minerals of Canada to recover nickel and cobalt from Caron Plant tailings from the Votorantim plant at Niquelandia in Brazil and from below cutoff grade saprolite samples from PT ANTAM's Pomalaa deposit in Indonesia. It consists of atmospheric tank leaching using a relatively small amount of sulfuric acid to selectively extract nickel and cobalt. In the case of the saprolite ore containing 1.33 % Ni, 0.02 % Co, 8.1 % Fe and 17.2 % Mg, the conducted tests indicated optimum conditions enabling the extraction of Ni in 57.9 %, Co in 65 %, Fe in 10 % and Mg in 46.5 % using sulfuric acid with a rate of 350 kg/t of dry ore. The overall extraction of Ni and Co as hydroxides was predicted to be greater than 99 %. Atmospheric leach process for the recovery of nickel and cobalt from limonite, saprolite and mixed ores has also been used in the Promeca/Norilsk Process (developer: Promeca Consulting and Norilsk Nickel in Finland) [110]. It is a two-stage countercurrent leach process using sulfur dioxide and concentrated sulfuric acid. The first step involves a reducing atmospheric tank leach with sulfur dioxide and sulfuric acid solution returning from the second stage. The sulfur dioxide reduces trivalent iron, manganese and cobalt coming from the second stage to bivalent soluble sulfates. After solid–liquid separation, the solids are treated in a kiln or pug mill-type second-stage reactor with hot concentrated sulfuric acid to complete the leaching.

In the Taganito HPAL Plant Project, the wet-screened laterite ores are preheated up to 200°C and then leached in autoclave using 98 wt.% H_2SO_4 [111]. After HPAL stage, the leach slurry is neutralized by using limestone slurry, and then sent to decantation circuit for residue washing. The washed leach liquor is again neutralized with limestone slurry, and this stage enables impurity removal (iron, aluminum and chromium). Zinc, as a major impurity, is precipitated ZnS (sulfidization with H_2S). The Zn-free pregnant liquor is sent to the sulfidization circuit for the complete precipitation of nickel and cobalt (final recovery over 99 % of nickel and cobalt). The project has also considered the waste liquor utilization. The liquor is again neutralized and recycled back to the plant to be used in the production of limestone slurry.

The Neomet Process has also been developed by Neomet Technologies to recover Ni and Co from limonite and saprolite ores [112, 113]. It is an atmospheric hydrochloric acid leaching process coupled with a pressure system to regenerate the HCl. Nickel and cobalt extractions are high (>90 %) and no secondary neutralization is required to eliminate residual iron. Nickel and cobalt are recovered as basic chlorides, which can be further processed to metals or calcined to form oxides. Iron and magnesium oxides are possible by-products.

Limonite-type laterite is also treated in the Caron Process which is more effective by involving reduction roasting of the ore followed by ammonia leaching [114]. In this process, the dried ore is fed to reduction roasting stage, where it is gradually heated to 700–900°C under a reducing atmosphere. After reduction, the resulting material is cooled at 120 °C–150 °C under inert atmosphere to prevent re-oxidation of the nickel and other desired metals. After cooling, the roasted ore is sent for ammoniacal leaching, where metallic nickel is oxidized and dissolved. In this technology, 80–85 % Ni and 20–30 % Co can be extracted. The remained leaching residues contain less than 0.5 % Ni and 0.1 % Co. The conditions and results of selected leaching stages of Ni-Co ores are presented in Table 7.4.

7.3.3.2 Leaching of low-grade copper containing ores

Up to the present time, smelting technology remains the only commercially viable copper recovery route from low-grade copper ores; however, since smelting remains relatively expensive, alternative, more economical and less environmentally harmful chalcopyrite treatment processes are still being sought. Hydrometallurgical processing of low-grade copper ores (constituting of combination of crushing, leaching, solvent extraction and electrowinning) contributes about 20 % of annual copper production. Especially, secondary copper sulfides are the most profitable copper ores and numerous lixiviants processes for the leaching of copper from chalcopyrite concentrates have been proposed and patented [119]. Using acidic sulfate-based chemical systems, the main processes enabling copper extraction from chalcopyrite have been developed and patented. The main example is the process developed for treatment of Hudson Bay Mining and Smelting copper concentrates for possible replacement of the existing smelting operation at Flin Flon (Manitoba, Canada). This process involves oxidative leaching of chalcopyrite in autoclaves at 150 °C (the reaction may also be conducted at 90 °C) with an oxygen overpressure above about 400 to 750 kPa, H_2SO_4 at concentrate of 120–150 g/L and using coal in an amount of 25 kg per ton of concentrate as an effective antiagglomerant [120]. It was shown that the extraction of copper in the presence of the antiagglomerant was 98 %, with calcium lignosulfonate 71 % and without the antiagglomerant 49 %. Moreover, the unleached sulfide is removed to the leach after the floating. The consortium of Anglo

Table 7.4: Leaching conditions and results of selected leaching stages of Ni-Co ores.

Plant, country	Ore grade and type of leaching	Leaching conditions	Ref.
Caldag pilot plant, Turkey	Ore grade 1.13 % Ni 0.07 % Co HPAL or HAL	Atmospheric heap leaching process (HL) with H_2SO_4 (528 kg/t) at ambient temperature and duration 548d Results: extraction Ni (79 %), Co (83 %), Fe (30 %), Mg (79 %), Al (37 %). High-pressure acid leaching with H_2SO_4 (528 kg/t) at 250°C and duration 1 h. Results: extraction Ni (94 %), Co (94 %), Fe (1.7 %), Mg (84 %), Al (51 %) and 82 % Mn	[106, 115]
Taganito HPAL Plant Project, Philippines	Ore grade: 1.26 % Ni 0.089 % Co HPAL	Wet-screened laterite ores are preheated up to 200°C and leached in autoclave using 98 wt.% H_2SO_4, T = 240–250°C. Pregnant leach solution containing about 6 g/L Ni, 0.5 g/L Co and 30–50 g/L H_2SO_4.	[111]
Murrin Murrin Plant Project, Australia	Ore grade: 1.01 % Ni 0.074 % Co HPAL and HL	Leaching with H_2SO_4 condition: concentration 50–65 g/L Results: Ni and Co extraction in 92–95 % after 70–90 min leaching (HL required 150–200 days).	[105]
Neomet Process, Canada	Ore grade 1.72 % Ni 0.03 % Co HPAL	Leaching with HCl (600 kg/t) at 100–110°C under atmospheric conditions. Results: Ni and Co extraction in 99 and 91 % after 30 min of leaching other ingredients in 96 % Fe, 50 % Al, 82 % Mg and in 7 % Ca. Fe and Al are removed during HCl regeneration	[113]
Yongtong Plant, Henan, China,	Ore grade: 0.8 % Ni 0.07 % Co NAPL	Leaching with HNO_3 condition: HNO_3 concentration:130 g/L; nitric acid/ore ratio: 0.38 g/g; particle size: 150 μm (100 %); pulp density: 32.5 wt.%; leaching temperature: 190 °C; leaching duration: 60 min. Results of extraction: Ni = 84.5 %, Co = 83.9 %, Fe = 1 %	[116]
Coral Bay Nickel HPAL Plant Palawan, Philippines	Ore grade: 1.26 % Ni 0.09 % Co HPAL	Ore slurry (1.26 % Ni, 0.09 % Co, 2.21 % Mg and 8.15 % Si) is preheated up to 200 °C. Using 99.8 % H_2SO_4 the leaching reaction is carried out on 245 °C and 4 MPa. Results of extraction: Ni = 55–58 %, Co = 4–5 %	[117]
Coron Processadopted by QNI Pty Ltd (QNI), Yabulu, Australia	Ore grade: 1.6 % Ni 0.15 % Co	Reduced ore (reduction roasting at ~750 °C) is cooled and then leached in ammonia/ammonium carbonate liquor at atmospheric pressure. Product liquor contains 10–12 g/L Ni and 0.6 g/L Co or after mixed nickel-cobalt hydroxide processing 23 g/L Ni and 1.1 g/L Co. Results of extraction: Ni = 80–82 %, Co = 50–60 %	[118]

American Corporation and the University of British Columbia has developed a medium pressure oxidation process, in which the molten sulfur was dispersed in the presence of a surfactant [121]. The Mt Gordon Copper Process treats chalcocite/pyrite ore or concentrate through two autoclaves operating at around 90°C [122]. The ore is leached with pressure oxygen in acidic raffinate from solvent extraction. After solid/

liquid separation and washing of leach residue, the pregnant leach solution is directed to solvent extraction. Peroxide as an oxidant for the sulfuric acid has also been tested as an effective route for copper and zinc recovery from Nigerian complex sulfide ore [123].

Recovery of value metals through bioleaching process is a very common technique and is gaining importance in the field of waste treatment [124]. In bioleaching process, minerals are biooxidized through the action of microbially generated inorganic acids such as nitric and nitrous acids, sulfuric and sulfurous acids and carbonic acid. Both chemolithotrophic and heterotrophic microorganisms play an active role in the solubilization of metal ions [125]; however, *Acidithiobacillus (At.) ferrooxidans* is the dominant bacterium responsible for metal sulfide solubilization and works mainly via enzymic oxidation of the ferrous irons. Especially microbial leaching of copper ores is becoming increasingly important in copper extraction and it is estimated that annually 5–10 % worlds copper output is recovered though bioleaching processes. In the case of polymetallic concentrates, hybrid processes comprised of chemical leaching and bioleaching seems to be adequate but limitations due to low extractions and low kinetics with mesophiles have limited its implementation on commercial scale.

GEOCOAT Process (GeoBiotics, Colorado) was initially developed for biooxidation of refractory sulfide gold concentrates and to the bioleaching of copper, nickel, cobalt, zinc and polymetallic base metal concentrates. Mesophilic bacteria are typically used, except for the processing of chalcopyrite concentrates, where the higher temperatures associated with the use of thermophile microorganisms have proven highly beneficial in increasing the rate and extent of copper [126].

The purpose of Alliance Copper (a joint venture company formed between BHP Billiton and Codelco) was to exploit the BioCOP™ technology, which was commercial demonstrated at the Chuquicamata Mine in Chile, with a design production rate of 20,000 t copper per annum [127]. The process utilizes thermophilic microorganisms operating at temperatures up to 80°C to leach copper sulfide mineral concentrates. Leaching of chalcopyrite concentrate is complete within 10 days. The copper is subsequently recovered by conventional solvent extraction and electrowinning, producing a high-value copper metal product. Example of microorganism participation in the leaching is also Brisa process (Biolixiviación Rápida Indirecta con Separación de Acciones: Fast Indirect Bioleaching with Actions Separation), which was developed for the copper recovery from chalcopyrite concentrates [128]. Two copper concentrates (with a copper content of 9 and 10 wt.%) with chalcopyrite as the dominant copper mineral have been leached with ferric sulfate at 12 g/L of ferric iron and pH 1.25 in agitated reactors using silver as a catalyst. Liquors generated in the chemical leaching were biooxidized for ferrous iron oxidation and ferric regeneration with a mixed culture of *ferrooxidant* bacteria. The participation of thermophilic microorganisms has also been proposed by Mintek and its partners, Industrias Peñoles S.A. de C.V. of Mexico and BacTech. The successfully demonstrated plant was commissioned and operated in Monterrey (Mexico). It was a bioleaching (at 45 °C) of a Cu-Zn concentrate containing high levels of Ag [129, 130]. The Ag was recovered from the bioleach residue using hot lime treatment followed by cyanidation. Recoveries of >93 % Cu and 96 % Zn was obtained at a concentrate feed rate of 2.7 t/day, provided the concentrate was milled to 100 %.

Example of the bioheap leach treatment of run-of-mine low-grade copper sulfide is the Escondida Sulfide Leach Project (Figure 7.6), which treats the material extracted as waste under the current mining operations [131]. Metallurgical testwork carried out to date has included a 300,000 t test dump and have confirmed very high extrability of copper. Finally, according to the production plan, the low-grade ore feed will be dumped onto an engineered pad located to the northeast of the Escondida pit and to the southeast of the future Escondida Norte pit, next placed onto the leach pad.

Figure 7.6: Escondida copper mine located in Northern Chile (the material is reproduced with the permission of BHP (Source: BHP)).

The use of hybrid sulfate–nitrate systems to copper recovery from low-grade ores has also been studied, but only nitrogen-species-catalyzed (NSC) sulfuric acid pressure leach has ever been built and operated successfully on an industrial scale [132]. The application of NSC oxidative pressure leaching was tested on a low-grade nickel–

copper sulfide concentrate composed of pyrrhotite, pentlandite and chalcopiryte and resulted in the Cu and Ni dissolution while elemental sulfur and iron oxide remained in the residue. Another application of the NSC oxidative pressure leaching was tested on chalcopyrite concentrate and an arsenic-bearing copper concentrate on a laboratory scale. Researchers at Kennecott Copper in Utah built a pilot plant for treating copper concentrate with a nitric–sulfuric acid mixture and concluded that the process was feasible but not economical because of the incomplete regeneration of HNO_3. Another example is the Kennecott Copper pilot plant in Utah which involved a mild oxidative pressure leach in sulfuric acid to which a small amount of HNO_3 was added [133]. The addition of chloride to sulfate has also been considered to improve the copper dissolution in leaching solution. In the chloride–sulfate system, the leaching rate of chalcopyrite is significantly faster as compared to the leaching in pure sulfate media and that effect has been attributed to changes in the morphology of the elemental sulfur produced, as well as changes the speciation of both copper and iron through the formation of Cu- and Fe-complex chloride ions in concentrations reflecting the overall solution composition [134].

Example of the chloride–sulfate leaching process is the Cuprochlor, in which secondary sulfide copper ores, mainly of the chalcocite type, are leached. In this process, dissolved calcium chloride is added in addition to water and sulfuric acid. The $CaCl_2$ immediately reacts with the acid to produce calcium sulfate acting to agglomerate, while chloride ions permit high cuprous ion concentration in the solution [135]. The PLATSOL™ process is applied to process mixed base and precious metal sulfide concentrates containing copper, nickel, cobalt, platinum, palladium, gold and silver [136]. The processing of a bulk concentrate enables to produce copper cathode, a precious metal precipitate (Au, Pt, Pd, Rh, others), as well as nickel and cobalt cathode. The PLATSOL™ process treats concentrate under total oxidation conditions at 225°C and at overpressure of 100 psi/g of oxygen and with the addition of NaCl in concentration of 5 g/L. The conditions and results of selected leaching stages of copper ores are presented in Table 7.5.

Treatment of copper ore and copper sulfide ores at low temperatures in chloride leaching systems is a promising method of dissolving. The majority of those processes targeted the processing of copper concentrates and were operated at atmospheric pressure and at temperatures approaching the boiling point of ferric chloride solutions. In chloride leaching system, acidic oxidizing solutions containing chloride salts are employed as lixiviants. Chloride is proposed to act as the active agent in the dissolution process, while ferric and cupric chloride act as oxidant for chalcopyrite. The strong interest in chloride systems is combined with much higher solubilities of iron and other metals than in sulfate system, enhanced redox properties and the faster leaching kinetics of sulfide ores. Very important is also the precipitation of sulfur in an elemental form and reduction of role of pyrite [139].

Sumitomo Metal Mining Co., Ltd has developed a new technology to recover copper from its ores. Concentrate leaching technology, in conjunction with a

Table 7.5: Leaching conditions and results of copper ores treatment.

Process	Concentrate	Leaching conditions	Ref.
Brisa Process	Chalcopiryte Cu content: 8.9–9.9 wt.%	Leaching with ferric sulfate at 12 g/L of ferric iron, at pH of 1.25 and at 70 °C with Ag as catalyst (0.5–2 mg Ag/g concentrate). Copper extractions >95 % after 10h. Ag elimination by leaching a leach residue with 200 g/L NaCl–0.5 M H_2SO_4 medium at 90 °C and 10 wt.% of pulp density.	[128]
Galvanox™	Chalcopyrite with addition of pyrite	Ferric sulfate leaching at T = 80 °C, P_{80} = 53–75 µm and at solution potential of 470 mV or higher. Process is enhanced when silver-treated pyrite is used in the ratio chalcopyrite: pyrite 2:1.	[137]
Sepon Copper Process	Chalcocite Cu content: 3.8 wt.%	Atmospheric acid ferric sulfate leaching of copper (T = 80°C, P_{80} = 100 µm) followed by residue washing and pyrite/elemental sulfur flotation. Copper recovery in 90 %	[138]
AAC/UBC Copper Process	Chalcopyrite Cu content: 3.1 wt.%	Medium temperature leaching process at 150 °C in an acid–sulfate system. The concentrate is reground to fine size (P80 5–20 µm) and leached under moderate pressure.	[121]
BioCOP™	Predominantly chalcopyrite and bornite	High-temperature sulfate bioleach uses thermophilic bacteria (75 °C). The oxidative bioleaching (O_2, air) consumes limestone during leaching and requires indirect cooling of the bioreactors. Other conditions: Pressure 1 atm, P_{80} = 37 µm.	[127]
Bactech/ Mintek	Polimetalic Cu concentrate containing (wt.%): Cu (21.7 %), Fe (23.2), Zn (7.53), Pb (5.77), SiO_2 (2.46)	Bioleaching with 5 g/L Fe at 45°C, pH of 1.4 and 1 atm. Nutrients concentrations: 1 g/L $(NH_4)_2SO_4$, 0.5 g/L $MgSO_4 \cdot 7H_2O$, 0.1 g/L KCl and 0.5 g/L K_2HPO_4 results a final Cu recovery in >93 % Cu (as cathode) and Zn in 96 % (as precipitate).	[130]
NSC leaching (partial chalcopyrite oxidation leach)	Chalcopiryte 25 % Cu, 31 % Fe, 36 % S (total).	Sulfate–nitric leach of chalcopyrite concentrate: Initial H_2SO_4 50 g/L, pressure 620 kPa/g, T = 125 °C, 2 g/L N and P_{80} = 10 µm. Copper dissolution after 20 min of the leaching: 99.5 % Cu and 99.6 % Fe; residue 0.5 % Cu and 0.4 % Fe.	[132]

(continued)

Table 7.5: *(continued)*

Process	Concentrate	Leaching conditions	Ref.
PLATSOL™	Concentrate from NorthMet Ore Cu (15.5 %), Ni (3.7 %), Co (0.2 %), Fe (28.7 %), S^{2-} (26 %), Au (2.8 %), Pt (2.5 %), Pd (11 %)	Leaching involves a one-step autoclave oxidation to dissolve base and precious metals. Autoclave conditions: moderately high temperature (225°C), 30–60 g/L of H_2SO_4 and 10 g/L NaCl. Recovery: Cu (94 %), Ni (77 %), Co (46 %), Au (77 %), Pt (76 %), Pd (76 %).	[136, 138]

conventional milling and flotation concentrator, allows primary and secondary copper sulfide minerals to be transformed to copper cathodes through the efficient pressure leaching and elextrowinning process. The leaching process with chlorine as lixiviant enables dissolving more than 90 % of copper from chalcopyrite concentrate [140]. The Cuprex™ process leaches chalcopyrite concentrate at atmospheric pressure with $FeCl_3$ as the leaching agent. The plant at the Tecnicas Reunidas Research Center in Torrejon (Spain) employed a single-stage leaching operation comprising five reactors arranged in series. The pilot operations were carried out using a chalcopyrite concentrate with efficiencies of greater than 97 %, which were obtained with a residence time of 9–10 h at 95–100 °C. The leaching stage resulted in an overall copper recovery in the process of greater than 96 %. After leaching, the pregnant liquor containing copper, iron and minor amounts of impurities mainly Zn, Pb and Ag was sent to the extraction stage of the SX circuit. After the three-stage selective extraction, the aqueous solution of copper chloride was then sent to the electrolysis section, which is fed to the cathode compartment of an electrowinning cell to produce granular copper [141].

Intec Copper Process involves a chloride/bromide solution leach at 85 °C and atmospheric pressure. This hydrometallurgical process was developed for the extraction of pure copper and precious metal from sulfide concentrates [142]. Patented technology is able to extract copper, gold, silver and other metals from a range of mineral feedstocks, including chalcopyrite, chalcocite, covellite, bornite, enargite as well as a range of other minerals. Following purification, the copper can then be directly electrowon from the chloride/bromide electrolyte without the formation of corrosive chlorine [143]. Additionally, Ag is precipitated with mercury, and then produced as a chloride for further processing. Au is recovered onto carbon for further processing. Innovation of this process is reagent Halex™ being source of oxidizing agent ($BrCl_2^-$) formed at anode.

The HydroCopper Process is a new advanced technology to produce copper from copper concentrate. In this process, cuprous oxide is precipitated with sodium hydroxide from the purified pregnant leach solution. Sodium hydroxide is regenerated in a chlor-alkali cell. The other products of the electrolyzer are also recirculated, for example chlorine gas to concentrate leaching and hydrogen gas for the reduction of cuprous oxide to metallic copper. Pilot plant testing has been undertaken, and a demonstration plant is being built at Pori, Finland, to produce 1 t/day copper [144].

The CESL Process employ a chloride-enhanced oxidative pressure leach process in which high amounts of sulfur remain in the elemental form in leach residue [145]. The leaching process enables conversion of the copper to copper sulfate, the iron to hematite and sulfur to elemental sulfur. The process is continuing to develop precious metal recovery from the leach residue and tested in the pilot plant in Vancouver, Canada. Table 7.6 presents the examples of the chloride leaching copper ores which have been tested at least in a pilot scale.

Table 7.6: Chloride leaching conditions and results of copper ores treatment.

Process	Concentrate	Leaching conditions	Ref.
Cuprex™	Copper concentrate: Cu (27.5 %), Fe (26 %), S_T (28.5 %), SiO_2 (12 %)	Leaching with $FeCl_3$ in single-stage reactor at T = 95°C and duration 9–10 h. Composition of pregnant leach solution: 25 g/L Cu(II), 120 g/L Fe and 6 M Cl^-.	[141]
Intec Copper Process	Low-grade copper ore containing 10 % Cu with As, Sb, Bi, ug	Oxidizing chloride leaching with 0.47 M Cu(II), 4.9 M NaCl and 0.27 M NaBr at T = 85 °C and at oxidative redox-potential 950–1,000 mV (Ag/AgCl). Halex™ is oxidizing agent ($BrCl_2^-$) formed at anode, P_{80} of concentrate 40 µm.	[142]
HydroCopper™	Copper concentrate: Cu (25.7 %),Fe (28.1 %), Zn (2.1 %), S (32.3 %), Pb (0.7 %)	Oxidizing chloride leaching with 0.3 M $CuCl_2$ in 4.8 M NaCl at pH 2–2.4 (2nd stage) (HCl); T = 85–100°C, P_{80} = 100 µm leaching at three countercurrent stages with retention time 15–25 h. Results: 98 % leaching of Cu. Leaching of Au at redox potential 650 m/V (Ag/AgCl). Copper was precipitate as cupric hydroxychloride at pH 4.	[144]
CESL	Copper concentrate: Cu (31.6 %), Fe(28 %), S (26.6 %)	The concentrate slurry is fed to the autoclave, where it is combined with oxygen and recycled process liquor containing 12 g/L chloride and acid. Results: 96 % leaching of Cu (residue 1.5 %)	[145]
Sumitomo	Copper concentrate: Cu (30.3 %), Fe (25 %), S (31.5 %)	First stage: reduction with Cu(II) and Fe(III) and total chloride 5.7 M. Second stage: partially leaching with Cl_2 gas.	[140]

7.3.4 Chemical treatment of slag and tailings

Demand for Ni, Co, Cu, Zn and Ag is growing in the world, which makes it increasingly necessary to process not only low-grade ores but also tailings and slags that so far have been considered as wastes. Peet et al. have pointed to a possibility to recover valuable metals from nickeliferous pyrrhotite tailings containing 0.27 % Ni, 0.57 % Zn and 0.14 % Cu (Sudbury basin Xstrata Nickel) [63]. The processing has been confirmed by the conducted economic analysis, which has shown that 70 % recovery of Ni, Fe and S can be profitable. Previously, several pyrometallurgical processes were investigated for the treatment of the pyrrhotite tailings [146]; however, most were abandoned due to the production of SO_2 gas during the roasting process and the introduction of additional Fe in the slag. The treatment of nickelliferous tailings is likely to result in a deficiency of washwater when distilling pregnant liquors with at least 1 % Ni and as a result ammonia losses may be increased to such an extent that steaming of the tailing will be economically justified. The presence of ferric hydroxide in the tailing may also cause losses of adsorbed NH_3. It may be expected that the values of the pregnant liquors may vary between 1 and 2 % (both Ni and Co). The microbial-mediated recovery of nickel from nickeliferous pyrrhotite tailings (0.6 wt.% Ni) produced by Vale base metals in the Sudbury Basin of Ontario has been studied [147]. The results showed very high nickel recovery in the pH-controlled (pH = 1.5) biotic (93 %) and abiotic experiments (90 %). Moreover, it was shown that a high concentration of Fe(III) promotes the dissolution of Ni, and the conversion of sulfide into elemental sulfur was found to be high in the abiotic experiments conducted at pH = 1.5 (32 %).

The leaching of the Cuban nickel tailings, residues produced in the ammonia carbonate leaching process at the Ernesto Che Guevara nickel plant, Moa (North-East of Cuba), was investigated by using organic acids (tartaric and oxalic acids). The maximum extraction of about 80 % Co, 70 % Ni and 40 % Fe was achieved after 5 days of leaching from residues containing 0.34 % Ni, 0.08 % Co and 44.2 % Fe. The metals as oxalates were then precipitated with 80 % yield [148]. The proposed technology, being a combination of percolation leaching, metal salt precipitation and heating of organic salts at 200 °C, can be used to concentrate of nickel and cobalt. In addition, the magnetic separation of leaching residues would lead to an iron concentrate. The acidophilic chemolithotrophic microorganisms have also been used in leaching of flotation and acid-leaching tailings obtained from Zambia sulfide ores [149]. The result showed that 62.7 % and 61.7 % copper were dissolved by bioleaching of flotation and acid-leaching tailings, respectively. Fosso-Kankeu et al. have tested the heterotrophic microorganism (*Bacillus subtilis*) for biological recovery of nickel from tailing obtained from the Inkhomatsi Mining Company SA (0.16 w% Ni) [150]. The obtained results have not shown so spectacular efficiency as expected, but the uninterrupted bioleaching over 24 h allows dissolution of 28.5 % nickel from tailings and represents better performance over the conventional method. Bioleaching of arsenic was also conducted from highly concentrated arsenic-bearing tailings using the iron-oxidizing bacteria (*Acidithiobacillus ferrooxidans*). It was shown that at reaction temperatures 15–30 °C the As leached amount exceeded 85 %; however, the pH range should be controlled to prevent the formation of jarosite [151].

The column bioleaching of copper flotation tailings obtained from the Yulong Copper Mine in Xizang of China was also investigated. The main chemical composition of tailings was 0.19 % Cu, 19.73 % Fe, 2.25 % S, 3.27 % Ca, 1.25 % Na, 4.6 % Al and 1.4 % K. The microorganisms were isolated from the Dexing Copper Mine in Jiangxi (China). The results indicated that the tailings copper extractions achieved 54.61 %, 60.09 % and 43.93 %, respectively in a layered heap construction method, agglomerate heap construction method and a pelletized sintering heap construction method [152]. Studies were conducted to establish the possible application of bioleaching to recover valuable metals from the sulfidic tailing of Golgohar Iron Mine in Kerman (Iran) in the presence of a mixed culture of moderately thermophilic microorganisms at a stirring rate of 150 rpm, temperature of 45 °C and with the addition of yeast extract (0.02 % (w/w)). The results have shown that 55 % of copper, 98.2 % of nickel and 59.5 % of cobalt are extracted from the tailing through the bioleaching process after 30 days at 5 % (w/v) pulp density [153]. Xie et al. have proposed mixed nitric–sulfuric acid solutions as leaching medium for the recovery of Ni, Cu and Co from low-grade Ni–Cu sulfide tailings (0.23 wt.% Ni, 0.25 wt.% Cu, 0.01 wt.% Co and 10.43 wt.% Fe) [154]. The oxidative leaching of Ni–Cu sulfide tailings was performed at room temperature and atmospheric pressure, which resulted in the effective dissolution of some of the valuable metals contained in the tailings. The optimal weight ratio of tailings:nitric acid: sulfuric acid was 1:0.42:0.18 and the overall recovery yield of Ni, Cu and Co was 81.7 %, 75.7 % and 52.4 %, respectively, as their enriched sulfides (ca. 7–8 % for Ni and Cu). The method described may be applicable for the recoveries of Ni, Cu from low-grade Ni–Cu sulfide tailings on an industrial scale. The roasting (480–500 °C) and sulfuric acid leaching (0.3 M H_2SO_4) at 85°C was also established as the optimal parameters for extraction of Co from pyrite waste; however, the technology efficiency is very low (35 %) [155]. The pre-enrichment via sulfation roasting method has been proposed as a stage improving the leaching efficiencies of copper and cobalt from pyrite ashes formed during the calcination of concentrated pyrite. The conducted procedure (H_2SO_4 leaching tests performed following sulfation roasting at 250°C and with a duration of 1 h) enables 80 % Cu and 70 % Co recovery. The recovery parameters were found to be appropriate for pilot-plant tests [156]. The treatment of slag with chlorine and chloride as lixiviant also enabled efficient dissolving of copper (75–80 %) and the most significant parameters were the particle size and the initial chlorine concentration [87]. However, this technology has not paid more attention.

Lv et al. [157] pointed out that more than 2.45 million tons of cyanidation tailings are discharged into tailing ponds every year in China. Relatively high concentrations of residual Au can be found in these wastes. These tailings could be a source of cheap gold, since the mining and comminution costs have already been incurred. Gold that was not recovered during the initial treatment tends to be either extremely fine or refractory. Examples of large-scale remining of tailings are the ERGO project in South Africa. In this process, the flotation concentrates together with the flotation tails are treated by the conventional carbon-in-leach process (39–40 % extraction or after the concentrate milling 75–76 % extraction). After the cyanidation, the dissolved gold is adsorbed onto activated carbon. Once the gold has been eluted from the carbon, the gold is precipitated, calcined and smelted in the existing smelthouse.

Removing Ni, Co and Cu from slag and introducing them into the aqueous phase have been considered in many more research studies. Anand et al. [158] reported that over 95 % of Ni and Co could be recovered from a converter slag using a mixture of ferric chloride and diluted sulfuric acid. Jia et al. [159] have investigated the leaching behavior of nickel smelter slags in dilute sulfuric acid at room temperature. Over 65 % of Ni and 75 % of Co along with a comparable amount of iron to cobalt were extracted from the slags in a 0.1 M H_2SO_4 acid after 20 h. Higher acid concentrations accelerated the leaching process; however, no selective leaching of Ni and Co was observed in the acid at concentrations of 0.05 M or above. Gbor et al. [160] from the same group attempted aqueous sulfur dioxide leaching of a nickel smelter slag from a Canadian mining company INCO Limited. Maximum extractions of 77 % for Co and 35 % for Ni were reached after 3 h of leaching with an aqueous 1 M SO_2 solution at 35 °C. Iron with extraction of 70 % displayed similar leaching characteristics to cobalt. High-pressure oxidative acid leaching was successfully applied to slow-cooled converter slags from Vale's operations in Sudbury (Ontario, Canada). The conducted tests at 250 °C, at overpressure 0.62 MPa O_2 and using 70 g/L H_2SO_4 indicated that extractions of Ni, Co and Cu exceeded 95 % after 45 min [161]. In another work, Sobol [162] demonstrated the advantages of using elevated temperatures and oxygen overpressures in terms of metal recoveries to solution, reaction rates and selectivity versus iron. It was reported that by pressure leaching with 0.35 N H_2SO_4 (1.6 times the stoichiometric requirement), about 92 % of Cu and more than 95 % of Ni and Co could be extracted with only 0.8 % extraction of Fe from a copper converter slag containing 4.03 % Cu, 1.98 % Ni and 0.48 % Co for 10 % solids in slurry at 130°C and 0.6 MPa O_2 overpressure in 4 h. At higher temperatures of 150–190 °C and a solid loading of 20 %, Sobol achieved extractions of more than 90 % for Ni, Cu and Co from a converter slag in less than 60 min [162]. In previous works, a pressure oxidative acid leaching was also proposed to clean up a waste smelter slag with low base metal contents of 1 % Ni and Cu, 0.2 % Co from a nickel/cobalt smelter and economically recover the base metals at mild conditions with lower oxygen overpressure and lower acid addition [163].

Altundogan et al. [164] have conducted the extraction of Cu, Co, Zn and Fe from copper converter slag generated in Ergani Copper Plant (Maden, Elazig, Turkey) by oxidative leaching with potassium dichromate–sulfuric acid. It was shown that the leaching of copper converter slag in the presence of dichromate is more advantageous with respect to the leaching with sulfuric acid alone [165]. Rudnik et al. [166] have proposed a hydrometallurgical method for selective recovery of copper and cobalt from industrial copper converter slag. In this process, the alloy (composed of Cu (27 %),Co (6 %), Fe (64 %),Pb (1.5 %)) obtained after reductive roasting of the slag was dissolved in the ammonia–ammonium chloride solution. It resulted in the distribution of copper and cobalt with iron remaining in the slime. Cu and Co were separated at electrowinning stage. Altundogan and Tumen [84] have also reported that the high-temperature roasting of slag before leaching causes an increase in extraction of copper. Basir and Rabah [89] showed that copper, zinc and lead are successfully recovered from brass smelting slag by treatment using acid or alkali leachants. The results obtained showed that the mineral acids leach almost all metals in hot conditions, while ammonia selectively leaches copper and zinc with high efficiency [89]. A combination of sulfuric acid and sodium chlorate as effective leaching mixture of Cu, Zn and Co directly from the copper smelter slag has been proposed by Yang et al. [167]. According to this method, the target metals are extracted with high efficiency, while Si and Fe in 3.2 and 0.02 %, respectively. The obtained leachate solution was then neutralized with calcium hydroxide to remove Si and Fe as oxide. Finally, the overall recovery of Cu, Zn and Co was 98 %.

The biological leaching of Cu, Co and Ni from a converter slag was also proposed with promising result [168]. The bacterial action was attributed to iron oxidation and simultaneously acidification of the solution (generation of sulfuric acid). The acidity of the aqueous solution has an influence on the copper dissolution (99 % at pH of 2), while Ni and Co recoveries were 30 %. Carranza et al. have obtained a similar efficiency of dissolution by using moderately thermophilic microorganisms with ferric sulfate solutions to leach a converter slag [169]. Cu dissolution reached 90 % at 2 % pulp density at 60 °C. Muravyov et al. [170] also used thermophilic microorganisms for dissolution of Cu from a copper smelter slag. However, in this study, the ferric sulfate–sulfuric acid mixture as a lixiviant was used. An inhibitory effect of Fe(III) on zinc and iron leaching was indicated, and finally the process resulted in solubilization of 82 % copper and 38 % zinc with relatively low iron recovery (26 %). Metal content in the leach residue was 0.13 % Cu and 3.69 % Zn. Kaksonen et al. conducted and compared the results of the dissolution of Si, Fe, Cu and Zn from a smelter slag under acidic chemical and bacterial leaching conditions [171]. In the case of bioleaching, the acidophilic iron- and sulfur-oxidizing bacteria enriched from the slag were used. Approximately, 80 % Cu and 25 % Zn were dissolved from the slag (10 % pulp); however, the bioleaching required the addition of S. Chemical leaching was much less effective; moreover, at pH 2.3–4.0 there was no substantial dissolution of valuable metals.

7.4 Summary

Several technologies have been or are currently being developed for removal of metals from ore. Most of them co-produced large amounts of unprofitable materials that so far have been treated as waste (mining, processing and metallurgical wastes) storage. The produced waste must be responsibly processed and disposed of to prevent environmental damage. Only *in situ* leaching enables the recovery of mineral from the ore with a minimal surface disturbance of the landscape. The rest of the mining processes require additional steps to increase product purity and to make the mining technology more economic, as well as, environment-friendly. This chapter presented various aspects of the extraction and mining processes with the waste material production and further treatment to recover valuable metals, especially from a hazardous waste.

References

[1] Kippenberger C. Materials Flow and energy required for the production of selected mineral commodities—summary and conclusions (Geologisches Jahrbuch). Stuttgart, Germany: Schweizerbart, 2001: 13–16.

[2] Krausmann F, Gingrich S, Eisenmenger N, Erb K, Haberl H, Fischer-Kowalski M. Growth in global materials use, GDP and population during the twentieth century. Ecol Econ. 2009;68:2696–705.

[3] U.S. Geological Survey. Mineral Commodity Summaries 2017; San Antonio, TX, USA: United States Geological Service, 2017.

[4] Zhang FY, Zhang YH. Study on the recovery of gold and silver from cyanide residue. Rare Met Mater Eng. 2007;37:335–37.

[5] Zhang S, Xue X, Liu X, Duan P, Yang H, Jiang T, et al. Current situation and comprehensive utilization of iron ore tailing resources. J Min Sci. 2006;42:403–8.

[6] Li C, Sun HH, Bai J, Li LT. Innovative methodology for comprehensive utilization of iron ore tailings. Part 1. The recovery of iron from iron ore tailings using magnetic separation after magnetizing roasting. J Hazard Mater. 2010;174:71–77.

[7] International Minerals Statistics and Information U.S. Geological Survey. Available at: https:// https://minerals.usgs.gov/minerals/pubs/country/#links. Accessed August 2, 2017.

[8] Nuclear Energy Agency and the International Atomic Energy Agency. Uranium 2016: Resources, Production and Demand, NEA N°. 7301, Paris: OECD Publishing, 2016.

[9] Management of mining, quarrying and ore-processing waste in the European Union, BRGM/RP-50319-F, 2001. DG Environment, European Commission Co-ordination by P. Charbonnier.

[10] Study on the review of the list of Critical Raw Materials Critical Raw Materials Factsheets, Publications Office of the European Union. 2017. Available at: https://ec.europa.eu/growth/content/study-review-list-critical-raw-materials_en. Accessed August 13, 2017. Deloitte Sustainability; TNO; British Geological Survey; Bureau de Recherches Géologiques et Minières.

[11] Ashraf MA, Sarfraz M, Naureen R, Gharibreza M. Environmental impacts of metallic elements, Environmental Earth Sciences. Singapore: Springer Science+Business Media, 2015: 19–49.

[12] Swart P, Dewulf J. Quantifying the impacts of primary metal resource use in life cycle assessment based on recent mining data. Resour Conserv Recycl. 2013;73:180–87.

[13] Valero A, Arauzo I. Exergy as an indicator for resources scarcity. The exergy loss of Australian mineral capita, A case study. In: Proceedings of ASME IMECE2006, Chicago, USA, 2016.

[14] Valero A, Arauzo I. Evolution of the decrease in mineral exergy throughout the twentieth century. The case of copper in the US. Energy. 2008;33:107–15.
[15] Dixon CJ. Atlas of economic mineral deposits. Netherlands: Springer Science & Business Media, 2012.
[16] Wills BA, Napier-Munn T. Wills' mineral processing technology-an introduction to the practical aspects of ore treatment and mineral recovery seventh edition. Amsterdam: Elsevier Science and Technology Books, 2005.
[17] Technologies in Exploration, Mining, and Processing. In Evolutionary and Revolutionary Technologies for Mining, National Research Council. Washington, USA: The National Academies Press, 2002. (Committee on Technologies for the Mining Industries Board on Manufacturing and Engineering Design Board on Earth Sciences and Resources National Research Council.)
[18] Nagaraj DR, Farinato RS. Evolution of flotation chemistry and chemicals: a century of innovations and the lingering challenges. Miner Eng. 2016;96–97:2–14.
[19] Matis KA. An overview of the process. In: Matis KA, editor. Flotation science and engineering. New York, USA: Marcel Dekker, 1995: 1–44.
[20] Matis KA, Zouboulis AI. The role of bubble/particle size. In: Matis KA, editor. Flotation science and engineering. New York, USA: Marcel Dekker, 1995: 63–87.
[21] Peleka EN, Matis KA. Water separation processes and sustainability. Ind Eng Chem Res. 2011;50:421–30.
[22] Rao SR. The collector mechanism in flotation. Sep Sci. 1969;4:357–411.
[23] Gupta CK. Chemical metallurgy. Principles and practice. Germany: Wiley-VCH GmbH and Co, 2003: 194.
[24] Zhou Y, Tong X, Song S, Wang X, Deng Z, Xie X. Beneficiation of cassiterite fines from a tin tailing slime by froth flotation. Sep Sci Technol. 2014;49(3):458–63.
[25] Mulligan CN, Yong RN, Gibbs BF. Heavy metal removal from sediments by biosurfactants. J Hazard Mater. 2001;85:111–25.
[26] Crundwell FK. How do bacteria interact with minerals? Hydrometallurgy. 2003;71:75–81.
[27] Sanwani E, Chaerun S, Mirahati RZ, Wahyuningsih T. Bioflotation: bacteria-mineral interaction for eco-friendly and sustainable mineral processing. Procedia Chem. 2016;19:666–72.
[28] Nagaoka T, Ohmura N, Saiki H. A Novel mineral flotation process using Thiobacillus ferrooxidans. Appl Environ Microbiol. 1999;65(8):3588–93.
[29] Ohmura N, Kitamura K, Saiki H. Mechanism of microbial flotation using Thiobacillus ferrooxidans for pyrite suppression. Biotechnol Bioeng. 1993;41:671–76.
[30] Fowler TA, Crundwell FK. Leaching of zinc sulfide by Thiobacillus ferrooxidans: bacterial oxidation of the sulfur product layer increases the rate of zinc sulfide dissolution at high concentrations of ferrous ions. Appl Environ Microbiol. 1999;65(12):5285–92.
[31] Townsley CC, Atkins AS. Comparative coal fines desulphurisation using the iron-oxidising bacterium Thiobacillus ferrooxidans and the yeast Saccharomyces cerevisiae during simulated froth flotation. Proc Biochem. 1986;21(6):188–91.
[32] Pecina TE, Rodriguez M, Castillo P, Diaz V, Orrantia E. Effect of Leptospirillum ferrooxidans on the flotation kinetics of sulphide ores. Miner Engin. 2009;22:462–68.
[33] Fowler TA, Holmes PR, Crundwell FK. Mechanism of pyrite dissolution in the presence of Thiobacillus ferrooxidans. Appl Environ Microbiol. 1999;65:2987–93.
[34] Vasanthakumar B, Ravishankar H, Subramanian S. Microbially induced selective flotation of sphalerite from galena using mineral-adapted strains of Bacillus megaterium. Colloids Surf B. 2013;112:279–86.
[35] Gupta A, Yan DS. Mineral processing design and operation an introduction. Amsterdam: Elsevier B.V., 2006: 693.

[36] Olubambi PA, Potgieter JH. Effectiveness of gravity concentration for the beneficiation of Itakpe (Nigeria) iron ore achieved through Jigging Operation. J Miner Mater Character Engin. 2005;4:21–30.

[37] Sreenivas T, Venkatkrishnan RR. Preconcentration of molybdenum from a low-grade primary Mo ore by physical beneficition. Int J Metal Eng. 2012;1:96–101.

[38] Satoshi F. Magnetic separation method for International Journal of Metallurgical Engineering 2012, 1(5):96–101 continuous separation process. J Cryog Soc Jpn. 2002;37:321–27.

[39] Prakash S, Das B, Mohanty JK, Venugopal R. The recovery of fine iron minerals from quartz and corundum mixtures using selective magnetic coating. Int J Miner Process. 1999;57:87–103.

[40] Lu J, Yuan Z, Wang N, Lu S, Meng Q, Liu J. Selective surface magnetization of pentlandite with magnetite and magnetic separation. Powder Techno. 2017;317:162–70.

[41] Zhou Y, Tong X, Song S, Deng Z, Wang X, Xie X. Feng Xie Beneficiation of cassiterite and iron minerals from a tin tailing with magnetizing roasting-magnetic separation process. Sep Sci Technol. 2013;48:1426–32.

[42] Li Y, Zhu T. Recovery of low grade haematite via fluidised bed magnetising roasting: investigation of magnetic properties and liberation characteristics. Ironmaking Steelmaking. 2012; 39(2):112–20.

[43] Subba Rao DV. Mineral beneficiation: a concise basic course. London, UK: CRC Press, 2011:124.

[44] Song S, Lu S, Lopez-Valdivieso A. Magnetic separation of hematite and limonite fines as hydrophobic flocs from iron ores. Miner Eng. 2002;15:415–22.

[45] Huang RZ, Yang LY, Tang ZQ. Comparison of combination of flotation-magnetic-gravity separation process with bluk flotation for recovering fine tin slime. Adv Mater Res. 2014; 1010–1012:1598–602.

[46] Ferron CJ, Bulatovic SM, Salter RS. Beneficiation of rare earth oxide minerals. Materials Science Forum. 1991;70-72:251–70.

[47] Stromberg B, Banwart S. Kinetic modelling of geochemical processes at the Aitik mining waste rock site in northern Sweden. Appl Geochem. 1994;9:583–95.

[48] Clemente CA. Assessment of waste rock weathering characteristics at the Antamina mine based on field cells experiment, T. Columbia: University of British Columbia, 2010.

[49] Marescotti P, Azzali E, Servida D, Carbone C, Grieco G, De Capitani L, et al. Mineralogical and geochemical spatial analyses of a waste-rock dump at the Libiola Fe–cu sulphide mine (Eastern Liguria, Italy). Environ Earth Sci. 2010;61:187–99.

[50] Fortin S, Lamontagne A, Poulin R, Tassé N. The use of basic additives to tailings in layered co-mingling to improve AMD control. In: Singhal RK, Mehrotra AK, editors. Environmental issues and management of waste in energy and mineral production. Proceedings of the sixth international conference (SWEMP 2000, May 30-June 2, Calgary, Alberta). Rotterdam: A.A. Balkema, 2000: 549–56.

[51] Trópia De Abreu A, Faria EM, Chaves CTF, Leite AL, Carvalho De Lena J. Assessment of two kinetic tests to predict the acid mine drainage in waste rock samples of a uranium mine REM. R Esc Minas, Ouro Preto. 2014;67:107–13.

[52] Lèbre E, Corder G. Integrating industrial ecology thinking into the management of mining waste. Resources. 2015;4:765–86.

[53] Campbell G. Acid-forming characteristics of massive ore tailings and disseminated ore tailings. Company Report, August 1996.

[54] Clyde W Black Swan Nickel Project: preliminary waste characterization and tailings storage conceptual design. 1996, Project No. A3300164/003.

[55] Van Den Berg JAM, Yang Y, Nauta HHK. Comprehensive processing of low grade sulphidic molybdenum ores. Miner Eng. 2002;15:879–83.

[56] Liu Y, Zhang Y, Chen F, Zhang Y. The alkaline leaching of molybdenite flotation tailings associated with galena. Hydrometallurgy. 2012;129–130:30–34.
[57] Sirkeci AA, Gul A, Bulut G, Arslan F, Onal G, Yuce AE. Recovery of Co, Ni, and Cu from the tailings of Divrigi Iron Ore Concentrator. Miner Process Extr Metall Rev. 2006;27:131–41.
[58] Das SK, Kumar S, Ramachandrarao P. Exploitation of iron ore tailing for the development of ceramic tiles. Waste Manage. 2000;20:725–29.
[59] Gleisner M, Herbert Jr RB. Sulfide mineral oxidation in freshly processed tailings: batch experiments. J Geochem Explor. 2002;76:139–53.
[60] Wang HL, Shang JQ, Kovac V, Ho KS. Utilization of Atikokan coal fly ash in acid rock drainage control from Musselwhite mine tailings. Can Geotech J. 2006;43:229–43.
[61] Lutandula MS, Maloba B. Recovery of cobalt and copper through reprocessing of tailings from flotation of oxidised ores. J Environ Chem Engin. 2013;1:1085–90.
[62] Zhang Y, Li H, Yu X. Recovery of iron from cyanide tailings with reduction roasting–water leaching followed by magnetic separation. J Hazard Mat. 2012;213–214:167–74.
[63] Peek E, Barnes A, Tuzun A. Nickeliferous pyrrhotite – "Waste or resource?". Miner Engin. 2011;24:625–37.
[64] Kim KK, Kim KW, Kim JY, Cheong YW, Min JS. Characteristics of tailings from the closed metal mines as potential contamination source in South Korea. Environ Geol. 2001;41:358–64.
[65] Ghose MK, Sen PK. Characteristics of iron ore tailings lime in India and its test for required pond size. Environ Monit Assess. 2001;68:51–61.
[66] Gzogyan TN, Gubin SL, Gzogyan SR, Melnikova ND. Iron losses in processing tailings. J Min Sci. 2005;41:583–87.
[67] Licsko I, Lois L, Szebenyi G. Tailings as a source of environmental pollution. Water Sci Technol. 1999;39:333–36.
[68] Gupta CK. Chemical Metallurgy. Principles and Practice. Germany: Wiley-VCH GmbH and Co, 2003: 353.
[69] Gorai B, Jana RK. Characteristics and utilisation of copper slag—a review. Resour Conserv Recycl. 2003;39:299–313.
[70] Kierczak J, Neel C, Puziewicz J, Bril H. The mineralogy and weathering of slag produced by the smelting of lateritic Ni ores, Szklary, southwestern Poland. Can Mineral. 2009;47:557–72.
[71] Gbor PK, Mokri V, Jia CQ. Characterization of smelter slags. J Environ Sci Health A. 2000;35:147–67.
[72] Tang CB, Li Y, Chen YM, Yang SH, Ye LG, Xue HT. Distribution behaviours of cu, Co and Fe during Cu smelter slag cleaning process, 6th International Symposium on High-Temperature Metallurgical Processing 2016, 429–36.
[73] Maweja K, Mukongo T, Mutombo I. Cleaning of a copper matte smelting slag from a water-jacket furnace by direct reduction of heavy metals. J Hazard Mater. 2009;164:856–62.
[74] Kim TJ. Flash smelting in Korea. In: George DB, Taylor JC, editors. Copper smelting—an update. New York: AIME, 1981: 33–39.
[75] Suziki T, Yanagida T, Goto M, Echigoya T, Kikumoto N. Recent operation of Mitsubishi continuous copper smelting and conProcess at Naoshima. In: George DB, Taylor JC, editors. Copper smelting—an update. New York: AIME, 1981: 51–76.
[76] Arslan C, Arslan F. Recovery of copper, cobalt, and zinc from copper smelter and converter slags. Hydrometallurgy. 2002;67:1–7.
[77] Banza AN, Gock E, Kongolo K. Base metals recovery from granulated copper slag by digestion method. Proceedings of the R'02, 6th World Congress on Integrated Resources Management, Geneva, Switzerland, 2002, Paper No. 327.

[78] Tumen F, Bailey NT. Recovery of metal values from copper smelter slags by roasting with pyrite. Hydrometallurgy. 1990;25:317–28.

[79] Yücel O, Sahin FC, Sirin B, Addemir O. A reduction study of copper slag in a DC arc furnace. Scan J Metal. 1999;28:93–99.

[80] Yin F, Xing P, Qiang L, Wang C, Wang Z. Magnetic separation-sulphuric acid leaching of Cu–co–Fe matte obtained from copper converter slag for recovering Cu and Co. Hydrometallurgy. 2014;149:189–94.

[81] Anand S, Kanta P, Rao PK. Jena recovery of metal values from copper converter and smelter slags by ferric chloride leaching. Hydrometallurgy. 1980;5:355–65.

[82] Rao GV, Nayak BD. Flotation of copper from converter slags. J Mines Met Fuels. 1992; 40(3-4):131–34.

[83] Bulut G. Recovery of copper and cobalt from ancient slag. Waste Manag Res. 2006;24:118–24.

[84] Li Y, Chen Y, Tang C, Yang S, He J, Tang M. Co-treatment of waste smelting slags and gypsum wastes viareductive-sulfurizing smelting for valuable metals recovery. J Hazard Mater. 2017;322:402–12.

[85] Altundogan HS, Tumen F. Metal recovery from copper converter slag by roasting with ferric sulphate. Hydrometallurgy. 1997;44:261–67.

[86] Barnes CD, Lumsdaine J, O'Hare SM. Copper converter slag treatment at Mount Isa Mines Limited, Mount Isa, Qld. AusIMM Proceedings. 1993;298:31–37.

[87] Herreros O, Quiroz R, Manzano E, Bou C, Vinals J. Copper extraction from reverberatory and flash furnace slags by chlorine leaching. Hydrometallurgy. 1998;49:87–101.

[88] Aghajania H, Khakib JV, Goudarzic M, Hosseinia N. Sulfuric acid leaching of sulfide slag of Sarcheshmeh copper reverberatory furnace. SCIREA J Metal Engin. 2016;1:12–27.

[89] Basir SMA, Rabah MA. Hydrometallurgical recovery of metal values from brass melting slag. Hydrometallurgy. 1999;53:31–44.

[90] Gupta CK. Chemical metallurgy. Principles and practice. Germany: Wiley-VCH GmbH and Co, 2003: 462–84.

[91] Seredkin M, Zabolotsky A, Jeffress G. In situ recovery, an alternative to conventional methods of mining: exploration, resource estimation, environmental issues, project evaluation and economics. Ore Geol Rev. 2016;79:500–14.

[92] Roberts PA, Zhang H, Prommer H, Benvie B, Jeffrey MI, Johnson CD, et al. Ore characterization, hydrometallurgical and reactive transport studies for in place leaching of oxidized gold deposits. Miner Metall Process. 2010;27(2):72–80.

[93] Organisation for Economic Co-operation and Development (OECD). Uranium 2005: Resources, Production and Demand. OECD Nuclear Energy Agency. 2005. NEA N°. 6098.

[94] CA Pat. 880708.

[95] US Pat. 3834760.

[96] US Pat. 3640579.

[97] US Pat. 3063499.

[98] Yannopoulos JC. The extractive metallurgy of gold. USA: Van Nostrand Reinhol, 1991.

[99] Berezowsky RM, Xue T, Collins MJ, Makwana M, Barton-Jones I, Southgate M, et al. Pressure leaching las cruces copper ore. JOM. 1999;51:36–40.

[100] Whittington BI, Muir D. Pressure acid leaching of nickel laterites: a review. Miner Proces Extract Metal Rev: Inter J. 2000;21(6):527–99.

[101] Chen J, Huang K. A new technique for extraction of platinum group metals by pressure cyanidation. Hydrometallurgy. 2006;82:164–71.

[102] Mpinga CN, Bradshaw SM, Akdogan G, Snyders CA, Eksteen JJ. Evaluation of the Merrill–crowe process for the simultaneous removal of platinum, palladium and gold from cyanide leach solutions. Hydrometallurgy. 2014;142:36–46.

[103] Warner AEM, Diaz CM, Dalvi AD, Mackey PJ, Tarasov AV. World nonferrous smelter survey, Part III: nickel: laterite. JOM. 2006;58:11–20.
[104] Crundwell FK, Moats MS, Ramachandran V, Robinson TG, Davenport WG. Overview of the hydrometallurgical processing of laterite ores. In: Extractive metallurgy of nickel, cobalt and platinum-group metals. Great Britain: Elsevier Ltd, 2011: 117–22.
[105] Brown M. Murrin Murrin operations – past, present and future, ALTA 2011 Conference, Perth, Australia, 2011.
[106] Oxley A, Sirvanci N, Purkiss S. Çaldağ nickel laterite atmospheric heap leach project. Assoc Metall Eng Serbia, AMES. 2007;13:5–20.
[107] Readett D, Sullivan J GME's NiWest Nickel Project – ongoing development of Ni laterite heap leach, sx-ew process, ALTA 2015 Conference, Perth, Australia, 2015.
[108] Oxley A. The Piauí Nickel Heap Leach Project, Brazil, Paper from the Proceedings of the Third International Conference on Heap Leach Solutions held in Reno, Nevada, US, 2015.
[109] Dreisinger D. The Starved Acid Leaching Technology (SALT) for nickel recovery from saprolites, ALTA 2016 Conference, Perth, Australia, 2016.
[110] Rastas J, Saikkonen P, Snåre L, Lindell E, Jääskeläinen E. An atmospheric dissolving process for laterites, ALTA 2015 Conference, Perth, Australia, May 2015: 79–93.
[111] Shibayama K, Yokogawa T, Sato H, Enomoto M, Nakai O, Ito T, et al. Plant, project. Miner Eng. 2015;88:61–65.
[112] Taylor A. Technical & cost comparison of laterite treatment processes, Part 1, ALTA 2014 Conference, Perth, Australia, 2014: 35–46.
[113] Harris B, White C. Recent developments in the chloride processing of nickel laterites, Paper presented at ALTA 2011, Perth, Australia, 2011: 12–24.
[114] Rhamdhani MA, Chen J, Hidayat T, Jak E, Hayes P. Advances in research on nickel production through the Caron process. Proceedings of EMC 2009. 2009;899–913.
[115] Önal MAR, Topkaya YA. Pressure acid leaching of Çaldağ lateritic nickel ore: an alternative to heap leaching. Hydrometallurgy. 2014;142:98–107.
[116] Ma B, Yang W, Yang B, Wang C, Chen Y, Zhang Y. Pilot-scale plant study on the innovative nitric acid pressure leaching technology for laterite ores. Hydrometallurgy. 2015;155: 88–94.
[117] Llerin EJ, Nishikawa I, Kawata M. Coral bay nickel HPAL plant expansion project, Paper presented at ALTA 2011, Perth, Australia, 2011: 511–44.
[118] Fittock JE. Nickel and cobalt refining at QNI Pty Ltd Yabulu Queensland. AUSIMM Monograph 19, 2nd ed. Western Australia: Australasian Institute of Mining and Metallurgy in Melbourne. Ed. 1992:1–47.
[119] Ghorbani Y, J-P F, Petersen J. Heap leaching technology – current state, innovations and future directions: a review. Miner Proces Extract M Rev. 2015;37(2):73–119.
[120] U.S. Patent 5,730,776.
[121] Dreisinger DB, Steyl JDT, Sole KC, Gnoinski J, Dempsey P. The Anglo American Corporation/University of British Columbia (AAC/UBC) chalcopyrite process: an integrated pilot–plant evaluation. In: Rivieros A, Dixon DG, Dreisinger DB, Menacho J, editors. Copper 2003-Cobre 2003, Vol VI, Hydrometallurgy of copper. Montreal, 2003: 223–37.
[122] Reisinger DB, Richmond GD, Hess F, Lancaster T. The competitive position of the Mt Gordon Copper Process in the copper industry, Proceedings of ALTA Cu Conference 2002, ALTA Metallurgical Services Australia, 2002.
[123] Olubambi PA, Borod JO, Ndlovu S. Sulphuric acid leaching of zinc and copper from Nigerian Complex Sulphide Ore in the presence of hydrogen peroxide. J South African Inst Min Metal. 2006;106(11):765–70.
[124] Watling HR. Review of biohydrometallurgical metals extraction from polymetallic mineral resources. Minerals. 2015;5:1–60.

[125] Groudev SN, Genchev FN, Guidarjiev SS. Observations on the microflora in an industrial copper dump leaching operation. In: Murr LE, Torma AE, Brierley JA, editors. Metallurgical applications of bacterial leaching and related microbiological phenomenon. New York: Academic Press, 1978: 253–73.
[126] Johansson C, Shrader V, Suissa J, Adutwum K, Kohr W. Use of the GEOCOAT™ process for the recovery of copper from chalcopyrite. Process Metal. 1999;9:569–76.
[127] Batty JD, Rorke GV. Development and commercial demonstration of the BioCOP™ thermophile process. Hydrometallurgy. 2006;83(1-4):83–88.
[128] Romero R, Mazuelos A, Palencia I, Carranza F. Copper recovery from chalcopyrite concentrates by the BRISA process. Hydrometallurgy. 2003;70(1-3):205–15.
[129] Gericke M, Muller HH, Van Staden PJ, Pinches A. Development of a tank bioleaching process for the treatment of complex Cu-polymetallic concentrates. Hydrometallurgy. 2008;94: 23–28.
[130] Van Staden PJ. The Mintek/Bactech copper bioleach process, ALTA Copper Hydrometallurgy Forum. Brisbane, 1998.
[131] BHP-Billiton annual report April 2004. Available at: http://www.bhp.com/-/media/bhp/documents/investors/news/sulphideleach.pdf?la=en. Accessed September 10, 2017.
[132] Anderson CG. The treatment of chalcopyrite concentrates with nitrogen species catalysed oxidative pressure leaching. In: Young SK, Dreisinger DB, Hackl RP, Dixon DG, editors. Proceedings of Copper99–cobre99 International Conference. Hydrometallurgy of Copper, vol. IV. Warrendale: TMS, 1999: 139–49.
[133] Habashi F. Nitric acid in the hydrometallurgy of sulfides. In: Mishra B, editor. Proceedings of the EPD Congress 1999. Warrendale, PA: TMS–AIME, 1999: 25–32.
[134] Ruiz MC, Montes KS, Padilla R. Chalcopyrite leaching in sulfate–chloride media at ambient pressure. Hydrometallurgy. 2011;109:37–42.
[135] Herreros O, Quiroz R, Longueira H, Fuentes G, Viñals J. Leaching of djurleite in Cu2+/Cl– media. Hydrometallurgy. 2006;82:32–39.
[136] Ferron CJ, Fleming CA, O'Kane PT, Dreisinger D. Pilot plant demonstration of the PLATSOL™ process for the treatment of the NorthMet copper-nickel-PGM deposit. Min Engin. 2002; 54(12):33–39.
[137] Dixon DG, Mayne DD, Baxter KG. Galvanox™ — a novel galvanically-assisted atmospheric leaching technology for copper concentrates. Can Metall Q. 2008;47:327–36.
[138] Baxter K, Dreisinger DB, Pratt G. The Sepon copper project: development of a flowsheet. In: Young CA, Alfantazi AM, Anderson CG, Dreisinger DB, Harris B, James A, editors. Hydrometallurgy. Warrendale: TMS, 2003: 1487–502.
[139] Watling HR. Chalcopyrite hydrometallurgy at atmospheric pressure: 2. Review of acidic chloride process options. Hydrometallurgy. 2014;146:96–110.
[140] Imamura M, Takeda K, Ando K, Nagase N, Ojima Y. Iron recovery in the Sumitomo chlorine leach process for copper concentrates. In: Dutrizac JE, Riveros PA, editors. Iron Control Technologies, Proceedings of the Third International Symposium on Iron Control in Hydrometallurgy (Montreal). Montreal: CIM, 2006: 191–204.
[141] Dalton RF, Diaz G, Price R, Zunkel AD. The cuprex metal extraction process: recovering copper from sulfide ores. JOM: J Min Met Mater Soc. 2006;43(8):51–56.
[142] Everett PK. Development of Intec Copper Process by an international consortium. In: Hydrometallurgy'94. Springer, Dordrecht, 1994. Available at: http://www.intecip.com.au/intec-copper-process.pdf. Accessed September 12, 2017.
[143] Haavanlammi L, Hyvarinen O, Yllo E, Hydrocopper™ – a new method for treating copper concentrates, Presented at Extractive Metallurgy Operators' Conference 2005, Brisbane, Australia, 2005: 1–14.

[144] Defreyne J, Barr G, McCunn G, De Souza Costa R, Goncalves LRL. CESL copper process – moving from pilot plant to production scale operation. Available at: http://infrastructure.companyoung.com/cesl_copper_process_moving_from_pilot_plant_to.pdf. Accessed September 12, 2017.
[145] Garg S, Papangelakis VG, Mahadevan R, Edwards E, Krause E. Bioleaching of pyrrhotite tailings for Ni extraction—initial scoping tests. In: Towards Clean metallurgical processing for profit, social and environmental stewardship, Proceedings of the 51st Conference of Metallurgists COM12, Niagara Falls, ON, Canada, 2012: 315–29.
[146] Shoonewille RH, Rioux D, Kashani-Nejad S, Kreuh M, Muinonen MES, editors. Canadian Institute of Mining Metallurgy and Petroleum: Montreal, QC, Canada, 2012.
[147] Garg S, Papangelakis V, Mahadevan R, Edwards E. Bioleaching of pyrrhotite tailings for Ni extraction- insights into an adaptive evolution study, COM 2015, The Conference of Metallurgists, Toronto, Australia, 2015.
[148] Hernández CM, Banza AN, Gock E. Recovery of metals from Cuban nickel tailings by leaching with organic acids followed by precipitation and magnetic separation. J Hazard Mater. 2007;139(1):25–30.
[149] Lianga Y, Yin H, Xiao Y, Tang M, Feng X, Xie Z, et al. Microbial leaching of copper from tailings of low grade sulphide ores in Zambia. Adv Mat Res. 2015;1130:473–76.
[150] Fosso-Kankeu E, Mulaba-Bafubiandi AF, Mamba BB, Barnard TG. Assessing the effectiveness of a biological recovery of nickel from tailing dumps. Min Eng. 2011;24(5):470–72.
[151] Park J, Han Y, Lee E, Choi U, Yoo K, Song Y, et al. Bioleaching of highly concentrated arsenic mine tailings by Acidithiobacillus ferrooxidans. Sep Purif Technol. 2014;133:291–96.
[152] Hao X, Liang Y, Yin H, Ma L, Xiao Y, Liu Y, et al. The effect of potential heap construction methods on column bioleaching of copper flotation tailings containing high levels of fines by mixed cultures. Min Engin. 2016;98:279–85.
[153] Ahmadi A, Khezri M, Abdollahzadeh AA, Askari M. Bioleaching of copper, nickel and cobalt from the low grade sulfidic tailing of Golgohar Iron Mine, Iran. Hydrometallurgy. 2015;154:1–8.
[154] Xie Y, Xu Y, Yan L, Yang R. Recovery of nickel, copper and cobalt from low-grade Ni–Cu sulfide tailings. Hydrometallurgy. 2005;80(1–2):54–58.
[155] Akdag M. Investigation of cobalt solubilities from pyrite cinder in sulphuric acid solution. IBSU Sci J. 2008;2:137–42.
[156] Erust C, Akcil A. Copper and cobalt recovery from pyrite ashes of a sulphuric acid plant. Waste Manag Res. 2016;34(6):1–7.
[157] Lv CC, Ding J, Qian P, Li QC, Ye SF, Chen YF. Comprehensive recovery of metals from cyanidation tailing. Min Engin. 2015;70:141–47.
[158] Anand S, Rao KS, Jena PK. Pressure leaching of copper converter slag using dilute sulphuric acid for the extraction of cobalt, nickel and copper values. Hydrometallurgy. 1983;10:305–12.
[159] Jia CQ, Xiao JZ, Orr RG. Behavior of metal in discard nickel smelter slag upon reacting with sulphuric acid. J Environ Sci Health A. 1999;34(5):1013–34.
[160] Gbor PK, Ahmed IB, Jia CQ. Behavior of Co and Ni during aqueous sulphur dioxide leaching of nickel smelter slag. Hydrometallurgy. 2000;57:13–22.
[161] Perederiy I, Papangelakis VG, Buarzaiga M, Mihaylov I. Co-treatment of converter slag and pyrrhotite tailings via high pressure oxidative leaching. J Hazard Mater. 2011;194:399–406.
[162] Sobol SI Chemistry and kinetics of oxidative sulphuric acid leaching of cobalt-bearing converter slags. Extractive metallurgy of copper, nickel and cobalt, vol. 1. Fundamental Aspects, Proceedings of the Paul E. Queneau International Symposium, TMS, Denver, CO, USA, 1993: 803–11.
[163] Li Y, Perederiy I, Papangelakis VG. Cleaning of waste smelter slags and recovery of valuable metals by pressure oxidative leaching. J Hazard Mater. 2008;152:607–15.

[164] Altundogan HS, Boyrazli M, Tumen F. A study on the sulfuric acid leaching of copper converter slag in the presence of dichromate. Min Engin. 2004;17:465–67.

[165] Ahmed IM, Nayl AA, Daoud JA. Leaching and recovery of zinc and copper from brass slag by sulfuric acid. J Saudi Chem Soc. 2016;20(1):280–85.

[166] Rudnik E, Burzyska L, Gumowska W. Hydrometallurgical recovery of copper and cobalt from reduction-roasted copper converter slag. Min Engin. 2009;22:88–95.

[167] Yang Z, Rui-Lin M, Wang-Dong N, Hui W. Selective leaching of base metals from copper smelter slag. Hydrometallurgy. 2010;103:25–29.

[168] Mehta KD, Pandley BD. Premchand, Bio-assisted leaching of copper, nickel and cobalt from copper converter slag. Mater Trans JIM. 1999;40:214–20.

[169] Carranza F, Romero R, Mazuelos A, Iglesias N, Forcat O. Biorecovery of copper from converter slags: slags characterization and exploratory ferric leaching tests. Hydrometallurgy. 2009;97:39–45.

[170] Muravyov MI, Fomchenko NV, Usoltsev AV, Vasilyev EA, Kondrat'eva TF. Leaching of copper and zinc from copper converter slag flotation tailings using H_2SO_4 and biologically generated $Fe_2(SO_4)_3$. Hydrometallurgy. 2012;119–120:40–46.

[171] Kaksonen AH, Särkijärvi S, Puhakka JA, Peuraniemi E, Junnikkala S, Tuovinen OH. Chemical and bacterial leaching of metals from a smelter slag in acid solutions. Hydrometallurgy. 2016;159:46–53.

Daria Wieczorek and Dobrawa Kwaśniewska

8 Economic aspects of metals recover

Abstract: One of the modern economy models is circular economy in which wastes should be considered as resource and used in an efficient and sustainable way. This also concerns to metals included in scraps. However, the need for metal recovery from waste is not only the result of the latest economic trends but also the result of large and constantly changing demand for metals. Shrinking natural sources of metals, concentrations of ores in small number of countries in the world and resulting from this dependence on import, geopolitical situation, new technologies demands are only a few most important determinants that have been changing the structure of the metal market over years. In this chapter, authors focused on the presentation of economic aspects of metal recovery from various sources. The chapter presents the characteristic of metal market elements (supply, demand and price) and changes that took place over decades, underlining the structure of precious and highly desirable metal market elements. Balance between the demand and supply ensures price stability and rationalizes inflation. However, growing demand on many means that secure supply chains, such as recycling and material recovery, are essential to ensure continuity in the supply chain and guarantee unrestricted technological progress and innovation. The data included in this chapter presents also the concentration of different metals and group of metals in wastes pointing that recycling of waste can become one of the possibilities of acquiring missing and critical metals. Metal-laden wastes include a few groups: waste electrical and electronic equipments, catalysts of different application, introduced on chemical, petrochemical or automotive market, galvanic wastes and wastewaters. The profitability assessment of recycling processes is very complicated. Nevertheless cited data shows that profitability of recovery depends on the metal analyzed and the type of waste.

It must be underline that an optimized management of wastes is of a great importance for the global economy and allow achieving not only economic but also environmental and social benefits.

Keywords: circular economy, metal market elements, metal recovery

---

This article has previously been published in the journal *Physical Sciences Reviews*. Please cite as: Wieczorek, D., Kwásniewska, D. Economic aspects of metals recovery. *Physical Sciences Reviews* [Online] **2018**, *3* (4). DOI: 10.1515/psr-2018-0027

https://doi.org/10.1515/9783110547061-008

8.1 Introduction

Amount of metal-containing products placed on the market every year is increasing both in industrialized and industrializing countries. Rapid industrialization, human activities and shortening of product life cycle generate large amount of metal-laden wastes [1–3]. In traditional economic systems characterized by linear material flows growing economies result in overuse of natural capacities mainly due to increasing consumption of resources, generation of potentially harmful effluents and overuse of final disposal reservoirs [4]. The international scientific community agrees that an optimized management of wastes can allow achieving economic, environmental and social benefits [5]. Alsheyab and Kusch (2013) suggest that decoupling can be an efficient strategy for the conservation of natural resources and a step forward for sustainable development. Authors points that decoupling strategies comprise all means to break linear dependencies between economic development and requirement of resources was well as generation of final waste [4]. Metals are well suited to sustainable development goals. They are not biodegradable with virtually an unlimited life span. As metals are considered to have potential for unlimited recyclability they are considered as renewable materials. However, the sources of primary metals, mineral resources are "non-renewable" and their supply is finite. Presented in Table 8.1 number of years to exhaust mineral metals resources highlights the importance of the problem. If assuming consumption rate closely balanced to total production rate and no recycling the supply of iron will end in 77 years, aluminum in 162 years, copper in 22 years, lead in 10 years, zinc in 16 years and nickel in 43 years [6]. These data does not correspond with these presented in table and much more alarming.

Table 8.1: The estimated time to exhaustion of resources selected metals in the absence of recycling [7].

Metal	Years to exhaust mineral metal resources
Iron	190
Chrome	370
Manganese	90
Nickel	105
Vanadium	570
Molybdenum	95
Cobalt	130
Tungsten	55
Aluminum	375
Magnesium	4503
Copper	56
Zinc	38
Lead	29
Platinum	186
Tin	14

This does not necessarily mean scarcity while recycling of metals from wastes can become a solution of such worldwide problem. Due to increasing demand for raw materials (mainly metals) and environmental protection requirements, waste processing mainly of electronic and electrical as well as car catalysts scrap is becoming an economic necessity and an imperative of the modern economy. Furthermore, recycling of waste became a valuable source of raw materials, especially for countries where there are no natural sources of metals, increasing their independence. European Union countries are in 100% dependent on import from other countries such metals like: antimony, beryllium, cobalt, germanium, indium, magnesium, niobiumt [7].

Table 8.1 shows the approximate years to exhaust mineral metal resources. However it should be distinguish between resources and reserves. A resource is a concentration of naturally occurring material (metal) in the earth's crust, but it such form and amount that extraction of a metal from the concentration is currently or potentially feasible. While a reserve is that part of an identified resource which could be economically extracted or produced at the time of determination [6]. Thus, the year in which metal reserves will run out changes constantly. Changes in metal prices and/or new processing technologies may change the world reserves of metals. However very important source of metal that can extent the reserves is "metal in use". This rapidly growing stock of metals represents a virtually permanent asset for society. Metals can be recycled indefinitely and passed along from one generation to the next only if they are used in non-dissipative applications where the metal can be economically recovered. Figure 8.1 shows how the "years of supply" increases as the recycling rate increases for the various metals [6].

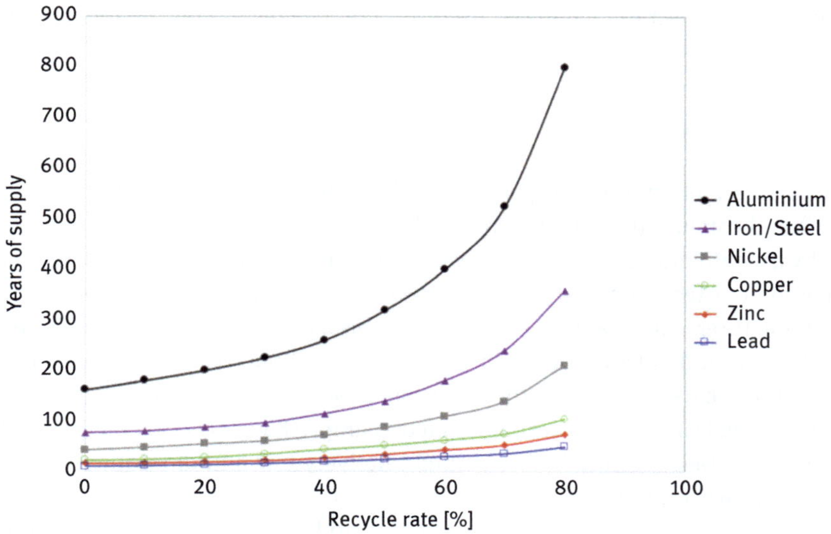

Figure 8.1: The effect of recycle rate on the supply of metals [6].

8.2 Elements of metal market

Demand, supply and prices are key elements of each market. This also applies to the metal-market; however, the considerations usually cover the world market. The relationships between these elements and their dependencies refer to a demand–supply–price mechanism.

Over the last decades there has been a significant increase in the availability of electrical and electronic equipment (EEE). Consumption of these goods and the ever-increasing demand are associated with an increase in the amount of generated electrowaste. However, it seems that recycling of these wastes is insufficient, and for example in 2009 only 13% of e-waste was recycled [8]. At the same time it is estimated that between 20 and 50 million tons of wastes from EEEs are generated each year and the growth rate is from 3% to 5% [9]. Recycling of EEEs is rational for economic reasons but also for the protection of the environment. Increasing environmental awareness has led some countries to adopt a strategy of Extended Producer Responsibility (EPR). It assumes that the manufacturer is responsible for the entire life cycle of the product [10].

E-waste are not chemically a homogeneous group, these differences are due to age and type of equipment. However, e-wastes are a mixture of metals, plastic, glass and ceramics. In economic terms, it is very important to recover the metals present in e-waste. In the waste from EEEs, except for basic and special metals (e.g. Fe, Al, Ni, Zn, Se, In and Ga), the presence of precious (Au, Ag, Pt, Pd) and dangerous (Hg, Be, Pb, As, Cd and Sb) metals can be noted [8, 11]. However, not only e-waste is a rich source of precious metals. Platinum group metals (PGMs) can be recovered from used car catalysts, jewel scrap, catalysts used in the chemical and petrochemical industries.

8.2.1 Platinum groups metal

Six metals namely platinum (Pt), palladium (Pd), rhodium (Rh), ruthenium (Ru), iridium (Ir) and osmium (Os) enter into so-colled PGMs [12]. These metals are characterized by their unique chemical and physical properties, so it is not surprising that they find numerous technological applications. It is estimated that the world's largest PGMs' manufacturers are located in South Africa, Russia, Canada, Zimbabwe, United States, Columbia and Australia [13]. In general, the volume of PGMs mining has changed over the years. Essentially, the last hundred years have brought a significant increase in mining PGMs what is best illustrated in graph (Figure 8.2). By the 1940s of the last century, PGMs' mining had remained relatively low and did not exceed 20 tons per year. However, the third industrial revolution resulted in an increase in the demand for these metals. So far, the highest PGMs' production was observed in 2006, at 515 tons [14].

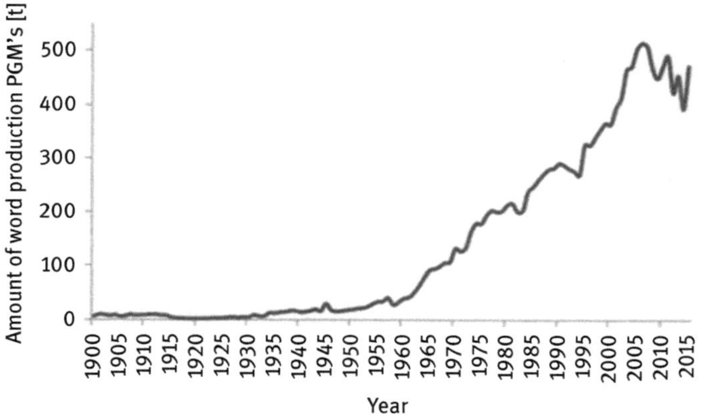

Figure 8.2: PGM's production volumes [14].

In 2015, the total production of PGMs was 473 tons, of which platinum output was 189 tons (Figure 8.3) and palladium 216 tons. At present, the largest platinum mining is definitely taking place in South Africa and Russia. In 2015, 139 and 22 tons of platinum were mined in these countries, respectively. In the same year, palladium production in Russia and South Africa was at a similar level and amounted, respectively, to 81 and 83 tons [14].

The greatest importance in the global economy of platinum, especially platinum, palladium and rhodium, has the production of catalysts for cars. This industry consumed 90 tons of platinum in 2012, which accounted for 40% of world consumption, 208 tons of palladium and 25 tons of rhodium. In the years 2008–2009, the automotive market witnessed a crisis, which led to a 20% decrease in demand

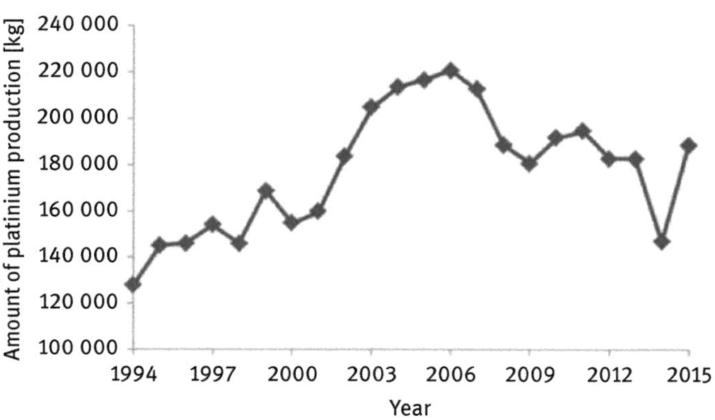

Figure 8.3: Platinum production volume [14].

for PGMs, as early as 2010, the consumption of these metals increased again. It is estimated that jewellery is a branch dealing in second place in terms of the demand for platinum and palladium. So far, the biggest palladium consumption was recorded in 2005, it was 46 tons, while in 2012 and reached the level of 14 tons. The use of platinum in jewellery has been in recent years at 75–85 tons per annum. Since the 1980s there has an increase in demand for platinum and palladium in the electronics industry. However, the development of technology most likely and the replacement of PGMs with other materials will drop demand for these raw materials. Nevertheless, half of the world's use of ruthenium is in the electrical and electronic industries. It is estimated that the global recovery of platinum from autocatalysts and jewellery in 2014 was 34.3 tons, while waste EEE recovered 1.48 tons. In the same period, 14 tons of palladium were recovered from the e-waste and 61.3 tons of autocatalysts [15, 16].

8.2.2 Lithium

Lithium is electrochemically active metal and resources are estimated at 11 million tons. At present there are 100 fossil mines in 20 countries around the world. However, the largest tonnage mining is located in Portugal, Zimbabwe, Argentina, Brazil, Chile, Canada, USA, China and Australia. Lithium in nature do not appear in the form of a pure element can be found in about 20 minerals. It is estimated that, in terms of industrial use, the most important are Lepidolite ($KLi_{1.5}Al_{1.5}[Si_3O_{10}][F,OH]_2$), Spodumene ($LiO_2$ Al_2O_3 $4SiO_2$), Petalite (LiO_2 Al_2O_3 $8SiO_2$) and Amblygonite ($LiAl[PO_4][OH,F]$) [16]. The volume of production of lithium calculated on Li_2O in years 2008–2012 is presented in Table 8.2.

Between 2002 and 2006, the world market have been observed an increase in supply of about 60%, which was 52.7 thousand ton Li_2O. This trend was due to

Table 8.2: The volume of production of lithium calculated on Li_2O [16].

	Production of lithium calculated on Li_2O[ton]				
	2008	**2009**	**2010**	**2011**	**2012**
Portugal	1395.5	1494.4	1604.4	1501.4	1500.0
Zimbabwe	2150.0	2150.0	2021.0	2064.0	2279.0
Argentina	6246.4	4707.2	6438.4	5324.8	5478.4
Brazil	1041.1	1146.9	1132.8	563.0	540.0
Chile	21,320.2	11,250.7	19,534.7	26,054.6	28,295.1
Canada	1584.0	720.0	–	–	–
China	7000.0	8000.0	8400.0	8800.0	9600.0
Australia	17,246.0	14,218.7	21,240.0	30,340.5	32,898.3
World	**50,983.2**	**43,687.9**	**60,371.3**	**74,648.3**	**80,590.8**

increased demand for lithium resulting from favorable economic situation in the US and Western Europe also not without significance was the economic growth of China and India. In subsequent years (2007–2008) there were no longer growth trends in supply, but the demand for lithium continues to grow. This situation led to a strong price increase. The financial crisis in 2009 contributed to a significant decrease in the demand for lithium. However, in 2010–2012, the improving situation on the raw materials market lithium meant that it is possible to see an increase in total supply of as much as 84%. Available data indicate that lithium production over the years 2011–2015 remained at an even level. In 2016, the worldwide lithium market in various sectors had the percentage structure shown in the Figure 8.4 [17].

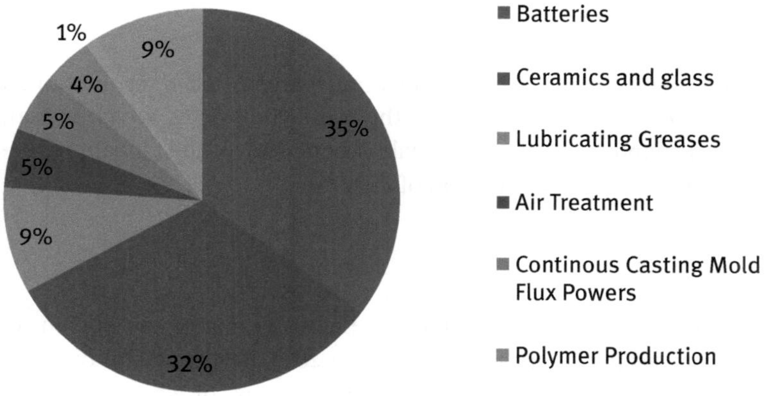

Figure 8.4: Distribution of global lithium end-uses at various sectors for various applications [17].

8.2.3 Indium

World indium resources are estimated at about 11,000 tons. Indium is usually found in zinc ore, and tin, lead, tungsten, iron and pyrites. With the participation of hydrometallurgical methods, a highly polluted indium is obtained, so-called black indium, which requires further processing by electrochemical methods for metallic indium [16]. It is estimated that over the past 30 years consumption grew by over 2000% [18]. Over the past few years, the largest indium producers were China and Republic of Korea. Essentially, in the last decade the production of indium has been dominated by Asian countries, in this period as many as 70–80% of world production of indium came from Asia. According to a United States Geological Survey (USGS) report, worldwide total primary production reported at 655 tons in the year 2016. Detailed data on indium production, including the largest producers, is provided in the Table 8.3 [14, 15].

Table 8.3: The volume of production of indium [14].

	Refinery production (*t*)						
	2010	**2011**	**2012**	**2013**	**2014**	**2015**	**2016**
Belgium	30	30	30	30	25	20	25
Canada	67	75	62	65	65	70	65
China	340	380	405	415	460	350	290
France	–	–	–	33	43	41	
Japan	70	70	71	72	72	70	70
Republic of Korea	70	70	165	150	150	195	195
Peru	–	–	11	11	14	9	5
Russia	NA	5	13	13	5	4	5
World	**609**	**662**	**782**	**799**	**844**	**759**	**655**

Production of modern LCDs and LEDs as well as transparent diodes resulted to an increase in In-Sn oxide (ITO) consumption in the years 2008–2012, which in turn stimulated a rise in demand for indium materials. About 80% of the global production of indium is dedicated to the production of ITO [16].

8.2.4 Gold and silver

Gold is one of the more useful precious metals. Eighty to eighty-five percent of its output comes from residual deposits, another 15–20% of other metal ore. It is estimated that the size of gold deposits is 51 thousand tons, of which 14% is in Australia, 12% in South Africa, 10% in Russia, 7% in Chile, 6% in USA, 6% in Indonesia, 5% in Brazil, 4% in China and another 4% in Peru. In addition, gold deposits can be found in 100 other countries [16]. However, the volume of mining is slightly different in comparison to the volume of resources. Over the past few years, China was largest producer of gold, in 2015 has produced 450 tons of this ore. Detailed data on the volume of extraction in each country are provided in the Table 8.4 [14]. The supply of gold in recent years has remained stable at 3.8–4.5 thousand tons, 60–65% of this comes from mining, another 21–39% is scrap, and some share is also sold reserves of central banks [16].

So far, jewellery remains the most important direction for the use of this ore.

Over the years 2000–2012, as much as 75–89% of gold was used in the arts and jewellery making. In the same period, the consumption of gold for electrical products increased from 4% to 6%. It is estimated that 3175 tons of gold were used to make jewellery only in 1 year 2000 [15, 16].

The amount of silver on the earth is estimated at 545 thousand tons, of which the largest deposits are in Peru, Poland, Chile, Australia, China, Mexico, USA and Bolivia. However, this is not exactly reflected in the volume of extraction. The largest

Table 8.4: The volume of production of gold [14].

	Mine production of gold (ton)					
	2010	2011	2012	2013	2014	2015
United States	231	234	235	230	210	214
Australia	261	258	250	265	274	278
Brazil	58	62	65	71	80	81
Canada	91	97	104	124	152	153
China	345	362	403	430	450	450
Ghana	82	80	87	90	91	88
Indonesia	120	96	59	61	69	97
Mexico	73	84	97	98	118	135
Papua New Guinea	68	66	53	57	53	60
Peru	164	164	161	151	140	145
Russia	192	200	218	230	247	252
South Africa	189	181	160	160	152	145
Uzbekistan	90	91	93	98	100	102
World	**2560**	**2660**	**2690**	**2800**	**2990**	**3100**

silver mining is currently in Mexico [14, 16]. The volume of production of silver in years 2010–2015 is presented in Table 8.5.

In 2005, the global market observed a record supply of 29.6 thousand tons, the years 2006–2009 brought a decrease in supply by 4%, but this trend did not hold. Silver supply in 2010–2012 reached 32.5–33.5 tons. Silver available on the market in 75% comes from mining extraction in 24% from scrap, only 1% is silver from government reserves [16].

Table 8.5: The volume of production of silver [14].

	Mine production of silver					
	2010	2011	2012	2013	2014	2015
United States	1270	1120	1060	1040	1180	1090
Australia	1860	1730	1730	1840	1720	1430
Bolivia	1260	1210	1210	1290	1340	1190
Chile	1280	1290	1190	1170	1570	1370
China	3500	3700	3900	4100	4060	3100
Mexico	4410	4150	5360	4860	5000	5370
Peru	3640	3410	3480	3670	3780	3850
Poland	1180	1170	1150	1200	1260	1180
Russia	1150	1350	1500	1720	1330	1430
World	**23,100**	**23,300**	**25,500**	**26,000**	**26,800**	**25,100**

Over the years, the structure of the use of silver has declined. The limitation of the use of silver in the photography industry was the most visible, which directly connected with the widespread use of digital photography. In turn, the consumption of silver in the electrical and electronic industries increased and in 2010 it reached the value of 15.6 thousand tons. In the next few years, the demand for this industry was lower than in 2010. Another big user of silver is jewellery, the volume of demand in this industry grew until 2010, and in 2011–2012 the demand slightly decreased in 2013 again to grow [16].

8.2.5 Copper

Most of the copper available on the market comes from primary sources, from which concentrates are produced which can be further processed into anode copper and refined copper. In case of copper available on the market as much as 33% comes from secondary sources [16, 19]. In recent years the largest copper mining carried out in Chile, Peru, China and the US. Detailed data on copper production in 2010–2015 can be found in the Table 8.6. Copper production in the past 110 years has shown exponential growth [20].

Table 8.6: The volume of production of copper [14].

	Mine production of copper [kt]					
	2010	2011	2012	2013	2014	2015
United States	1110	1110	1170	1250	1360	1380
Australia	870	958	958	990	970	971
Canada	525	566	579	632	696	697
Chile	5420	5260	5430	5780	5750	5760
China	1190	1310	1630	1600	1760	1710
Congo	343	520	600	970	1030	1020
Mexico	260	443	440	480	515	594
Peru	1250	1240	1300	1380	1380	1700
Russia	703	713	883	833	742	732
Zambia	690	668	690	760	708	712
World	**15,900**	**16,100**	**16,900**	**18,300**	**18,500**	**19,100**

It has been estimated that the economy of copper extraction has changed drastically over the past decade, at that time, sector operating costs increased. Irrespective of the downturn in 2007–2009, global copper consumption has increased since 1970. China, which through intensive purchases and increased consumption (by 3.6 million tons in the period 2008–2012), contributed the most to the rise in demand and the occurrence of metal shortages in 2010–2012. Forecasts show that the global supply of refined copper from all sources may rise to 23 million tons in 2017 [16, 21].

8.2.6 Aluminum

The importance of aluminum in the economy is evidenced by the fact that in terms of consumption takes second place behind the iron. In high industrialized countries, the increase in secondary aluminum production has been observed in recent years. In turn, in developing countries where energy is cheaper, primary aluminum production is predominant. Hence, it is not surprising that the largest production of primary aluminum is in Asia, the leading producer is China. The world aluminum supply consists of the production of primary aluminum also secondary aluminum production [16]. Available date indicate that recycling of aluminum including manufacturing scarp is on level 40% in Europe and 37% on the World [22].

Literature indicates that over the 25 years it is from 1990 to 2015 aluminum production grew on average by 2.5% per year [23]. Growing production over the past decades has been accompanied by demand, amounting to 2.4% a year [24]. However, they are average figures, because 2009, as a result of global recession, aluminum production decreased by 6%. However, in the years 2010–2012, the world supply of aluminum has recovered growth tendencies (see Table 8.7). The multiplicity and variety of aluminum applications, as well as the growth of industrialization (mainly in China, India and Brazil) has led to an estimated double consumption of aluminum products in 2020 compared to 2005 [24, 25].

Table 8.7: The volume of production of aluminum [16].

	World production of primary aluminum [kt]				
	2008	**2009**	**2010**	**2011**	**2012**
Europe	9847	8362	8510	8582	8277
Asia	17,402	17,695	21,632	24,404	26,855
South America	2668	2511	2282	2236	2056
North and Central America	5778	4757	4690	4974	4851
Africa	1717	1705	1820	1849	1767
World	**37,412**	**35,030**	**38,934**	**42,045**	**43,806**

8.2.7 Rare earth metals

Rare earth elements (REEs) are a group of 17 elements of the key technological applications (the Table 8.8 contains examples of REE applications). This group of metals includes lanthanides, scandium and yttrium.

Table 8.8: Rare earths usage by application, in % [26].

Application	La	Ce	Pr	Nd	Sm	Eu	Gd	Tb		DyY
Magnets	–	–	23.4	69.4	–	–	–	0.2	5	–
Battery alloy	50	33.4	3.3	10	3.3	–	–	–	–	–
Metallurgy	26	52	5.5	16.5	–	–	–	–	–	–
Autocatalysts	5	90	2	3	–	–	–	–	–	–
Polishing powders	31.5	65	3.5	–	–	–	–	–	–	–
Glass additives	24	66	1	3	–	–	–	–	–	2
Phosphors	8.5	11	–	–	–	4.9	1.8	4.6	–	69.2
Ceramics	17	12	6	12	–	–	–	–	–	53

The size of the penultimate column is not right.

Dissemination of rare earths in the earth's crust is relatively high, although their concentrations of economic importance are not frequent. Not often, REEs appear as native elemental metals in nature, commonly REEs occur as part of the host mineral's chemistry. All this makes REEs are recovered using sophisticated chemical processing methods. REEs are included in more than 200 minerals, but it is estimated that the significant REE ores are bastnäsite, xenotime and monazite [15, 27]. For several years, the largest producer of REEs has been China. In addition, significant quantities of rare earths are also produced in Russia, Australia and India. Table 8.9 contains

Table 8.9: The volume of production of rare earth elements [14].

	Mine production of REE [t]					
	2010	2011	2012	2013	2014	2015
United States	–	–	800	5500	5400	5900
Australia	–	2200	3200	2000	8000	12,000
Brazil	550	250	140	330	–	880
China	130,000	105,000	100,000	95,000	105,000	105,000
India	2800	2800	2900	2900	–	1700
Malaysia	30	280	100	180	240	500
Russia	–	–	2400	2500	2500	2800
Thailand	–	–		800	2100	760
Vietnam	–	–	220	220	–	250
World	133,000	111,000	110,000	110,000	123,000	130,000

available data on rare earth production; many data are not published because they are confidential.

Worldwide REE production has risen sharply by around 7% annually between 1994 and 2005. Only in 2006 was this trend retarded and production dropped. In spite of the economic crisis in 2007–2010 there was a regrowth of REE production. China in 2010 reduced the supply of REE due to the protection of limited resources. In addition, 90% of all rare earths are produced by China even though the country has less than 40% reserves. Currently it is believed that the REEs as the most critical raw materials group, with the highest supply risk. For example, due to the wide application of rare earths, global demand is estimated at 136,000 tons per year, with world production in 2010 of 136,000 tons. The forecast supply and demand of rare earths till 2014 is presented in Table 8.10. Numerous uses of rare earths allow you to predict that over the next 25 years the demand for neodymium and dysprosium will rise by 700% and 2600%, respectively [15, 26].

Table 8.10: Estimated supply and demand of rare earth elements in 2014 [28].

Elements	Supply (REO in tons)	Demand (REO in tons)
Lanthanum	54,750	51,050
Cerium	81,750	65,750
Praseodymium	10,000	7900
Neodymium	33,000	34,900
Samarium	4000	1390
Europium	850	840
Gadolinium	3000	2300
Terbium	350	590
Dysprosium	1750	2040
Erbium	1000	940
Yttrium	11,750	12,100
Ho-Tm-Yb-Lu	1300	200
Total	**203,500**	**180,000**

REE's market analyzes predict that by 2020 the demand for these elements will still continue to grow at a rate of 6% per year. It is assumed that with a low recycling and substitution rates of REE, supply security will be crucial for the producers and consumers of REEs all around the world [29].

Metal market is specific, however governed by the market law, law of supply and demand. Above-presented demand and supply of metals is differentiated. To the most important market and nonmarket demand factors belong prices of substitute and complementary goods and demographic factors. While to the supply determinants belong to non-price change (increase) in production costs, technological progress.

The dynamics of prices changes of metals is significant. Table 8.11 presents prices of metals in 2017 compare to 1995. Only platinum, molybdenum and nickel are cheaper nowadays in comparison to the price in 1995.

Table 8.11: Prices of metals in 2017 with compare to 1995 [14, 30].

	Price	
	2017	1995
Metal	USD/kg	
Aluminum	2.12	1.5
Chromium	2.48	1.7
Cobalt	58.99	25
Copper	6.43	5.2
Gold	41,370	34,070
Lead	2.46	0.92
Mercury	1.85	1.7
Molybdenum	16.00	21
Nickel	10.50	14
Palladium	29,930	23,214
Platinum	29,640	37,607
Rhodium	37,290	16,396.88
Silver	540	514
Steel/Iron	0.3	0.12
Tin	20.85	17
Zinc	3.15	1.7

8.3 Metals content in wastes and their recovery rate

Since the last two decades it was tried to put the bases for the development of a circular economy. In such economy wastes should be considered as resource and used in an efficient and sustainable way [9]. On the other hand development of new materials and technologies, growth in the defense, energy and transport sectors, will ensure that the demand for rare earth metals and their compounds will continue to grow. This demand means that secure supply chains, such as recycling and material recovery, are essential to ensure that technological innovation is not hampered [7]. Recycling of waste is also becoming one of the possibilities of acquiring missing and critical metals. Metal-laden wastes include a few groups: Waste EEE [31], catalysts of different application, introduced on chemical, petrochemical or automotive market [32] and galvanic wastes [33]. From metal-laden wastes many valuable metals can be recovered. The data presented in Table 8.12 show estimated world waste production of metals by different types of industries in 1990s. It is observed that the electronic industries play an important role in the emission of metals to the environment [33].

Table 8.12: Estimated world waste production of metals by different types of industries [33].

Industry	Metals	Quantities [t/year]
Electronic	As, Cr, Hg, Se, Ni, Cu	1,200,000
Oil and Coal	As, Pb, V, Cd, Ni, Zn	1,200,000
Mining and Metallurgy	Hg, Cr, Cu, As, Zn, Pb	390,000
Agriculture	Mg, As, Cu	1,400,000
Manufacturing	Cr, Co, Ni, Fe	240,000
Others		720,000

In the last decades, the electronics industry has revolutionized the world. The consumption of EEE is growing year by year. During the past decades, a huge upsurge in the consumption of high-tech products has resulted from the rapid technological progress [15]. These wastes content a various type of metal as is presented in Table 8.13.

In parallel to growing demand, the e-waste stream is increasing too and is the fastest growing waste stream on the planet [8]. It is estimated that the total amount of EEE put on the market in 2012 was about 56.5 million tons [15], while in 2014 around 42 million tons of waste electrical and electronic equipment (WEEE) was created. Within the EU, 9.1 million tons were generated in 2014 with a projected increase to 12 million tons per year by 2020 [34]. The estimated annual growth rate of WEEE equals to 3–5% [9]. Such e-waste production represents 3–5% of the global municipal waste production [34, 35].

There are ten WEEE directive categories: (1) Large household appliances; (2) Small household appliances; (3) IT and telecommunications equipment; (4) Consumer equipment; (5) Lighting equipment; (6) Electrical and electronic tools; (7) Toys, leisure and sports equipment; (8) Medical devices; (9) Monitoring and control instruments; (10) Automatic dispensers [2]. Only in UK by the year 2000, 392 tons of large household appliances become an e-waste. UK Status Report on Waste from Electrical and Electronic Equipment Industry Council for Equipment Recycling reports also that waste of IT equipment was about 357 tons what constitutes 22 million units [36].

Waste electronics are a mixture of various basic and special metals (including Fe, Al, Ni, Zn, Se, In and Ga), precious metals (Cu, Au, Ag, Pd and Pt), hazardous substances (Hg, Be, Pb, As, Cd and Sb), plastics and glass and ceramics [8].

Depending on the type of equipment, the age of the device and the producer, the content of the individual components in the electronic scrap is varied. It is important to approach the subject of e-waste. The weight domination is not equivalent to value. In most WEEE plastics and steel tend to dominate in terms of weight, but in terms of value, gold and the other precious metals dominate. In DVD-layer Fe weight-share is the highest while value share is the highest for Au similarly for PC boards, cell phones

Table 8.13: Materials composition of some EEE [9].

Materials	LCD Notebooks	LED Notebooks	CRT TVs	LCD TVs	LED TVs	CRT Monitors	LCD Monitors	LED Monitors	Cell Phones	Smart phones	PV Panels	HDDs	SSDs	Tablets
														g/unit
Aluminum			67			242	130	130	12	2.9	1370	441	441	
Antimony	0.77	0.77	14	0.71	0.71					0.084				0.154
Arsenic	0.01	0.01												0.002
Barium	2.5	2.5				1								0.049
Beryllium										0.003				
Cadmium			0.2											
Cerium	<0.001	<0.001		0.005	<0.001		<0.001	<0.001						<0.001
Chromium	0.07	0.07	0.03											0.014
Cobalt	0.065	0.065							3.8	6.3				0.013
Copper	135	135	656	824	824	952			26	14	78	15	15	27
Dysprosium	0.06	0.06										0.06		0.012
Europium	<0.001	<0.001		0.008	<0.001		0.001	<0.001						<0.001
Ferrite						483								
Gadolinium	<0.001	<0.001		<0.001	0.002		<0.001	0.002						<0.001
Gallium		0.0016			0.005		0.003	0.003			0.119			
Gold	0.22	0.22		0.11	0.11	0.31	0.2	0.2	0.024	0.038		0.005	0.005	0.044
Indium	0.04	0.04		0.003	0.003		0.079	0.082			0.119			0.008
Lanthanum	<0.001			0.007			<0.001							<0.001
Lead	5.3	5.3	1319			464	16		1	0.6				1.1
Mercury	<0.001	<0.001					<0.001	0.004	1					<0.001
Molybdenum	0.04	0.04					0.633	0.633			0.295			0.008

(continued)

Table 8.13: (*continued*)

Materials	LCD Notebooks	LED Notebooks	CRT TVs	LCD TVs	LED TVs	CRT Monitors	LCD Monitors	LED Monitors	Cell Phones	Smartphones	PV Panels	HDDs	SSDs	Tablets
														g/unit
Neodymium	2.1	2.1								0.05		1		0.427
Nickel	3.6	3.6				199			1	1.5				0.722
Palladium	0.04	0.04		0.044	0.044		0.04	0.04	0.009	0.015		0.003	0.003	0.008
Platinum	0.004	0.004								0.004				
Praseodymium	0.274	0.274		<0.001			<0.001			0.01		0.145		0.055
Selenium											0.119			
Silicon									5		226			
Silver	0.25	0.25		0.45	0.45	1.25	0.52	0.52	1	0.244		0.031	0.031	0.05
Steel/Iron			2088			3322	2530	2530	11	8		62	62	
Tantalum	1.7	1.7									0.406			
Tellurium														
Terbium	<0.001			0.002			<0.001							<0.001
Tin			32	18	18	20	24	24	1	1	0.116			
Titanium							0.633	0.633						
Tungsten							0.633	0.633						
Vanadium						1								
Yttrium	0.002	0.002		0.11	0.005	1	0.016	<0.001						<0.001
Zinc	0.004	0.004	8.6						4	1	0.4			<0.001
Number of precious metals	4	4	0	3	3	2	3	3	3	4	0	3	3	3

and calculator, gold and the other precious metals make up more than 80% of the value, while for TV boards and DVDs, they still contribute around 50% of the value. Copper also contributes next in the value ratings [37, 38]. Table 8.13 presents examples of the metal composition of different electronic scraps. There is no average scrap composition, even the values given as typical averages actually only represent scrap of a certain age and manufacturer. Additionally, non-ferrous metals and precious metals contents have gradually decreased in concentration in scrap [38].

Recycling of such basic metals like aluminum, iron, magnesium, zinc, manganese, nickel, copper, silver, tin, antimony, iridium, gold and lead and bismuth from the scrap of electrotechnical and electronic confirms the profitability of the project [7]. Therefore, WEEE is a significant source of such metals, however only approximately 35% were recycled [39].

Innovations make EEE more functional by using an increasing number of metals, particularly the critical metals (e.g. rare and precious metals) with specialized properties, which means that more and more critical metals will be stored in these products [40]. In 2014 the consumption of metals palladium, platinum, gold, silver, REE, cobalt, lithium, indium in EEE ranged from 3.65% to 218% of the total mine production. The consumptions in EEE of silver, REE, cobalt and lithium were about one third of world mine production, precious metals were about 10% while indium over 200%. It is claimed that if the consumption of metal is over 10% the recycling of WEEE is necessary to relieve the supply risk. Except for huge demand of critical metals, their deficiency in the Earth's crust as well as existence in a few regions lead to the serious depletion and supply risks in the future [15].

REEs are essential ingredients for the development of modern industry as well as designing and developing high technology products used in our daily lives. Consequently, the worldwide demand of rare earth metals is rising quickly and predicted to exceed the supply by 40,000 tons annually. However, their availability is declining, mainly due to the export quotas imposed by the Chinese government and actions taken against illegal mining operations. China is the exclusive supplier of many metals from REE group. It is estimated that China has over 95% of world production of rare earths and holds 36% of proven reserves. REEs have become a strategically important area (critical and deficient material) due to the continuity of undistorted, independent supply. This has laid emphasis to exploit and expand technologies to meet the future necessities of rare earth metals. Considering, moreover, the limited access and shrinking mining resources of rare earths, including lanthanum, cerium, praseodymium, neodymium, samarium or gadolinium, whose high price is associated not only with low concentration in ore extraction but also with the high cost of their enrichment, confirms the purpose of recovery. These elements are present in increasing quantities in the construction of modern means of transport, e.g. electronics, motors and batteries for hybrid cars, cells, catalytic converters, controls and active safety. In addition, rare earth metals are widely used in high technology industries, wind turbines, energy saving light bulbs, screens and displays (LEDs, LCDs, plasmas), hard disks, camera lenses, glass, batteries, industrial accumulators, medical equipment and water treatment [7, 28, 41].

The ever-increasing demand for rare earth metals, mainly in the metallurgical, electronic, defense or medical industries and the strategic nature of the economic importance of rare earths requires the search for alternative sources of their extraction. It is justified to recycle them from post-operational scrap from machinery components and from EEE [15]. Such technologies are applied now.

The common element of electronic devices and the main carrier of most precious and special metals are printed circuit boards (PCBs) in which metals and their alloys in combination with ceramics are combined with resin layers of a peripheral plate laminate. In general, PCBs contain approximately 28–40% metals (including precious metals), 23–30% plastics, with the remainder being ceramics and glass materials [42]. Such a combination of metals and plastics requires the right processing technology for this waste, providing maximum levels of recovery of valuable components while eliminating dangerous substances while preserving the principles of protection [8]. PCBs account for about 6% of the total weight of WEEE. They contain the highest precious metal values (Au, Ag), and may be considered as a high-value waste. Also PCB may contain PGMs (Pd, Pt, Rh, Ir and Ru), base metals (Cu, Al, Ni, Sn, Zn and Fe), hazardous metals (Hg, Be, In, Pb, Cd, As and Sb), scarce or trace metals (Te, Ga, Se, Ta and Ge) [43]. The concentration of metals in PCB scrap is differentiated (Table 8.14) and the weight percentage depends on the method of determination and the author. Material recovery from PCB scrap may contribute to reduce the environmental impacts caused by the extraction of high-valued and/or highly toxic materials from nature [1]. The concentration of precious metals in waste PCBs is higher than the concentration of precious metals in their ores. As it is reported currently mined ores for the extraction of gold and palladium include less than 10 g/t of precious metals compared to the concentrations of the similar metals in PCBs of personal computers (250 g/t of gold

Table 8.14: The concertation of metals in PCB scrap [44–50].

Metal	Weight [%]
Aluminum	0.26–5.5
Iron	4.97–8.0
Copper	16.0–26.8
Gold	0.03–0.1
Silver	0.05–1.5
Palladium	0.005–0.022
Nickel	0.47–1.0
Lead	0.8–2.0
Tin	1.7–3.0
Zinc	1.5–1.8

and of 110 g/t of palladium) not to mention negative environmental impacts of the extraction of precious metals through mining [35]. Furthermore such metals like Cd, Cu, Ni, Pb and Zn released into the environment would pose a risk to the ecosystems.

Recovery metals from WEEE are particularly profitable when considering such metals like Au, Ag, Pd. It was estimated, on the basis of average weight of mobile phone (125g) and of PC/laptop (2.5kg) and number of 1,200 million units of mobile phones and 255 million units PCs and Laptops that value of potential recovery of Au, Ag, Pd will reach $1,891.6; $238.7 and $356.1 million, respectively. Also value of copper and cobalt recovered from such devices is very high and was $968.1 and $385.5 million, respectively [4]. Also the content of gold silver and palladium metals in mobile phones and computers was compared to total demand from these appliances and world mine production. Investigation showed that Au, Ag, Pd recovered from mobile phones and PC/laptops represents 3%, 4% and 16% of world mine production, respectively. Moreover cobalt recovered for these devices is 23% of world mine production [37]. Above examples show only the value of recovered metals. The full profitability assessment of recycling processes is very complicated and not always verified. However such calculations made for UE WEEE divided into a few categories of devices. Net present value (NPV) and discounted payback time (DPBT) are used as reference indexes for the evaluation of investments. NPV was positive for WPCBs coming from IT and telecommunication equipment, while negative for such devices as fridges, washing machines, air conditioners, vacuum cleaners, toasters, fryers and other big and small household appliances. Investigation showed that DPBT varies from 1 to 2 years with interesting influences on returns from investments. That led to the conclusion that profitability of metal recycling strongly depends on WEEE streams as well as on individual device. However, the profitability of WEEE treated as a whole obtained positive economic values [51]. Waste recycling processes represent not only an economic point of view but also an environmental protection action. It was estimated that energy saved by recycling metals when compared to primary resources is at the level of 95% for aluminum, 85% for copper, 74% for steel, 65% for lead and 60% for zinc [46]. A good example of environmental aspect is the process of obtaining 1 kg of primary Mg by thermal reduction with silicon or electrolysis, where the energy demand is 20–35 kWh, depending on the method of processing and the purity of the material. Recovering magnesium from scrap requires less than ten times the energy consumption (only about 3.5 kWh) [7].

Catalyst is used in all chemical industries. In basic chemistry, where it used in synthesis of some acids, ammonia, methanol and aromatics, in petrochemistry, in polymerization, in refining or in reactions of fluid catalytic cracking (FCC) residue cracking catalyst (RFCC), hydrodesulfurization (HDS) catalyst [52]. However the best know are automotive catalysts. The recovery of metals and precious metals from catalysts was an important topic from the economic point of view. The reduction of catalytic waste can prevent from the environmental pollution.

PGMs (platinum, palladium, rhodium and iridium, ruthenium and osmium) are widely applied not only in electronics sector (hard disks, PCBs) or jewellery but also in catalysts in chemical industry.

The worldwide market for catalysts used in chemical, pharmaceutical and polymers is estimated at about $1.5 billion and global demand will grow 9–10 % per year [53]. However the sector of automotive is still leading one demand for PGMs (especially Pt, Pd and Rh). In 2010 the demand of platinum, palladium and rhodium in autocatalyst reached 84.6, 146 and 20.6tons, respectively. The recovery level of these metals from all industry branches (autocatalysts, electro, invest, jewellery, glass, medicine, chemistry and others) was ca. 24% for platinum, 21% for palladium and 27% for rhodium [32].

Waste effluents can be also a rich source of metals. Contaminated metallic wastewater poses a significant risk to health and the environment, but also provides a chance for metal recovery [54]. The metal included in contaminated water can post health and environmental concerns. Heavy metals included in wastewater are directly or indirectly discharged into the environment. The level of this phenomenon is varied, however bigger in developing countries. Due to the fact that most metals are not biodegradable and may be accumulated in living tissues of animals, plants and human bodies, causing diseases and disorders the recovery of metals is highly justified. Ecological point of view underline the necessity of recovery, especially toxic heavy metals (zinc, copper, nickel, mercury, cadmium, lead and chromium), which are of particular concern in treatment of mainly industrial wastewaters [55]. Table 8.15 shows the concentration of major metals in different wastewater. It is clear that industrial wastewater contains far more metals than municipal waste, including those with higher market values such as silver, gold, copper, zinc, etc. [3]

The US Water Environmental Research Foundation (WERF) reported in 2011 that approximately 37,854 m^3 per day of wastewater could have a silver recovery potential of $8,849–3,904,664 per year. Also it was reported that from wastewater, gold can be recovered at the concentrating of metal to 165 μg/l, which translates into about $4,600 gold per 10 m^3 of water. Although estimation of the value of metals recovered from wastewater is promising, concentrations of metals in most wastewater are relatively low. They are mainly in the range of μg/l to μg/l, thus the technology of recovery of such materials must be developed to treat a large amount of water, effectively focusing on targeted metals [3].

The worldwide production of cars grew steadily over the past 30 years despite a dip in 2008–2009 due to the economic crisis [56]. High demand for cars also generates large amounts of auto scrub. Statistical data shows that approximately 6.5 million tons of end-of-life vehicles (ELVs) were produced in Europe in 2011 [57]. ELV is composed (by weight) of ferrous metal (65.4–71%), non-ferrous metal (7–10%), plastics (7–9.3%), tires and rubber (4–5.6%), glass (2.9–3%), fluids (0.9–6%), battery (1–1.1%), polymers (1–1.1%), electronics (0.4–1%) [57]. The current ELV recycling rate is around 80%. The remaining waste is called automotive shredder residue (ASR), or car fluff [58].

Table 8.15: Material concentration in wastewater [3].

Metal	Municipal treatment plant µg/l	Road wash water	Tannery	Mining	Battery factory
Aluminum		0.467–26.1		0.161	0.2–7.3
Arsenic	0–0.0019				
Cadmium	0–0.0033		0.056	0.004	0.02–0.12
Calcium			255	548	83–255
Chromium	0.04–0.56	0.004–0.107	391		<0.0044–0.08
Cobalt			1.55	0.0126	
Copper	0.079–0.58	0.0111–0.177		0.244	<0.0033–0.38
Iron	0.48–3.9	2.59–26.8	4.4	0.033	0.02–20
Lead	0–0.039	<0.018–0.053	0.872		4.0–13
Magnesium			268	29.52	15–26
Manganese	0.067–1.16		0.396		0.04–0.6
Mercury	0–0.0002	2.1			
Molybdenum		0.0154–0.021			
Nickel	0.0067–0.77	<0.006–0.0525	0.179	0.142	0.07–0.38
Potassium			183	34.9	
Silver	0–0.0014				
Sodium			25,519	100.05	
Tin	0–0.028				
Zinc	0.26–0.75	0.105–1.56	0.684	0.023	0.6–17
Estimated value (USD)*	0.02	0.06	121.35	61.56	28.29

Note: *Suppose recovering 1000l wastewater with 100% recovery efficiency.

Automobile recycling generates approximately 5 million tons of ASR annually, which equals to about 5% of the world's industrial waste worldwide [42]. Worldwide, ASR is considered an increasingly problematic mixture of materials that needs the development of a processing solution [59]. Average composition of ASR material includes aluminum, copper, iron, zinc, lead, chromium, manganese, antimony, cadmium and mercury, which are accounted for 5–23% [60, 61]; however, recent studies reported that ASR residues include significant amounts of Au, Cu, Pt, Ag and Pd representing 4.8, 3.7, 2.2, 0.8 and 0.8 wt%, respectively. Values of metals included in ASR are significantly higher, which confirms the legitimacy of extraction [61].

8.4 Summary

In times of strong civilization development there is a need to save energy and raw materials as well as the evolution of safe, lightweight structures easy to recycle. Such actions lead to further technical progress in the level of structural reliability and safety of use and protection of the environment. It can be achieved by reducing fuel consumption and reducing the emissions of harmful substances into the atmosphere from means of transport, among others, by using modern electronic systems. Great progress in the design of modern vehicles and driving safety was achieved not only by the rapid development of microelectronics, including closely linked to the achievements of material engineering. Such fast development of electronics is also inseparably linked with new technologies and new materials. The dynamics of economic development is determined by the need for the development of waste recycling techniques and the decommissioning and recovery of valuable raw materials, especially metals. Recycling of metals is also associated with reduction of greenhouse gas emissions, because the primary mining of metal (iron, aluminum, magnesium) requires high energy and financial inputs while the recovery and secondary melt demand for energy is generally much lower.

Due to the depletion of natural resources as well as the increasing ecological awareness of society relevant legislation in the EU was implemented. In June 2010, the Raw Materials Supply Group meeting in Brussels presented a report on critical mineral resources to the EU. The report listed 14 critical minerals of economic importance to the EU. These materials are an increased risk of supply disruptions for which there are no primary and secondary sources in EU countries. This list includes 12 metals with different levels of production and demand, different origins, and importance for EU and non-EU economies, is as follows: antimony, beryllium, gallium, germanium, indium, cobalt, magnesium, PGMs, tungsten, niobium, tantalum, rare earth metals.

Critical, the risk of shortage and impact of this shortage on the EU economy is higher than for other raw materials. The high risk of their shortage is mainly due to the fact that a significant portion of world production is in a small number of countries such as China (antimony, fluorspar, gallium, germanium, graphite, indium, magnesium, rare earths, tungsten), Russia (platinum, tungsten), the Democratic Republic of the Congo (cobalt, tantalum) and Brazil (niobium and tantalum).

The justification for the high demand for metal recovery is not only their high price, shrinking deposits of natural resources, increasing new technologies demand for precious metals, but also the territorial dependence of natural ores.

References

[1] Rubin RS, de Castro MA, Brandão D, Schalch V, Ometto AR. Utilization of life cycle assessment methodology to compare two strategies for recovery of copper from printed circuit board scrap. J Clean Prod. 2014;64:297–305.

[2] Bigum M, Brogaard L, Christensen TH. Metal recovery from high-grade WEEE: a life cycle assessment. J Hazard Mater. 2012;207:8–14.

[3] Wang H, Ren ZJ. Bioelectrochemical metal recovery from wastewater: a review. Water Res. 2014;66:219–32.

[4] Alsheyab M, Kusch S. End of life of electronic communication devices in the context of strategies to decouple resources use from economic growth. In: Conference of Informatics and Management Sciences, 2013:133–37.

[5] Cucchiella F, D'Adamo I, Gastaldi M. Sustainable management of waste-to-energy facilities. Renew Sust Energ Rev. 2014;33:719–28.
[6] Norgate TE, Rankin WJ. The role of metals in sustainable development. Green Process. 2002;49–55.
[7] Wojciechowski A, Dyduch J, Lankiewicz K. Odzysk metali z elektroniki samochodowej i sprzętu AGD. Logistyka. 2014;6:11259–67.
[8] Willner J, Fornalczyk A. Złom elektroniczny jako źródło metali szlachetnych. Przem Chem. 2012;91(4):517–22.
[9] Cucchiella F, D'Adamo I, Koh SL, Rosa P. Recycling of WEEEs: an economic assessment of present and future e-waste streams. Renew Sust Energ Rev. 2015;51:263–72.
[10] Li J, Liu L, Zhao N, Yu K, Zheng L. Regional or global WEEE recycling. Where to go? Waste Manage. 2013;33(4):923–34.
[11] Robinson BH. E-waste: an assessment of global production and environmental impacts. Sci Total Environ. 2009;408(2):183–91.
[12] Rao CR, Reddi GS. Platinum group metals (PGM); occurrence, use and recent trends in their determination. TrAC-Trend Anal Chem. 2000;19(9):565–86.
[13] Glaister BJ, Mudd GM. The environmental costs of platinum–PGM mining and sustainability: is the glass half-full or half-empty? Miner Eng. 2010;23(5):438–50.
[14] Commodity statistics and information (Accessed September 28. 2017 at https://minerals.usgs.gov/minerals/pubs/commodity/)
[15] Zhang S, Ding Y, Liu B, Chang CC. Supply and demand of some critical metals and present status of their recycling in WEEE. Waste Manage. 2017;66:113–27.
[16] Smakowski T, Ney R, Galos K, Eds. Bilans gospodarki surowcami mineralnymi Polski i świata 2012. Instytut Gospodarki Surowcami Mineralnymi i Energią Polskiej Akademii Nauk. Pracownia Polityki Surowcowej. 2014. (in polish).
[17] Swain B. Recovery and recycling of lithium: a review. Sep Purif Technol. 2017;172:388–403.
[18] Boundy T, Boyton M, Taylor P. Attrition scrubbing for recovery of indium from waste liquid crystal display glass via selective comminution. J Clean Prod. 2017;154:436–44.
[19] Fu X, Ueland SM, Olivetti E. Econometric modeling of recycled copper supply. Resour Conserv Recy. 2017;122:219–26.
[20] Singer DA. Future copper resources. Ore Geol Rev. 2017;86:271–79.
[21] Kabwe E, Yiming W. Analysis of Copper's market and price-focus on the last decade's change and its future trend. Inter J Sci Tech Res. 2015;4(10):54–61.
[22] Rombach G. Raw material supply by aluminium recycling–efficiency evaluation and long-term availability. Acta Mater. 2013;61(3):1012–20.
[23] Sverdrup HU, Ragnarsdottir KV, Koca D. Aluminium for the future: modelling the global production, market supply, demand, price and long term development of the global reserves. Resour Conserv Recy. 2015;103:139–54.
[24] Grimaud G, Perry N, Laratte B. Life cycle assessment of aluminium recycling process: case of shredder cables. Procedia CIRP. 2016;48:212–18.
[25] Jirang CU, Roven HJ. Recycling of automotive aluminum. T Nonferr Metals Soc China. 2010; 20(11):2057–63.
[26] Binnemans K, Jones PT, Blanpain B, Van Gerven T, Yang Y, Walton A, et al. Recycling of rare earths: a critical review. J Clean Prod. 2013;51:1–22.
[27] Charalampides G, Vatalis KI, Apostoplos B, Ploutarch-Nikolas B. Rare earth elements: industrial applications and economic dependency of Europe. Procedia Econ Financ. 2015;24:126–35.
[28] Jha MK, Kumari A, Panda R, Kumar JR, Yoo K, Lee JY. Review on hydrometallurgical recovery of rare earth metals. Hydrometallurgy. 2016;165:2–26.

[29] Klossek P, Kullik J, van den Boogaart KG. A systemic approach to the problems of the rare earth market. Resour Policy. 2016;50:131–40.

[30] Current and historical commodity and metal prices. Available at: http://www.infomine.com/investment/metal-prices/. Accessed: 28 Sept 2017.

[31] Lehner T E&HS aspects on metal recovery from electronic scrap. In: IEEE International Symposium on Electronics and the Environment, 2003:318–22.

[32] Fornalczyk A. Industrial catalysts as a source of valuable metals. JAMME. 2012;55(2):864–68.

[33] Veglio F, Quaresima R, Fornari P, Ubaldini S. Recovery of valuable metals from electronic and galvanic industrial wastes by leaching and electrowinning. Waste Manage. 2003;23(3):245–52.

[34] Charles RG, Douglas P, Hallin IL, Matthews I, Liversage G. An investigation of trends in precious metal and copper content of RAM modules in WEEE: implications for long term recycling potential. Waste Manage. 2017;60:505–20.

[35] Ghodrat M, Rhamdhani MA, Brooks G, Rashidi M, Samali B. A thermodynamic-based life cycle assessment of precious metal recycling out of waste printed circuit board through secondary copper smelting. Environ Develop. 2017;24:36–49.

[36] UK Report on Waste from Electrical and Electronic Equipment. http://icer.org.uk/. Accessed: 28 Sept 2017.

[37] Hagelüken C, Corti CW. Recycling of gold from electronics: cost-effective use through 'design for recycling'. Gold Bull. 2010;43(3):209–20.

[38] Cui J, Zhang L. Metallurgical recovery of metals from electronic waste: a review. J Hazard Mater. 2008;158(2):228–56.

[39] Cieszyńska A. Waste Electronic and Electrical Equipment (WEEE)-scraps or valuable source of precious metals. Pol J Comm Sci. 2016;49 :43–54.

[40] Diaz LA, Lister TE, Parkman JA, Clark GG. Comprehensive process for the recovery of value and critical materials from electronic waste. J Clean Prod. 2016;125:236–44.

[41] Tunsu C, Petranikova M, Gergorić M, Ekberg C, Retegan T. Reclaiming rare earth elements from end-of-life products: a review of the perspectives for urban mining using hydrometallurgical unit operations. Hydrometallurgy. 2015;156:239–58.

[42] Mallampati SR, Lee BH, Mitoma Y, Simion C. Sustainable recovery of precious metals from end-of-life vehicles shredder residue by a novel hybrid ball-milling and nanoparticles enabled froth flotation process. J Clean Prod. 2018:171:66–75.

[43] Khaliq A, Rhamdhani MA, Brooks G, Masood S. Metal extraction processes for electronic waste and existing industrial routes: a review and Australian perspective. Resources. 2014; 3(1):152–79.

[44] Yamane LH, de Moraes VT, Espinosa DC, Tenório JA. Recycling of WEEE: characterization of spent printed circuit boards from mobile phones and computers. Waste Manage. 2011; 31(12):2553–58.

[45] Petter PM, Veit HM, Bernardes AM. Evaluation of gold and silver leaching from printed circuit board of cellphones. Waste Manage. 2014;34(2):475–82.

[46] Akcil A, Erust C, Gahan CS, Ozgun M, Sahin M, Tuncuk A. Precious metal recovery from waste printed circuit boards using cyanide and non-cyanide lixiviants – a review. Waste Manage. 2015;45:258–71.

[47] Behnamfard A, Salarirad MM, Veglio F. Process development for recovery of copper and precious metals from waste printed circuit boards with emphasize on palladium and gold leaching and precipitation. Waste Manage. 2013;33(11):2354–63.

[48] Ghosh B, Ghosh MK, Parhi P, Mukherjee PS, Mishra BK. Waste printed circuit boards recycling: an extensive assessment of current status. J Clean Prod. 2015;94:5–19.

[49] Vats MC, Singh SK. Assessment of gold and silver in assorted mobile phone printed circuit boards (PCBs). Waste Manage. 2015;45:280–88.

[50] Ikhlayel M. Environmental impacts and benefits of state-of-the-art technologies for E-waste management. Waste Manage. 2017;68:458–74.
[51] Cucchiella F, D'Adamo I, Koh SL, Rosa P. A profitability assessment of European recycling processes treating printed circuit boards from waste electrical and electronic equipments. Renew Sust Energ Rev. 2016;64:749–60.
[52] Yoo JS. Metal recovery and rejuvenation of metal-loaded spent catalysts. Catal Today. 1998; 44(1):27–46.
[53] Thayer AM. Making metathesis work. Chem Eng News. 2007;85:37–47.
[54] Lingamdinne LP, Koduru JR, Roh H, Choi YL, Chang YY, Yang JK. Adsorption removal of Co (II) from waste-water using graphene oxide. Hydrometallurgy. 2016;165:90–96.
[55] Fu F, Wang Q. Removal of heavy metal ions from wastewaters: a review. J Environ Manage. 2011;92(3):407–18.
[56] Vermeulen I, Van Caneghem J, Block C, Baeyens J, Vandecasteele C. Automotive shredder residue (ASR): reviewing its production from end-of-life vehicles (ELVs) and its recycling, energy or chemicals' valorisation. J Hazard Mater. 2011;190(1):8–27.
[57] Cossu R, Lai T. Automotive shredder residue (ASR) management: an overview. Waste Manage. 2015;45:143–51.
[58] Santini A, Passarini F, Vassura I, Serrano D, Dufour J, Morselli L. Auto shredder residue recycling: mechanical separation and pyrolysis. Waste Manage. 2012;32(5):852–58.
[59] Harder MK, Forton OT. A critical review of developments in the pyrolysis of automotive shredder residue. J Anal Appl Pyrol. 2007;79(1):387–94.
[60] Passarini F, Ciacci L, Santini A, Vassura I, Morselli L. Auto shredder residue LCA: implications of ASR composition evolution. J Clean Prod. 2012;23(1):28–36.
[61] Mallampati SR, Lee CH, Truc NT, Lee BK Quantitative analysis of precious metals in automotive shredder residue/combustion residue via EDX fluorescence spectrometry. Int J Environ Anal Chem. 2015;95(12):1081–89.

Index

https://doi.org/10.1515/9783110547061-009

Bei Fragen zur Produktsicherheit wenden Sie sich bitte an:
If you have any questions regarding product safety,
please contact:

Walter de Gruyter GmbH
Genthiner Straße 13
10785 Berlin
productsafety@degruyterbrill.com